Flow-induced alignment in composite materials

# Flow-induced alignment in composite materials

Edited by

T D Papathanasiou and D C Guell

WOODHEAD PUBLISHING LIMITED

Cambridge England

Published by Woodhead Publishing Limited
Abington Hall, Abington
Cambridge CB1 6AH, England

First published 1997 Woodhead Publishing Ltd

British Library Cataloguing in Publication Data
A catalogue record for this book is available from the British Library.

ISBN 1 85573 254 8

Designed by Geoff Green
Typeset by Techset Composition Limited, Salisbury, Wiltshire.
Printed by St Edmundsbury Press, Suffolk, England.

# CONTENTS

# PREFACE

The purpose of this book is to present, within a single volume, detailed information concerning the science, processing, applications, characterization and properties of composite materials reinforced with short fibres that have been oriented in a preferred direction by flows arising during processing. This book is intended to be useful not only to materials scientists and other specialists in the field, but also to researchers, postgraduate students and particularly, engineers, designers and producers of composites reinforced with short fibres. To our knowledge, this is the first time information pertaining to this important class of engineering materials has been systematically presented within a single volume.

The purpose of aligning short fibres in a fibre-reinforced material is to improve the mechanical properties of the resulting composite. Aligning the fibres, generally in a preferred direction, allows them to contribute as much as possible to reinforcing the material. In some instances, the mechanical properties of these aligned, short-fibre composites can approach those of continuous-fibre composites, with the added advantages of lower production costs and greater ease of production.

The scope of the book is precisely what its title implies: composite materials whose short fibres have been aligned by flow processing techniques. The book covers theory, processing, characterization and properties. More specifically, the topics considered in the following chapters include fibre alignment and material rheology; processes that can produce fibre alignment in polymeric, liquid crystal polymeric and metallic composites; materials characterization and mechanical properties; and modelling of processes and material properties. On the other hand, continuous-fibre composites are not considered, nor are processes that can cause fibre alignment by other means, such as by electrical or magnetic fields. Since this book focuses on aligned short-fibre composites, the general subject of composite materials is not covered in depth. Instead, the reader is encouraged to complement this book with other sources of general information regarding composite materials.

Each chapter of the book is a complete and independent treatise focusing on one subject concerning aligned-fibre composites. Numerous references are provided in each chapter where more extensive information is needed. The chapters are organized in a logical progression, beginning with fundamentals, moving to processing and closing with material characterization and properties. However, this organizational framework should not discourage the reader from exploring the chapters in the order directed by his or her interest and expertise. The opening chapter introduces the topic of aligned-fibre composites, the state of the art, their benefits, their history, applications and processing. The next two chapters set forth the fundamental principles underlying how fibres align during fluid flow. Chapter 2 deals with fibre suspension rheology, with a focus on effects arising from fibre–fibre and fibre–wall interactions. Chapter 3 considers theories for the constitutive behaviour of fibre suspensions and for the development of local fibre orientations and gives examples of numerical implementation of these theories for complex flows.

The next three chapters deal with processing and materials. Chapter 4 is a review of the technical literature on the subject of fibre orientation during injection molding of fibre-reinforced composites. Experimental and theoretical works are reviewed and the factors affecting fibre orientation are identified. Following this, Chapter 5 presents the principles and applications of shear controlled orientation technology for the control and manipulation of fibre orientation in large scale injection molded parts. Numerous examples are reported on parts of large thickness, thickness variations and parts with weldlines. The application of the technology to *in situ* formed fibres as well as to ceramic- or metal-matrix composites is also discussed. In the context of this book, liquid crystalline polymers are considered to be a form of aligned-fibre composite. Chapter 6 deals with the theory of the dynamics of orientation in this class of emerging materials.

The final three chapters deal with the morphological characterization and the mechanical properties of aligned-fibre composites. Chapter 7 is a comprehensive treatise on techniques used for the meso-structural characterization of aligned-fibre composites, including two-dimensional optical reflection and three-dimensional confocal laser scanning microscopy. Numerous examples, treating certain difficult issues such as the presence of curved or wavy fibres, are presented. Chapter 8 is a review of popular analytical models used for the prediction of the stiffness of aligned-fibre composites and their application in design. The use of state of the art computational techniques for the prediction of the mechanical behaviour of aligned fibre composites based on realistic descriptions of their microstructure is outlined in Chapter 9, where examples of structure-oriented modelling using the boundary integral method are presented.

We have attempted to keep the nomenclature as close as possible to accepted engineering practice. However, given the widely varying subjects covered in this book, some conflicts inevitably arise. In any event, symbols are introduced in each chapter as needed. Likewise, differences in spelling and usage may arise according to whether the authors have adopted US or UK conventions. Some minor overlap between chapters has been allowed to persist so that they remain independent and self contained, according to the intent of their authors. These issues should not detract from the content of each chapter.

Completing this project, we would like to extend our sincere appreciation to the authors for their meticulous work. They deserve the credit for this production. Dr. Papathanasiou wishes to acknowledge the support of Unilever Research during his tenure at Imperial College, as well as the support and encouragement of Mr. F.L. Matthews, Director of the Imperial College Centre for Composite Materials. Thanks are also due to Mrs. Patricia Morisson of Woodhead Publishing, first for suggesting this topic and second for her support and patience throughout this project. And finally, we would like to thank our wives, Connie and Ann, whose loving support and patience has made this an enjoyable experience.

# LIST OF AUTHORS

ADVANI S G   *Department of Mechanical Engineering, University of Delaware, Newark, DE 19716, USA*

ALLAN P S   *Wolfson Centre for Materials Processing, Brunel University, Uxbridge, Middlesex UB8 3PH, UK*

ARCHENHOLD G   *Molecular Physics and Instrumentation Group, Dept. of Physics and Astronomy, University of Leeds, Leeds LS2 9JT, UK*

BENARD A   *Department of Mechanical Engineering, Michigan State University, East Lansing, MI 48824, USA*

BEVIS M J   *Wolfson Centre for Materials Processing, Brunel University, Uxbridge, Middlesex UB8 3PH, UK*

BROOKS R   *Department of Mechanical Engineering, University of Nottingham, UK*

CLARKE A R   *Molecular Physics and Instrumentation Group, Dept. of Physics and Astronomy, University of Leeds, Leeds LS2 9JT, UK*

DAVIDSON N C   *Molecular Physics and Instrumentation Group, Dept. of Physics and Astronomy, University of Leeds, Leeds LS2 9JT, UK*

GUELL D C   *Los Alamos National Laboratory, Engineering Sciences and Applications, Energy and Process Engineering, Los Alamos, NM 87545, USA*

INGBER M S   *Department of Mechanical Engineering, University of New Mexico, Albuquerque, NM 87131, USA*

PHAN-THIEN N   *Department of Mechanical and Mechatronic Engineering, The University of Sydney, Sydney N.S.W. 2006, Australia*

PAPATHANASIOU T D   *Department of Chemical Engineering, Imperial College, London SW7 2BY UK*

RANGANATHAN S   *Kimberly-Clark Corporation, 2100 Winchester Road, Neenah, WI 54956, USA*

REY A D   *Department of Chemical Engineering, McGill University, Montreal, Quebec H3A 2A7, Canada*

ZHENG R   *Moldflow Pty. Ltd., 259–261 Colchester Road, Kilsyth, Melbourne, VIC. 3137, Australia*

# Flow-induced alignment in composite materials: current applications and future prospects

### DAVID GUELL AND ANDRÉ BÉNARD

This chapter introduces aligned-fiber composites and provides an overview of the processes used to manufacture them. Special attention is paid to flow processes that can be used to induce fiber alignment in composite parts because such processes appear to be the most likely route to widespread commercialization of composites containing short aligned fibers. The advantages of these materials and the issues arising during processing are also described. In short, this chapter provides readers of varying interests and backgrounds in composites with the information needed to explore the other chapters of this book.

## 1.1　A brief survey of composites

The idea of blending several materials together to retain the benefits of each constituent in the final structure is not new. Many examples of composite materials, including bone and wood, occur naturally. Concrete and carbon black in rubber are two examples of manmade composites that have been used for decades. The term 'composite material' has a broad meaning and refers to a structure made up of two or more discrete components which, when combined, enhance the behavior of the resulting material. This somewhat loose definition of a composite is often modified by the terms 'microscopic' and 'macroscopic' to describe the length scale on which internal heterogeneity exists. However, of all the possible definitions of composite material (see Schwartz,[1] for example), the type of composite that uses fiber reinforcement evokes the most interest for structural applications. Thus, for the purposes of this chapter, a 'fiber-reinforced composite' refers to a material containing fibers embedded in a matrix with well defined interfaces between the two constituent phases.

　The interest the engineering community has shown in fiber-based composites stems from the strength and stiffness of materials in fiber form. When fibers are combined with a matrix that provides cohesion, a composite can be made that has mechanical properties unmatched by its individual constituents. The fibers are generally the load bearing component of the composite, and the matrix generally holds the fibers in position and participates in the load transfer between the fibers. In most composites, the fibers are brittle, while the matrix (especially in polymer-based composites) is ductile. Composites are also attractive because the fibers can often be organized in different ways. This allows composites to be tailored for specific applications, providing added design flexibility.

Glass, carbon and aramid are the most commonly used types of commercial fibers. Other types of fibers such as boron, silicon carbide, various liquid crystal polymers and aluminum oxide are used in high performance applications but to a lesser extent. For low cost reinforcement, cellulosic fibers, in the form of jute, sisal or cotton, are also available, along with other natural fibers, such as wool and silk, which are not used as widely.[2] The low density and high strength of these fibers are usually exploited to provide a lightweight and stiff composite. Reinforcing fibers are also available in a wide variety of forms, as shown in Fig. 1.1. They range from short random fibers to braided preforms. The processing technique and the end application usually define which type of fiber is suitable and the arrangement of fibers. Different types of fibers can also be mixed to obtain hybrid composites. Thus, a nearly infinite range of composites can be imagined.

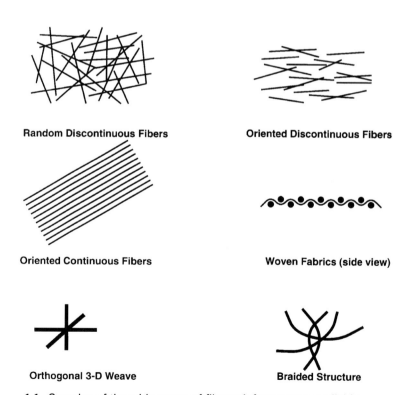

**Random Discontinuous Fibers**          **Oriented Discontinuous Fibers**

**Oriented Continuous Fibers**          **Woven Fabrics (side view)**

**Orthogonal 3-D Weave**          **Braided Structure**

*1.1* Samples of the wide range of fiber reinforcements available.

The desired part performance and the likely manufacturing method determine which materials are candidate constituents for a composite. A strong tie therefore exists between the part design, the materials used and the processing technique. Composites can be designed for high strength, light weight, fracture resistance, impact resistance or fatigue resistance.[3] A stadium seat, for example, might require randomly oriented short glass fibers in a polymer matrix, whereas a turbine blade in a high performance aircraft engine might require aligned silicon-carbide fibers in a metallic matrix.

In the past, short-fiber composites typically have been reserved for low performance parts. Yet such discontinuous-fiber composites can approach the performance of continuous-fiber composites if the fibers are aligned and if they are sufficiently long. The best properties of a load bearing composite are obtained when the fibers are aligned in the load bearing direction. Such alignment can be achieved by a variety of techniques that allow specific fiber orientations to be induced during processing. Since discontinuous-fiber parts are generally cheaper to manufacture than those containing continuous fibers, these techniques offer the potential for producing high performance parts at a reduced cost.

## 1.1.1  Brief history of composite materials

Composites, as they are known today, emerged almost immediately after the introduction of polymers during World War II. By 1942, an engineering monograph had been published presenting the processing and structural advantages of combining thermoplastic polymers with various fillers.[4] The first applications of composites involved fiber-reinforced plastics to protect radar equipment. The composites developed for these applications were well suited to this task because they were transparent to radio waves and were stiffer than the available polymers. At the same time, glass cloth reinforcements were used with the first low pressure laminating resins. Shortly afterward, thermosetting polymers similar to those used in today's composites became commercially available in the United States and Europe.

The aircraft industry continued to develop fiber-reinforced plastics and expanded their use during the 1950s, particularly in the United States. Three principal driving forces motivated interest in composite materials for high performance applications: theoretical predictions of extremely high potential crystal strength and elastic properties; the demand for low weight, high rigidity structures; and the flourishing US economy. Metal-matrix composites were also introduced during the 1950s, motivated by the need for high temperature materials for ballistic missile nosecaps.[5]

Composites also saw their first use in consumer goods in the early 1950s as chopped strand glass mats became available. These materials served as reinforcements for thermosetting polymers in boat hulls and sports car body moldings. These new materials were also employed in industrial applications such as pipes and chemical tanks. During this time, chopped-fiber-reinforced polyesters became familiar to the general public. However, the full structural potential of composites was not exploited in these applications because the new composite materials were used mainly as replacements for traditional materials such as wood. These applications of chopped-fiber thermoset composites became saturated because the first industrial composites exhibited limited performance and because the available molding technology was rudimentary.

The 1960s saw significant advances in the field of composites. High modulus whiskers and filaments were developed, and theories concerning the mechanics of laminated composite materials were discovered.[6,7] The first high performance composites (boron/epoxy composites) found applications in fighter aircraft. Advances were also made in press molding technology, extrusion, and injection molding of reinforced polymers. Improvements in graphite fibers led to significant improvements in mechanical properties and weight savings, advances that made them attractive for military and sporting goods applications.

As time passed, graphite/epoxy composites were successfully employed in military items and sporting goods such as golf clubs, tennis rackets and fishing rods. This success encouraged the development of high modulus organic fibers during the 1970s. Aramid and high density polyethylene fibers emerged as excellent candidates for various applications, including tethers for observation balloons. These fibers replaced steel in many other applications where high strength cables were required (for example, such as in 6 km lines for sonobuoy systems).[5]

Interest in composite materials reached a high point in the late 1980s, when the previous successes seemed to offer much more for the future. Yet, in recent years, the rate at which new composites have been applied commercially has not matched that explosive growth in interest. One reason for this slow growth has been the slow, costly and labor intensive processes available for composite manufacturing, a drawback that applies particularly to continuous-fiber composites. It now seems that a significant increase in the application of composites will occur only with the development of faster, less expensive processes capable of high production volumes. Thus, attention has focused on developing continuous processes for composites manufacturing. Most of these processes involve some kind of flow that tends to orient fibers in a preferred direction, an effect that can be exploited to enhance a composite's mechanical properties. Composite manufacturing processes that involve flow and fiber alignment are now, more than ever, of immediate practical importance.

## 1.1.2 An overview of aligned-fiber composites

Alignment of short, high-aspect-ratio fibers is not difficult to achieve. It often occurs unintentionally because fibers tend to align themselves naturally in the flow direction of a suspending fluid. This effect has been reported extensively and has been employed by a number of researchers to improve mechanical properties in composites. Examples of such processes include those employing injection molding,[8] extrusion,[9,10] compression molding[11] and converging flows.[12] Electric[13,14] and magnetic[15] fields have been proposed as alternatives to align fibers in a matrix precursor. Techniques have been designed in which fibers are aligned by flow and collected on the surface of a vacuum drum.[16] Vibration has been employed to align fibers[17] and aligned-fiber mats have been prepared by extruding a suspension of fibers onto a filter bed.[18] The common goal of these processes is to align the fibers in the direction of the load that the part sustains during service in order to increase tensile strength and stiffness in the principal stress direction. For example, greater hoop strength in a pipe or increased bending stiffness of a beam might be desired. In fact, the ease with which fiber alignment can be achieved through flow processing and the benefits fiber alignment can provide have contributed to the great variety of processes that have been used to align short fibers in composite materials.

In several of the processes that cause fiber alignment, short reinforcing fibers are mixed with the matrix material prior to processing or casting. Often, the resulting compound can be processed using the same techniques as those used to manufacture parts from the unreinforced matrix material. For example, injection molding and extrusion are frequently used to manufacture polymer parts and polymer matrix composites. Flow management techniques, described in Section 1.2, have been proposed to align fibers in chosen directions. The mechanical properties of the resulting parts can approach those of the corresponding continuous-fiber composite provided that they form a satisfactory bond with the matrix, that they are well aligned, and that the fiber length exceeds the critical length required for full development of the stress in the fiber. Short-fiber reinforced composites typically can be made more rapidly and at a lower cost than equivalent continuous-fiber composites. Figure 1.2 illustrates the qualitative trade-off between the performance and processability of a composite according to its fiber length.[14]

Aligned short-fiber composites provide advantages beyond those normally associated with random short-fiber composite materials, advantages such as light weight, improved mechanical properties, and resistance to chemical attack and high temperatures. They allow the properties of the composite to approach those of continuous-fiber composites by enabling every fiber to contribute fully to reinforcing the material. Also, aligned, short-fiber composites can be manufactured by continuous or semi-continuous processes that enjoy greater manufacturing output and lower costs than the hand lay-up processes commonly used to make continuous-fiber composites. Complex shapes can be manufactured, some of which are impossible to manufacture by hand lay-up due to the limited capacity of continuous fibers to conform to complex shapes. Thus, flow processes that induce fiber alignment are appealing because they are continuous,

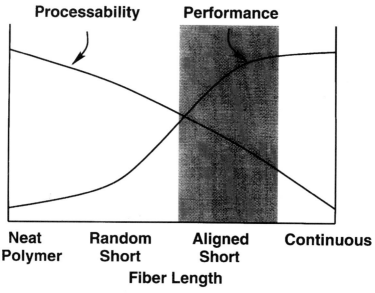

**Processability**   **Performance**

**Neat Polymer**   **Random Short**   **Aligned Short**   **Continuous**

**Fiber Length**

*1.2* The compromise often encountered between performance and processability of composites versus fiber length.

they can be incorporated into or adapted from existing processes, and they have the potential for lower cost and higher output rates.

Despite the advantages short-fiber composites offer, they do suffer some drawbacks. Short-fiber composites are weaker than continuous composites because stress concentrations form near the fiber ends and because the stress profile is not fully developed throughout the fiber length.[19] Generally, short fibers are less effective than longer fibers at bearing a load. Another problem that surfaces with discontinuous fibers is breakage during processing, which can significantly reduce the length of the fibers in the final product. For example, in injection molding, fibers are often mixed with the polymer before passing through the plasticizing screw, resulting in significant breakage. Thus, the advantages of aligning the fibers versus the loss of properties in the transverse directions must be weighed carefully.

### 1.1.3  Benefits of aligned short-fiber reinforcement

The potential that aligned fibers hold for improved mechanical properties was recognized years ago, as is evident from the statement by Dingle[19] that

composites made by conventional moulding techniques possess axial mechanical properties approaching those obtained from continuous-carbon-fiber composites of equal fiber volume fraction.... A composite made from perfectly aligned discontinuous carbon fibers can, theoretically, be expected to utilize the available fiber strength almost as efficiently as a unidirectional continuous-carbon-fiber composite. In practice, however, it has been difficult to realize.

Practical difficulties arise in using new or existing processes to align fibers with a high degree of precision because of the large number of factors affecting fiber orientation during processing, factors such as injection rate, aggregation of fibers, matrix viscosity, mold geometry, fiber length and fiber–fiber interactions. Furthermore, fiber breakage is difficult to reduce in processes such as injection molding, where high shear stresses are present and where significant mechanical interactions exist between the fibers and the machinery.

The reason that aligning fibers in a composite enhances tensile mechanical properties is intuitively clear. In a material in tension, a fiber aligned in the direction of tension will experience a strain along its entire length and, thus, will experience a greater total strain than a misaligned fiber. Such fibers will experience a greater stress and will bear a greater load. As long as the fiber–matrix bond is sufficiently strong to bear that load and the reinforcing fibers are stronger and stiffer than the matrix, the aligned-fiber composite will generally be stronger and stiffer in the direction of fiber alignment than an otherwise equivalent composite containing randomly oriented fibers. At the opposite extreme, a long thin fiber oriented perpendicular to the load direction will experience a strain only across its diameter, resulting in a much lower total stress in the fiber, and therefore a lesser load on it. Since tensile properties in the fiber alignment direction exhibit marked improvement over an equivalent random-fiber composite,[19] aligned-fiber composites are best suited to applications in which the fibers are aligned in the direction of principal stress in the part. The optimal fiber arrangement can be determined from equations derived from micromechanics. Numerous texts present models that allow the properties of a composite to be computed from the fiber arrangement of the composite.[1,3,20]

Aligned short-fiber composites offer benefits for processing and material properties, some of which have been noted above. These benefits have been pointed out in the early studies of composites.[21,22] Since then, numerous studies have reported improvements in the tensile properties of aligned-fiber composites. Improvements in fracture and wear properties have also been reported. A factor that makes short fibers especially desirable as reinforcements, whether they are aligned or not, is that some of the strongest materials known (whiskers and monocrystalline fibers) are available only in the form of short fibers.[12]

Perhaps the first application of flow-induced fiber-alignment techniques involved the manufacture of aligned-fiber prepregs in which fibers in dilute suspension were aligned by the flow field in a tapered slit.[21] In another early study of flow alignment induced by continuous processing methods, strength and stiffness measurements were made of random- and aligned-fiber epoxy composites reinforced with glass fibers.[23] In this study, aligned-fiber composites were made by extruding rods of epoxy, then compression molding the rods into thin plates. In the alignment direction, composites were found to be stiffer than (but not as strong as) the random fiber composite. Others have found that fiber alignment stiffens and strengthens a composite. Significant increases in the ultimate longitudinal strength and strain of aligned glass fiber materials have been observed in reinforced epoxy composites as the fiber length is increased.[24] In one study of injection molded glass fiber reinforced thermoplastics, the fiber alignment that occurred near the walls of the part significantly affected mechanical and fracture properties.[25] More recently, improved tensile strength and stiffness have been observed in carbon fiber reinforced epoxy composites whose fibers were aligned by the flow induced in the uncured material as a small cylindrical rod was passed through it.[26] This study found that aligned fibers were twice as effective as randomly oriented fibers at enhancing the stiffness and strength of the composites examined. Increased strength and stiffness have also been reported in zinc alloy matrix composites reinforced with alumina fibers aligned by extrusion.[27]

Fiber alignment enhances other mechanical properties as well. For example, impact toughness and shear modulus have been improved over conventional extruded polymeric composites.[28] Improved fracture properties in ceramics[29,30] have been reported, as have significant improvements in wear resistance.[27] Clearly, aligned short fibers in a composite can enhance mechanical properties in a variety of materials.

Processing materials containing short fibers also offers advantages over processes involving continuous fibers. A variety of processes exist by which parts can be made from short-fiber materials. In particular, composite parts can sometimes be made using the same processes that are used to produce parts from the unfilled matrix material. The best examples of these materials are polymer composites processed by injection molding and extrusion. Where such established processes as these are suitable, continuous, well developed and high volume processes can be employed directly, or can be adapted, to make composite parts.

Another advantage of aligned-fiber composites is that highly complex parts can be formed, particularly where three-dimensional shapes are concerned[12] or where specific alignment patterns are desired.[19] For example, short-fiber composites having different degrees of orientation can be prepared by flow molding techniques so that the fibers efficiently reinforce the material.[24] In a slightly different context, aligned, short-fiber prepregs formed by flow alignment methods can be molded or draped into complicated shapes, while retaining most of their desired properties.[24] A more orderly arrangement of fibers can, in principle, allow higher fiber loadings to be achieved and it has been argued that this is also true in practice.[19,31] In at least one instance, materials containing short fibers were found to be easier to mold than those containing continuous fibers.[28]

## 1.1.4   Flow-induced fiber alignment

Numerous experimental and theoretical studies have demonstrated that short fibers align themselves in laminar flows. In the classic theoretical study by Jeffery, the equations governing viscous (creeping) flow were solved around an ellipsoid of revolution in unbounded shear flow.[32] Jeffery's result showed that most of the time, high-aspect-ratio ellipsoids of revolution have their long axis nearly aligned in the plane of flow and spend only a small fraction of the time not so aligned. Subsequent theoretical work has employed slender body theory to derive similar results that apply to high-aspect-ratio cylinders, a geometry more representative of a typical cylindrical fiber.[33] Strong fiber alignment is also induced in pure extensional flows. In this case, fibers approach perfect alignment of their long axis with the direction of flow on the timescale of the fluid strain rate.[34] The group of S.G. Mason provided early experimental verification of the tendency of fibers to align themselves during flow, work which is reviewed by Ranganathan and Advani in Chapter 2. Whether in shear flow or extensional flow, high-aspect-ratio fibers tend to align themselves with the flow and it is this effect that underlies the flow alignment techniques that are the subject of this book.

The tendency of suspended high-aspect-ratio fibers to align themselves along the streamlines of a given flow is easily understood on intuitive grounds. Consider a high-aspect-ratio fiber aligned in a shear flow. In this case, only a relatively small mismatch exists between the velocity field that the fluid would adopt in the absence of the fiber and the velocity at each point on the surface of the fiber as it moves with the fluid. This 'mismatch' disturbs the surrounding fluid and causes the fiber to rotate. The thinner the fiber is, the smaller is the velocity mismatch and the slower the fiber rotates out of this aligned orientation. When a fiber is oriented with its long axis across the streamlines, the fiber rotates with the fluid, regardless of the fiber's diameter. Thus, a high-aspect-ratio fiber is slow to leave an aligned orientation and quick to realign itself once it is out of alignment. Therefore, high-aspect-ratio fibers spend a large fraction of the time aligned with the flow, with most of the fibers aligning themselves along streamlines. In shear flow, the fibers rotate periodically because no stable equilibrium orientation exists, whereas in pure extensional flow such an orientation is available, with fibers approaching that orientation with time. Even during processing of unreinforced polymers, these effects act to stretch and align polymer molecules.[35] The effect of flow on the alignment of the fibers is illustrated in Fig. 1.3 for three flow situations.

Fiber alignment is not merely a theoretical artifact but has been reported by many authors in the processing of real systems. For example, various microphotographs exist in the literature showing extensive fiber alignment in injection molded parts.[25,28,36] In these photographs, the skin–core structure common to injection molded parts is observed. Also, the degree of fiber

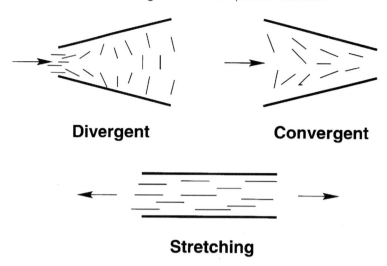

**Divergent**                    **Convergent**

**Stretching**

*1.3* Examples of flow induced orientation of short fibers.

orientation varies across the thickness of the part, with a greater degree of alignment occurring in the skin than in the core. Alignment of the fibers occurs through the thickness but also in the plane of the part. The high shear stresses found close to the walls, combined with the solidification process, result in the skin–core morphology often observed. The expansion of the flow immediately after the gate creates a weak in-plane orientation of the fibers in the core.

The theories of fiber alignment noted above pertain to dilute suspensions of fibers, that is, suspensions whose fibers do not interact with each other through the suspending fluid. As can be imagined, non-dilute suspensions of fibers can pose greater difficulties to achieving fiber alignment and such difficulties have been reported.[19,25] Steric interactions between fibers arise which can prevent the fibers from rotating freely into aligned orientations, causing the fibers to form an entangled mass instead. Nevertheless, fiber alignment occurs even when fibers interact. What constitutes a concentrated suspension of fibers, however, is not simply a matter of the volume fraction occupied by the fibers because high aspect ratio fibers are characterized by two lengths scales, a length $L$ and a diameter $a$. As a result, three different concentration regimes are often cited. Briefly, these are first, a dilute regime, in which the characteristic fiber separation distance is greater than the fiber length; second, a semi-concentrated regime in which this characteristic separation distance is less than the length of the fiber, but greater than the fiber diameter; and third, a concentrated regime in which fibers are separated by only a few fiber diameters. While early analyses of fiber motion pertained to dilute suspensions of particles, theories concerning non-dilute suspensions have also been developed to take account of the effects that occur in concentrated suspensions. These theories often employ a vector field that describes the average fiber orientation direction and accounts for the effect of local fiber alignment on the rheology of the suspension. In addition, models have been developed to describe the local fiber orientation in concentrated suspensions. The issues concerning fiber orientation and fiber suspension rheology are reviewed from different perspectives in Chapter 2 by Ranganathan and Advani and in Chapter 3 by Phan-Thien and Zheng.

As has been amply shown, fibers tend to align themselves in shear and extensional flows. These flows occur throughout materials processing, particularly in polymer processing. For example, the laminar, pressure driven flows that occur in injection molding appear as a local shear flow to fibers whose lengths are small compared with the length scale of the velocity gradient, typically a characteristic cavity thickness. Thus, fibers that are short compared with the thickness of an injection molded part often align themselves in the plane of the flow. The contractions that exist in polymer extrusion dies also give rise to extensional flows that cause fiber alignment. Thus, the mechanisms at work in simple shear and extensional flows are relevant to more complex processing operations.

## 1.1.5 Applications of aligned-fiber composites

Numerous applications have been proposed and demonstrated for producing aligned short-fiber composites and a few of these applications are highlighted in this section. Several processes have also been patented to produce aligned-fiber composites.[31,37,38] As noted before, the primary driving force for using composite materials is a combination of features such as reduced weight, improved mechanical properties and resistance to chemical attack. At first, their special properties led aligned-fiber composite materials to be considered primarily for specialty applications. However, the uses that have been found for these materials have broadened to include many kinds of structural applications. Although achievable fiber volume fractions may be lower than with continuous-fiber composites, short-fiber composites can be attractive when processing, part complexity and cost are important issues.[19]

The first uses envisioned for aligned-fiber composites were in aerospace applications and in extreme environments usually involving high temperatures. Examples of such applications include a thrust reverser cascade and a rocket motor nozzle.[19] Resistance to high temperature chemical attack has made ceramic composites good candidates for aerospace and turbine engine applications.[39] However, the brittleness of ceramics has prevented them from being applied directly in many applications. Fiber reinforcement is often pursued as a remedy to this problem and aligning the fibers in the load direction can provide needed improvement in mechanical properties. Other aerospace applications for polymer composites include vehicle body panels,[40] as well as missile covers and aircraft window frames.[16] Metallic composites are also candidates for aerospace and turbine engine applications[38,39] because they possess many of the advantages of ceramics, such as low density, high melting point, and resistance to oxidation.[41] However, the high price of the fibers required for use in metal matrix composites and difficulties in fabrication have inhibited commercial development of these materials.[41]

Other applications for aligned-fiber composites have been found. For example, in the automotive industry, ceramic composites have been proposed as the material of construction for engine valves in internal combustion engines,[38] and metal matrix composite connecting rods[31] have been proposed. In weight critical applications, graphite composite materials offer a desirable combination of structural properties, provided the orientation of the fibers is precisely controlled during the fabrication of the part. In the health care field, the specific properties of aligned-fiber composites have been employed to develop an artificial hip implant.[42] In this application, matching the stiffness of the implant with that of bone *in vivo* improves the likelihood of clinical success.

Every-day applications have also been cited for aligned-fiber composites. For example, by orienting fibers circumferentially, the hoop strength of extruded pipe can be increased.[9,10] In addition, extruded metal pipe and wire having enhanced tensile and wear properties have been reported.[27] Disks containing circumferentially aligned fibers have also been produced.[19] Liquid crystal polymers, while perhaps not a fiber-reinforced composite in the strictest sense, do bear strong similarities to the aligned-fiber composites considered here and, thus, they fall within the spirit and scope of this book. These materials have been developed under various trademarks for commercial use, and they have been proposed for structural and device applications.[43]

## 1.1.6  Types of fiber reinforcement

The high stiffness and strength found in fibrous reinforcements is at the heart of most high performance composites. Most materials attain their greatest strength in fiber form for several reasons. The principal reason is that flaws on the surface of a material significantly reduce the strength of the material because they serve as initiation sites for cracks that propagate into the rest of the structure, causing failure. A decrease in the volume and surface area of a material reduces the probability of occurrence of such flaws in a given fiber. The presence of only a few flaws has been observed to affect significantly the ultimate strength of brittle materials. The smaller the material, the less likely it is to contain a flaw. Slender fibers present a geometry which has a limited volume, a geometry that can be used effectively as a reinforcement. Materials produced in fiber form can thereby approach their maximum theoretical strength.

The choice of which kind of fiber or whisker is best suited to a specific application is strongly tied to the matrix material used and the processing route employed. Fibers vary in their mechanical, thermal and chemical properties and come in a variety of aspect ratios (e.g. chopped, continuous or whisker form). The type of fiber that is best suited to a given application also depends on the choice of polymer, metal or ceramic matrix.

The fiber used in a polymer-matrix composite generally controls the strength and stiffness of the material. Carbon/graphite, glass, aramid, polyethylene and boron are commonly used in polymer-matrix composites. Various types of fiber are available within each of these families, according to their composition and processing route, and they come in various cross-sections (e.g. circular, bilobal or star-shaped). The qualitative stress–strain relationships for several fibers commonly used as reinforcements are shown in Fig. 1.4. However, tensile behavior is only a starting point in choosing the best reinforcement for a given application. Heat resistance, cost, wear, density and chemical compatibility with the matrix must also be considered. In addition, batchwise variations in mechanical property values can be significant, so the qualitative relationships shown can only be considered estimates of what might be observed in practice.

Fibers used in metal-matrix composites must be suitable for processing at high temperatures. Metal-matrix composites are generally employed in applications requiring high specific stiffness at high temperatures. The fiber compositions that are typically used with metal-matrix composites are carbon, boron, silicon carbide and alumina–silica.[44] Zirconia-based fibers and silicon nitride whiskers are also available but are of lesser importance. As with fibers available for polymer-matrix composites, a range of products is available within each family, and several fiber morphologies are available. Fibers can be classified by diameter as whiskers ($<1$ $\mu$m), staple ($1$–$10$ $\mu$m), continuous multifilament yarns ($5$–$25$ $\mu$m) and continuous monofilaments ($>100$ $\mu$m). As with polymer composites, stiffness and strength are important properties, as are density, thermal expansion, flexibility and hardness.[44] Typical elastic moduli and tensile strengths of these fibers are shown in Fig. 1.5.

Metal-matrix composites can be processed somewhat inexpensively using whiskers and staple fibers. Fiber alignment can be expected to improve the properties of these composites because expensive continuous fibers have been used to make metallic composites that exhibit large increases in specific properties over the base alloy.

Ceramic-matrix composites typically include a polycrystalline ceramic, glass or a mixture of these, reinforced by particles, flakes, whiskers or fibers. Unlike polymer and metallic composites, which have relatively low modulus and high ultimate strain matrices, ceramic matrices have high moduli and very low elongation at failure. The elastic modulus of these matrices is often equal to, or greater than, the reinforcing fibers. The goal of reinforcing ceramics is, therefore, to obtain high strength and toughness at high temperatures. Toughness has been the most elusive of these goals because ceramics are notoriously brittle. Carbon and ceramic fibers appear to be promising reinforcements, but beyond this, numerous types of fibers are available for use with ceramic matrices. In the

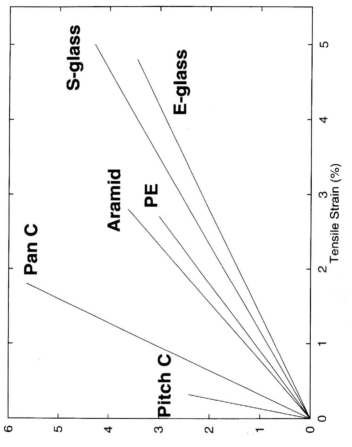

*1.4* Typical stress–strain relationships of various fibers, illustrating the relative moduli and strength.

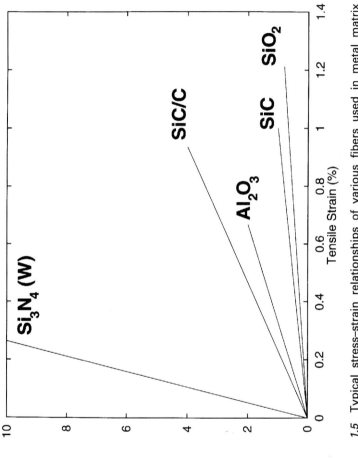

1.5 Typical stress–strain relationships of various fibers used in metal matrix composites.

selection of a fiber, chemical compatibility, high temperature resistance, thermal expansion and processability are all important issues for use with a ceramic matrix.

### 1.1.7  Critical fiber length

Fiber length is one of the most important factors influencing the mechanical properties of short-fiber composites. In continuous-fiber composites, the strain experienced by the fiber and the matrix is, on average, identical because both systems exist, in a sense, in parallel. In short-fiber composites however, the mechanism for stress transfer from fiber to fiber is more complicated. Figure 1.6 illustrates the deformation of the matrix surrounding a single fiber in tension.

The mechanism for stress transfer from fiber to fiber can be understood from what is known as shear lag theory. That is, fibers are assumed to be in tension only, while the matrix is assumed to bear only shear stresses. This implies that as the strain in the matrix increases, the strain in the fiber increases from the fiber ends. The stress increases from zero at the fiber tip and attains a plateau value along the middle of the fiber.[45]

Short-fiber composites are weaker than continuous composites because of the stress concentrations in the vicinity of the ends of the fiber and because the load bearing capacity of the fiber is not fully utilized as it is in a continuous-fiber composite. On the other hand, even though the stress profile is not fully developed along a portion of the fiber, the elastic properties of short-fiber composites containing well-aligned fibers can approach those of continuous-fiber materials, provided that the short fibers exceed the critical length for shear transfer through the matrix.[25]

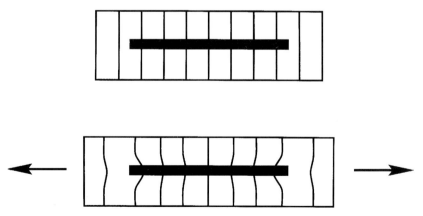

*1.6* Deformation of the matrix surrounding a single fiber under tensile loading.

A question that often arises in the composites literature is how long a fiber must be to provide the reinforcement efficiency of a continuous fiber. To achieve this reinforcement efficiency, the stress over the length of a fiber must be fully developed. The tensile stress in the fiber builds from a value of zero at each end of the fiber to a value of

$$\sigma_f = \frac{2x\tau}{r} \tag{1}$$

a result which is derived from a simple force balance over the fiber length. Here, $x=0$ refers to one end of the fiber (with $x$ extending to the middle of the fiber), $r$ is the fiber radius, $\sigma_f$ is the stress in the fiber and $\tau$ is the interfacial shear stress. For a given composite strain, $\varepsilon_c$, a critical fiber length $L_c$ exists for which the peak stress in the fiber just reaches the failure stress,[46] a length given by

$$L_c = \frac{\sigma_m r}{\tau_u}. \tag{2}$$

Here, $\sigma_m$ is the tensile strength of the fiber, $r$ is the radius of the fiber and $\tau_u$ is the shear strength of the matrix or of the interface. The critical fiber length, thus, depends on the fiber type (which determines $\sigma_m$), the matrix (which determines $\tau_u$) and the fiber diameter.

For fibers longer than $L_c$, tensile stress within the fiber increases along the length of the fiber until it reaches a maximum value at the center, the tensile stress that would be observed in a continuous fiber under equivalent conditions. The fiber breaks when the stress exceeds the ultimate fiber strength. The shear stress transferred through the matrix reaches a limiting value prescribed by the shear strength $\tau_u$ of the matrix or the interface. A region then exists at both fiber ends, where shear stress is uniform and equal to the yield strength of the matrix/interface. For fibers shorter than $L_c$, the stress in the fiber never reaches its maximum value because the shear stress at the interface ($\tau$) is higher than the stress in the fiber ($\sigma_f$). In this case, the fiber slips through the matrix when the composite breaks.

The notion of a critical fiber length is useful for determining the load-bearing capacity of a given fiber–matrix system. A fiber length smaller than $L_c$ implies that improvements can be achieved by increasing the fiber length in the end product and that a desirable stress distribution in the fibers has not been obtained. A desirable stress distribution is one in which the stress builds from zero to the maximum stress sustainable by the fiber/matrix interface, as shown in Fig. 1.7.[47,48] A fiber length of at least $5L_c$ has been observed to provide a suitable reinforcement efficiency in short fibers.[49] However, the fiber length in short-fiber composites processed by various techniques is often well below this desired value, indicating that significantly better mechanical properties could be obtained with small increases in the fiber length.[47,50] For example, the critical fiber length of 350–400 $\mu m$ for Nicalon fibers (SiC) in aluminum alloys is

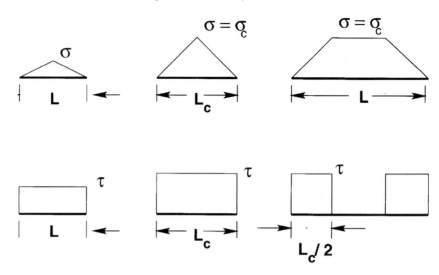

1.7  Idealized stress distributions in a single fiber for three fiber lengths. The first case considers $L < L_c$, the second $L = L_c$ and the third $L > L_c$.

greater than the fiber length observed in samples manufactured by compocasting and hot extrusion, both of which induce significant fiber breakage.[51] In this case, mechanical properties could be improved by employing processing methods that reduce fiber breakage.

## 1.1.8  Processing issues in aligned-fiber composites

Many processing issues arise in the context of aligned-fiber composites. These include fiber breakage, chemical compatibility and interfacial bonding between the fiber and the matrix, the rheology and processability of the fiber–matrix mixture and migration of the suspended fibers during flow. These factors will affect the suitability and the performance of the part. In this section, these issues are briefly described and their impact on aligned-fiber composites manufacturing is noted.

Fiber breakage during processing can be a significant problem if it reduces the ability of the fibers to bear the required load. As noted in Section 1.1.7, fibers shorter than a critical length cannot develop their maximum sustainable stress and are less efficient at reinforcing the material. In such a case, the mechanical properties of continuous-fiber composites are not approached. Mechanisms that induce fiber breakage during processing include the viscous fluid mechanical forces acting on the fibers and the interaction of the fibers with the processing equipment. Two studies of injection molding have found fiber breakage in the screw-barrel system, in the runners, and in the mold, and these studies have noted the possibility that part performance could thereby be degraded.[52,53]

Processes have been developed to limit this problem[52] and proper gate design can also help to reduce fiber breakage.[53] Fiber breakage can be an even greater issue in metal-matrix composites, where the large forces that act on fibers during solid-state deformation processing are often great enough to cause extensive breakage.[51] Processes have been developed for aligning whiskers and fibers that are too long to be flow oriented and extruded, yet too short and fragile to be processed by other methods.[54]

The rheology of fiber suspensions in viscous fluids has been studied intensively because of its importance to materials processing. The rheology of the fiber–matrix suspension determines whether the material flows smoothly and predictably during processing. If smooth predictable flows are not achieved, the quality of the cast part can be degraded. Consequently, Chapters 2 and 3 of this book consider different aspects of the rheology of fiber suspensions. In some cases, suspension rheology can significantly affect materials processing. In one study of injection molding, concentrated fiber suspensions did not always fill the mold and careful control of material rheology and mold design was required to achieve an adequately cast part.[25] The authors of this study concluded that 'the engineer must be continually alert ... to this interplay among rheology, local fiber alignment and mechanical behavior'.

A reinforcing fiber must be chemically compatible with the matrix and must form a suitable bond if it is to be an effective reinforcement. Generally, a strong fiber–matrix bond is desired. However, with ceramic composites, a stronger bond does not necessarily result in a better composite. Clearly, a fiber having a poor bond does not constitute an effective reinforcement, because any load born by the fiber must be transferred through that bond. Yet, a weakly bonded interface can reduce brittleness and increase fracture toughness in ceramics.[39] The reader is referred to the review by Piggott[55] for more information concerning fiber–matrix interfacial effects.

While interfacial bonding is a significant issue, chemical compatibility does not appear to be a significant problem in polymeric composites because commonly used carbon and glass fibers are relatively inert. However, the surface quality of the fibers can affect the nucleation rate of semi-crystalline polymer matrices.[56] Chemical compatibility is a larger issue in metal and ceramic matrix composites where typical processing temperatures can cause chemical degradation of the fiber surfaces. This can be a particular problem in metallic composites.[57] A mismatch in the coefficient of thermal expansion between the fiber and the matrix in ceramic composites can also lead to internal stresses induced by the thermal cycles that occur during processing.

Fiber migration during processing is another issue in fiber-reinforced composites processing. Particles in a suspension are well known to migrate in flowing suspensions, with the particles moving away from walls and toward the 'center' of the flow under many circumstances. Three different mechanisms have been identified as causing such migration. First, normal stress differences that arise from the viscoelasticity of the suspending fluid can cause particles to migrate toward the center of the flow.[58] This effect is relevant only to polymer matrix composites because ceramic and metallic composite precursors do not normally exhibit viscoelasticity. Second, the inertia of the surrounding fluid can cause suspended particles to move to a position midway between the wall and the center.[59] This inertial mechanism could occur in ceramic and metallic precursors where fluid viscosities are relatively low. However, if eddies form in the flow, the subsequent mixing will tend to randomize the fiber locations and orientations. In polymer processing, where Reynolds numbers are typically much less than one, these inertial mechanisms generally have little effect. Finally, fluid mechanical interactions between particles in a concentrated suspension also cause particles to migrate. In this case, fibers usually accumulate in the low shear rate region of the flow (for example, at the center of flow between flat plates or in a cylinder). This shear-induced particle migration has been extensively studied experimentally, theoretically and numerically, and is reviewed in greater detail in Chapter 3. Because this mechanism arises from interaction between particles, it occurs most often in concentrated viscous suspensions. In addition, the rate of migration is proportional to the square of the size of the suspended particles, so smaller particles migrate much more slowly than do larger particles.

Although particle migration during processing does not appear to be a serious problem, studies have reported the phenomenon in composites processing. Injection molded plates containing 75% glass fibers exhibited thin skin layers (consisting of about 5% of the part thickness) containing no fibers at all.[52] Similar results have been reported in injection molded parts reinforced with rigid and non-rigid particles, with a thin region near the surface of the part showing sharply reduced fiber concentration.[60] While particle migration is generally less important than the other issues noted, it does warrant awareness on the part of the practitioner.

The ability to recycle composite materials may not affect processing directly, but it can have an impact upon the decision concerning whether a given material is used in large quantities. The use of composites in automobiles serves as a particularly good example of the issues that arise. Today's automobiles have an average recyclability rate of 75% and maintaining or increasing that rate is desirable.[61] As more and more composites are used, new methods must be developed to dismantle, collect and recycle composite materials. The need to recycle will also likely affect part designs, as ease of dismantlement is built into parts at the design stage. Given the huge volume of material that could come from recycled composites in automobiles alone, processes must be developed, an infrastructure must be constructed and uses for the recycled materials must be found, if current recyclability rates are to be maintained.[61] As recycled materials are reused, the effect of repeated recycling on material properties and performance will also have to be determined. Since recycling affects the total life cycle cost of a part, these issues have an impact on the overall economics of employing large volumes of composite materials in such applications.

## 1.2    Flow processes for producing aligned-fiber polymer matrix composites

Polymer-based composites constitute the largest category among recently developed advanced materials. They offer a broad range of properties, from weak materials used in non-structural applications to lightweight, high performance composites that are well suited to military applications. Polymer composites have evolved rapidly over the years and significant progress is still being made, especially toward new manufacturing techniques. Polymer composites have also offered the most fertile ground for developing processes to produce aligned-fiber composites.

In the late 1970s, when high performance composites found widespread use in structural applications, fiber-reinforced plastics could be roughly divided into two categories: high performance, continuous-fiber-reinforced thermosets and low performance, short-fiber-reinforced thermoplastics.[62] Two factors were responsible for this division. First, the best mechanical properties were achieved with high loadings of long fibers, and thermosets are well suited to producing long fiber prepregs because of their low viscosity. Second, thermoplastics offered the ability to mold complex parts at high production rates with methods such as injection molding. Over the years, these two families of materials have moved into each other's territory, with thermoset products being produced by high productivity molding techniques and thermoplastics being applied with long fiber molding compounds and continuous reinforcements for structural components.

Among continuous-fiber polymer composites, thermosetting resins are the most popular matrix material. These resins start with low molecular weight species and undergo a chemical reaction, through the application of heat and pressure, to form a cross-linked network. Thermoplastic polymers are also used as the matrix material in continuous-fiber composites, but to a lesser extent because their higher viscosities make them somewhat more difficult to process. Normally, thermoplastics require higher temperatures and pressures to mold and shape than do thermosetting resins. However, they offer certain advantages, such as improved impact resistance, over thermosetting resins. In short-fiber-reinforced composites, the situation is reversed. Thermoplastic matrices are often used, principally because they can be easily injection molded and extruded. Such processes are well suited to large scale production.

Numerous processes are available for producing polymeric composites, depending on the matrix used and the desired fiber length. Table 1.1 provides a summary of the processes used for both thermosets and thermoplastic polymers and the resulting fiber lengths.

*Table 1.1.* Common processes used for manufacturing polymer composites and resulting average fiber lengths

| Process | Compound type | Fiber length (mm) |
|---------|---------------|-------------------|
| *Thermosetting* | | |
| Injection molding | Coating | 6–9 |
| Extrusion | Compounding | 0.1–1 |
| Compression molding | | |
| | | |
| *Thermoplastics* | | |
| Transfer molding | Dough blending | 3–6 |
| Compression molding | SMC | 6–25 |

Discontinuous fibers are currently the only type of reinforcement for which high productivity and low cost production methods are available for polymer-based composites. However, in order for these materials to find widespread use in load bearing applications, the processing methods used should be able to induce fiber alignment and reduce fiber breakage.

The part geometry and choice of materials determine the manufacturing method best suited to making a given part. For example, composites that employ thermosetting resins are processed differently from composites based on thermoplastics; thus a tubular structure might require a different manufacturing method from a part having a complex shape. The techniques based on flow-induced fiber alignment appear to be especially well suited to widely used processes such as injection molding, extrusion or compression molding. The next section introduces some of these processes and describes the impact they have on mechanical properties through fiber alignment.

## 1.2.1  Processing of thermoplastic polymer composites

Thermoplastic polymer molecules typically consist of linear chains whose molecular weights range from 5000 to 500 000. Typical families of thermo-plastic polymers include polyarylene ethers, polyimides, poly(phenylene sulfide), polybenzimidazole and aromatic liquid crystalline polymers. Thermo-plastic polymers can be broadly classified as amorphous or semi-crystalline, both of which are used in composite materials. Semi-crystalline polymer molecules are typically linear chains that organize themselves into an ordered state when brought below their melting point. Microscopic entities (termed spherulites) form during crystallization which consist of bundles of lamellae that branch regularly to form a sphere. The lamellae are simply the polymer crystals formed by a chain-folding mechanism.[63] Amorphous polymers, on the other hand, cannot crystallize because their molecular configuration does not allow regular, periodic organization of the chains. These polymers exhibit only a glass transition temperature, in contrast to semi-crystalline polymers which exhibit both a melting point and a glass transition temperature.

Long-fiber thermoplastic composites are often divided into three categories according to their fiber length. The family of composites that contain fibers with lengths of less than 0.5 cm is suitable for injection molding, while the family of composites that contains randomly oriented fibers longer than 1 cm is suitable for stamping. The last category employs continuous fibers and the products manufactured from them are similar to those manufactured with thermoset prepregs. For short-fiber composites, injection molding and extrusion have been used extensively to produce short-fiber composites.

### 1.2.1.1    Injection molding of thermoplastics

Injection molding of short-fiber-filled thermoplastics is probably the most attractive method for producing complex composite parts of relatively high stiffness economically. Injection molding has been used since the nineteenth century, when the first patent was awarded for what was then termed a 'stuffing machine'.[64] The number of thermoplastic products manufactured by injection molding today is practically endless.[65]

In injection molding of amorphous polymers, heating causes only a disentanglement of the polymer chains and the viscosity of the melt gradually decreases. For semi-crystalline polymers, heating causes both a melting of the crystalline phase and a disentanglement of the polymer chains. When fibers are added to these polymers, the material obtained is often called 'engineering compound' because the properties obtained by the addition of the fibers extend beyond those of the neat (fiber-free) polymer. These engineering compounds may include stabilizers, colorants, viscosity modifiers or secondary fillers.[66] E-glass fibers are often used as reinforcements because of their low cost. Mineral fillers, such as mica flakes or asbestos fibers, can also be used in lower performance applications. Carbon and aramid compounds can be prepared to obtain better performance, but at a higher cost. Various compounding techniques are discussed in the literature. The method chosen can result in significantly different fiber lengths in the end product.[28,47,66,67]

In injection molding of polymer-matrix composites, polymer pellets containing fibers are fed into a heated barrel where the polymer softens and becomes a viscous melt. In the barrel, one or two screws homogenize the polymer and melt it by friction. Supplementary heaters maintain a constant temperature in the barrel and assist in the melting process. When enough material has accumulated in front of the screw, the screw acts as a plunger and drives the polymer/fiber mixture into the mold. When the composite has solidified, the mold is opened and the part is ejected.

The properties of a polymer-matrix composite are highly dependent on the processing conditions. The aspect ratio and orientation of the fibers are important factors in determining the mechanical properties of the composite. Other factors such as the barrel and mold temperatures and the injection speed and pressure also influence the properties of the molded part.[65]

The fiber-aspect-ratio can be significantly reduced during the molding operation by breakage of the fibers in the screw barrel, the gates, the runners and the mold. The compounding technique can also affect the resulting fiber length. During molding, fiber orientation strongly depends on the fluid kinematics, which are determined by the runner, gate and mold geometries, the process conditions and the properties of the molding compound. The net result is that mold design for short-fiber-reinforced composite materials is strongly coupled to the manufacturing process. For example, the presence of fibers in the melt makes it more shear thinning than the neat polymer because the same macroscopic shear induces a higher local shear rate in the polymer occupying the space between fibers than it does in the neat polymer. Fibers also generally increase the thermal conductivity of a composite in comparison with that of the unfilled polymer.

In injection molding, an initial fiber orientation state is set up at the gate. As the material flows into the mold, deformation of the suspension affects the orientation of the fibers. The melt front progresses within the mold by the 'fountain flow' mechanism illustrated in Fig. 1.8. This flow leaves fibers in their oriented state, in a solidified layer, located close to the cold wall. Generally, three regions develop within injection molded composites, each with a different microstructure: a core, a shear zone and a skin.[68]

This skin–core effect is described in detail in Chapter 4 by Papathanasiou. Briefly, the skin is formed during mold filling by the extensional flow that occurs at the free surface of the melt front, which leaves a highly stretched layer of solidified polymer near the mold wall. Beneath the skin, a high shear region is formed as the polymer solidifies during filling. The size of this shear zone is

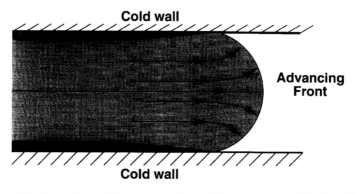

*1.8* Illustration of the fountain flow effect encountered during filling.

determined by the speed at which the solid/liquid interface moves toward the center of the part during mold filling. Once the mold is filled, flow ceases and the central core region retains the initial orientation that was created at the gate. The influence of flow-induced fiber orientation produced in the molding operation has been studied,[69–71] and the effect on fiber orientation of convergent and divergent flow near gates and of the shear flow within the mold, has been noted.

The alignment of short and long fibers induced during the injection molding process is quite different. In one study, long fibers were found to segregate significantly and an excess of fibers was observed, both in the core of the part and at the far end of the molding.[68] This phenomenon was explained by the lower volume fraction that resulted from less efficient packing of the long fibers in the more disordered zones. Other observations concerning the differences between the molding of short and long fibers include a higher in-plane orientation of the long-fiber compounds in the core than that of the short-fiber compound and similar fiber orientation and content for short- and long-fiber moldings in the shear zone, that is between the skin and the core.[68]

Several flow management techniques have been developed to orient the fibers in the mold and to reduce weld lines. One technique simply modifies the injection gate to favor a particular orientation in the core of the part. For example, the orientation produced by fan gates has been found to depend strongly on the injection rate, with the fibers more oriented transverse to the flow direction than when edge gates are used.[71] However, orientation can be controlled best by integrating converging and diverging sections in the design of the mold. The Scorim technique applies macroscopic shear to the material in the mold while it is solidifying.[72] In this process, pistons located near the injection gate apply the additional macroscopic shear needed to enhance fiber alignment in the molten polymer. The greater uniaxial fiber alignment that Scorim induces provides a significant increase in tensile modulus measured along the fiber alignment direction.[72]

### 1.2.1.2  Extrusion

Extrusion resembles injection molding in that a screw is used to plasticize polymer pellets. The main difference lies in the absence of a mold. In extrusion, a die is used instead of a mold to shape the polymer into a specific cross-section. Die swelling and melt fracture are significant issues in the processing of extruded structures, but when fibers are added other factors become important beyond fiber orientation. These factors include interfacial adhesion between the fibers and the melt and the increased melt viscosity. The addition of fibers enhances processability by increasing the melt consistency (especially for semicrystalline polymers), by increasing the thermal conductivity of the material and by reducing thermal expansion. Greater thermal conductivity reduces heating and cooling times, while a reduction in thermal expansion enhances the dimensional stability of the final product.

Short-fiber compounds are typically used when extrusion is employed. Extruded structures exhibit significant fiber orientation which results from the shear deformation near the surface and the elongation at the center of the extrudate. Die design can be modified to tailor the orientation of the fibers to specific applications. For example, hollow extrudates whose structures involve loading modes that require greater stiffness or strength in the circumferential direction can be made in this way.[9] Shear deformation that occurs in conventional composite pipe extrusion results in a high degree of orientation of the fibers near the surface. Furthermore, the common practice of using converging dies to produce these structures favors axial orientation in the central regions associated with elongational deformation. An expanding die geometry has been proposed to promote circumferential orientation in the core of the parts where elongational deformations are dominant, as is illustrated in Fig. 1.9. However, the expanding die has little effect in the shear dominated regions.[9,73-75] In order to orient the fibers in the hoop direction, extrusion through an expanding three-layer pipe die has been proposed.[76] This application employs unfilled thermoplastic in the inner and outer surface layers and a fiber-filled center.

Processing conditions also strongly affect the properties of parts made through the 'standard' extrusion process. In particular, fibers in the core region of a composite melt have been oriented perpendicular to the flow direction at high flow rates, but parallel to the flow at low flow rates.[77] This study reported that samples cut from different locations of a slit extrudate exhibited a distribution of mechanical properties. Specimens cut from the side of the extrudate showed higher yield strength and higher modulus (due to higher fiber content), longer fibers and low void content (<3%).

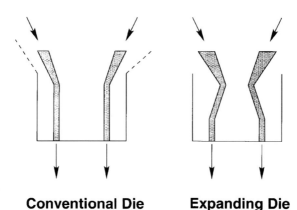

**Conventional Die    Expanding Die**

*1.9* Pipe extrusion dies, where the expanding die favors hoop orientation of the fibers.

The production of triextruded sheets using conventional dies has also been studied with the hope of obtaining the same results as those obtained with extruded pipes. These sheets provide a larger processing window than do neat materials. They also provide a better appearance and better 'stretchability' than the filled polymer.[9] Specialized die designs have been used to achieve a uniform axial flow rate across the die width and to reduce the anisotropy in the core material.[9]

The techniques discussed above for orienting fibers have primarily dealt with modifying mold and die designs so that polymer flow causes a desirable fiber orientation. The Scorex technique uses pistons located in the die to favor a specific orientation of the material while it is in the die. The molten material passes into the die chamber through a one way die and begins to solidify before exiting. A macroscopic shearing action is applied by the pistons to cause the alignment of the fibers in the extrudate. The pistons are arranged to orient the short fibers in extrudates, and proper sequencing of the pistons creates the desired degree of orientation. For example, when applied with a circular die, the Scorex method can produce a uniform structure in both the hoop and axial directions.[72]

### 1.2.1.3    Sheet forming

One attractive technique for producing finished thermoplastic composites is sheet stamping, a process based on the cold sheet stamping process widely used in the sheet metal forming industry.[78] Various attempts have been made to apply sheet stamping techniques to advanced thermoplastic composites. A significant difficulty in using thermoplastics for stamping is the limited draping capability of the fabrics. Forming of straight, continuous-fiber or woven-fiber composites results in wrinkling of the fibers and distortions. Only randomly oriented fibers have provided good formability, but without the advantages of the highly directional properties often desired in composite parts.[78,79] However, the formable sheets that consist of aligned, discontinuous fibers appear to have been used successfully.[78,79] These sheets permit extensive deformations while maintaining properties similar to those of continuous fibers, provided that the fibers are longer than the critical length. These sheets are made of aligned discontinuous fibers, commonly aramid, carbon or glass, with a fiber length ranging from 2–15 cm. During forming, drawing of the sheet is achieved by the sliding of the short fibers past one another. For a uniform drawing, axial draw as high as 50% is achievable, and mechanical properties are maintained.

The key to successful sheet forming (sometimes called stretch forming) is precise control over the fiber placement.[80] During the forming process, the fibers must not be compressed, but instead must be kept in tension to avoid the wrinkling caused by compression. If a fabric is used, interlocking of the fiber bundles must also be maintained to keep the fibers in their relative orientations.

Tensile deformation is therefore the preferred mode for the ordered short-fiber sheets during forming. Such deformation allows better control of the fiber placement and reduces fiber wrinkling. However, since the sheet must be clamped at the edges, the polymer at the interface of the clamp and the sheet acts as a lubricant. The clamp stress is thus determined by the shear modulus of the matrix. If the load needed to draw a sheet is higher than the shear yield stress of the composite, the sheet will slide from under the clamps. The forming of a complex three-dimensional part without loss of mechanical properties has also been achieved.[78] Precise placement of the fibers is possible and properties similar to those exhibited by parts that contain continuous fibers can be achieved.

## 1.2.2  Thermosets

Several classes of thermosetting resins are used to manufacture composite materials. Among these are unsaturated polyesters, vinyl esters, epoxides, phenolics and polyimides. A thermoset polymer is typically formed from oligomers which are cured to form a three-dimensional network of chains joined by covalent bonds. Catalysts and heat are generally used to induce curing during forming. Polyesters and vinylesters constitute the most popular matrix materials for short-fiber composites, mainly due to their low cost and rapid reaction rates.

Several processes using flows are used with thermosets. Reaction injection molding is probably the most popular technique. In reaction injection molding (RIM), polymeric parts are produced directly from low viscosity reactants which are injected into a mold. The part shape is determined by the shape of the mold and the fast polymerization of the polymer. Particulates or fiber fillers are often added to the polymer in order to reduce thermal expansion and enhance mechanical properties. This is the basis of the reinforced reaction injection molding process (RRIM). Disadvantages similar to those encountered with injection molding can arise (e.g. unintended orientation of the fibers, large increases in viscosity and heavy wear of the processing machinery). RIM is similar to injection molding from the fluid mechanical point of view; however, thermosetting and thermoplastic polymers exhibit different behavior during the solidification/curing phase. The same technique used with injection molding can therefore be used to manage fiber orientation during processing.

## 1.2.3  Aligned-fiber mats

Forming methods, such as the sheet forming process described above, require quality aligned short-fiber mats. Several processes have been developed over the years to produce such mats. Behind most of the techniques presented in the literature, hydrodynamic principles are used to give the fibers the desired orientation. One such process, the PERME filtration process, achieves fiber alignment by submitting a suspension of fibers to a shear flow field.[18,21] A fiber mat is formed by extruding the suspension through a slit in a V-shaped trough which is reciprocated over a filter bed. The suspending fluid is then removed rapidly by strong suction in a process which also maintains the orientation of the fibers. Some time is needed for filtration between each deposition. The quantity of fibers deposited simply depends on the number of repetitions of the process. But when thick mats are required, removing the carrier fluid rapidly can be difficult and can cause fiber misalignment.[18]

The centrifuge alignment process has been developed to overcome some of the problems associated with the filtration process described above.[18] Alignment in this process is achieved by ejecting a suspension of fibers through a tapered nozzle. As a result of the velocity gradient in the jet, the fibers are aligned in the direction of flow. The mat is formed by reciprocating the jet over a permeable cylinder. The cylinder rotates with an angular velocity that is sufficient to force the suspension through the surface and a layer of aligned fibers is thereby deposited on the surface of the cylinder. In order to keep the fibers aligned, the angular velocity must be increased as the thickness of the mat increases. The centrifugal alignment technique has been used to produce sheet material of finely divided and intimately mixed combinations of more than one type of fiber. In the study reporting this method, the degree of fiber alignment was greater than that achieved by the filtration process and the draping qualities of the original material were retained.[18]

The MBB-VTF method is a similar process which uses a rotating vacuum drum filter to align fibers in suspension. The flow of the suspension in a channel and the transfer from the channel to a drum causes the fibers to align as they are deposited on the drum in the direction of rotation of the drum. Suction is applied to remove the liquid and the fibers are kept aligned on the drum surface which is covered with a filter cloth.[16]

Finally, a recent technique involving electric fields has been proposed for the formation of oriented-fiber mats.[14] Briefly, the process involves three operations. The first employs an electric field to orient the fibers. Thereafter, the fibers are impregnated with dry powder. Finally, compression molding is used to form a flexible aligned-fiber mat into a composite.

## 1.2.4 Blends of liquid crystal polymers and thermoplastics

An alternative, low cost method for producing composites involves blending a liquid crystal polymer and a thermoplastic to create an 'in situ' composite. These polymer blends have been called self-reinforcing because the liquid crystal domains are elongated into fibrils that are oriented during processing, and once the blend solidifies, microdomains form. These microdomains act as the fiber or filler found in ordinary composites.[81] The processability of the thermoplastic matrix is also improved by adding small amounts of thermotropic liquid crystal polymers (TLCP) which have a relatively low viscosity.

The predominant processing method used to produce *in situ* composites is injection molding.[82] These composites can either be preblended in an extruder prior to producing the *in situ* composite, or the extrusion system can be fitted with a static mixer to provide additional mixing. Sheet extrusion has also been used, but without much success.[82] Novel methods, such as a dual extruder mixing technique,[83] have been proposed for producing *in situ* composites. Modifications of the methods noted previously, in the discussion on injection molding and extrusion in Sections 1.2.1.1 and 1.2.1.2, can also be utilized for producing *in situ* composites. These techniques include Scorim, Scorex and the use of converging/diverging dies in extrusion profiles.

*In situ* fiber formation has advantages that make liquid crystal polymers (LCP) an attractive alternative to short-fiber composites. First, the same equipment used for conventional polymer processing can be used to encourage the formation of *in situ* composites. Also, much less mechanical wear occurs with LCPs than occurs in short-fiber composite manufacturing. Products reinforced with liquid crystal polymers have shown mechanical properties similar to those of conventional composites.[81] Despite these advantages, LCPs do have a few disadvantages. First, the reinforcing TLCP must be chosen carefully for a given thermoplastic. Second, mixing and post-processing must also be carefully controlled in order to realize the desired reinforcement. Finally, poor interfacial adhesion is often observed because of the incompatibility of the polymers, even when compatibilizing agents are added.

## 1.3    Flow processes for producing aligned-fiber metal-matrix composites

While flow processing has been employed most extensively with polymer composites, such processes have also been used to align short fibers in metal-matrix composites. However, with metallic composites the concept of flow alignment must be expanded to include the alignment that occurs during deformation processing in the solid state. Fiber alignment during processing of a molten metal is generally not achieved because liquid metals have such a low viscosity that smooth laminar flows are much more difficult to exploit to align the short fibers than they are in the processing of polymeric composites. Instead, any eddies that form during processing or casting act to randomize fiber orientations in the molten metal. Nevertheless, the advantages of fiber alignment apply to metal-matrix composites, principally that mechanical properties can be improved. In particular, metal-matrix composites provide the potential for a high stiffness, lightweight material that can withstand high temperatures.[44] While processes for producing aligned-fiber metal-matrix composites have not been as extensively developed as those for polymers, uses for these materials have been envisioned, particularly in aerospace and turbine engine applications.

Aligned-fiber metallic composites can be produced by deformation processing in the solid state. In these processes, solid metals are deformed under large stresses. Examples of such processes include extrusion, rolling and forging. In extrusion, solid or semi-solid metal is forced through a die, generally at elevated temperatures. In this way, metal wire and pipe can be formed, for example, and a high degree of fiber alignment can be achieved in the extrusion direction in a variety of materials. The high stresses and temperatures present during the injection process allow a solid material to be formed. Forging involves shaping hot malleable metal through the forces applied by presses or hammers. In rolling, cold metal is repeatedly squeezed and/or bent between opposing rolls in order to form thin sheet or complex shapes from strip metal. Of these processes, extrusion appears to be the most commonly employed for making aligned-fiber composites.

Aligned-fiber metal-matrix composites can also be prepared by powder metallurgical techniques. For example, in powder metal injection molding, metal powder is mixed with a small amount of binder that allows the mixture to be injected into a mold in a manner somewhat similar to injection molding of plastics. The resulting part is heated to remove the binder, then sintered to allow empty interstitial spaces to be eliminated and a fully dense part to be formed. The resulting flow into the mold can cause the fibers contained in the mixture to align themselves. For example, in NiAl composites 'nearly perfect specimens were produced, indicating that this is a viable technique for the fabrication of aligned fibrous intermetallic matrix composites'.[84] Microphotographs reported as part of this study confirm the effectiveness of the alignment technique.

A combustion synthesis process has been described for making metallic composites. This process has been used to manufacture continuous fiber $Ni_3Al$ composites, and it is claimed that the process could be used to manufacture short-fiber composites as well.[85]

Aligning the reinforcing fibers in a metal-matrix composite increases tensile strength, stiffness and hardness and decreases brittleness. For example, extruded aluminum alloy composites reinforced with alumina fibers show increased strength, with higher extrusion temperatures resulting in higher strengths.[86] At higher temperatures, greater surface tearing is also reported, however steps can be taken to prevent this. Extruded SiC fiber-reinforced Al composites have been shown to exhibit fiber orientation and improved strength in the extrusion direction.[87] Mechanical properties and microstructure of short-fiber-reinforced Al composites manufactured by powder metallurgical methods and by extrusion have been investigated, with an emphasis on fiber alignment and the effect of processing on fiber length.[88] In a study of alumina fiber-reinforced $Al_3Ta$, the aligned-fiber material showed improved strength and was found to be brittle at ambient temperatures but ductile at higher temperatures.[41] The effect of fiber alignment on tensile and compression strength and modulus of fiber-reinforced Mg alloy composites formed by liquid metal preform infiltration has also been explored.[89] In this study, fiber alignment was shown to have a much greater effect on strength than on stiffness. Enhanced tensile strength and stiffness have also been reported in Mg–Li composites containing aligned carbon, alumina and silicon carbide whiskers.[57] Fiber alignment also increases the hardness of NiAl composites.[84] Superior wear properties have been reported in extruded Zn–Al alloys reinforced with alumina fibers.[27] Clearly, fiber alignment can enhance a variety of mechanical properties in metal-matrix composites.

A number of issues arise in the processing and manufacturing of metal-matrix composites. Among the most important are fiber breakage and fiber degradation. In deformation processing, fiber breakage occurs because of the high stresses acting on the fiber. When breakage reduces fiber length to less than the critical length, the potential of the reinforcement to enhance mechanical properties is not fully achieved. At least one method has been patented that claims to reduce fiber breakage in metallic aligned-fiber composites.[31] Quantitative measurements of fiber breakage have been performed which show that extrusion at higher temperatures can reduce fiber breakage.[86] On the other hand, in some applications it has been reported that, 'extrusion even at high temperature is very damaging for a fibre reinforced composite'.[51]

Beyond fiber breakage, the high temperatures at which metallic composites are processed lead to significant chemical degradation in many kinds of fibers, limiting the choice of fibers. In one study of a Mg–Li composite, the SiC whiskers were chemically stable, while other commonly available fiber reinforcements suffered significant chemical attack.[57] Fiber degradation has also been reported at the temperatures required for extrusion.[89] The presence of fibers in a metal matrix composite has even been reported to affect material microstructure.[51] Fiber prices and difficulties in fabrication technology have also inhibited the commercial development of metal-matrix composites.[44] Finally, effects that are known to exist in metal alloys of all kinds can have an impact on metal alloy-matrix composites. Because trace amounts of impurities such as carbon and oxygen can significantly affect material properties, material purity can be an issue.[41] Processing issues associated with producing aligned-fiber metal-matrix composites by reactive sintering, powder injection molding and hot isostatic compaction have been explored in more detail.[90]

## 1.4    Flow processes for producing aligned-fiber ceramic-matrix composites

Aligned short-fiber ceramic composites are reported less extensively in the literature than are polymeric or metallic composites. As with metal-matrix composites, the precursors of ceramic composites (often a dispersion of powder in water) can have very low viscosities. Thus, viscous laminar flows cannot generally be exploited to align fibers in an orderly way, as can be done in polymers. Nevertheless, aligned-fiber ceramic-matrix composites have been investigated and reported. These studies have been motivated by the well known need to reinforce ceramic materials, particularly with fibers, to reduce brittleness. Ceramics offer a unique combination of light weight, high stiffness and resistance to wear and chemical attack, particularly at high temperatures. These properties make them attractive candidates for aerospace and turbine engine applications. However, their extreme brittleness has so far prevented them from being employed in many applications, particularly where sudden, catastrophic failure cannot be tolerated. Fiber reinforcement holds the potential to reduce this problem greatly and much effort has gone into research in this direction.

The processes that have been reported for aligning fibers in ceramic composites reflect the special requirements for aligning short fibers in ceramic precursors by flow processes. For example, a tape casting and hot pressing process has been reported in which the flow between a blade and a carrier film tends to align whiskers in the casting direction.[29] Subsequent hot pressing can be applied to induce the whiskers to lie in the plane perpendicular to the pressing axis, but with a random orientation within this plane.[30] Hot pressing alone also has the same effect.

As with the other composites, the mechanical properties of ceramic composites are improved by fiber alignment. Since the most critical improvement needed in ceramics is in brittleness, most work has been focused in this direction. For example, 10–20% aligned, SiC whisker-reinforced alumina composites have shown greatly improved fracture toughness and reduced coefficient of thermal expansion.[29] The effect of interfacial shear strength on toughening mechanisms and crack propagation behavior of fiber-reinforced $Si_3N_4$ composites fabricated by hot pressing have also been investigated.[39] This study showed some degree of planar and axial alignment during hot pressing, as well as great sensitivity of fracture behavior to processing conditions and fiber–matrix interfacial properties. The fabrication, microstructure and mechanical and thermal properties of a series of ceramic composites reinforced with chopped carbon fibers and chopped zirconia fibers have been investigated.[30] In this case, aligning the fibers increased fractural strengths and work of fracture.

A complication that arises in ceramic composites that appears to have less impact on polymer- and metal-matrix composites is the difference in the coefficient of thermal expansion between the matrix and the fiber. Ceramics generally have a much lower coefficient of thermal expansion than do the reinforcing fibers. During the temperature cycle that occurs during sintering, internal stresses can develop that lead to matrix cracking[30] and reduced strength. Processing routes and fiber orientation have also been reported to have an impact on interfacial properties, which, in turn, influence mechanical properties.[39]

## 1.5    Future prospects

As this chapter shows, aligned-fiber composites offer numerous advantages. Tensile strength and stiffness can be improved, hardness can be increased and wear can be reduced, all by aligning the short fibers of a composite. Numerous processes are available, some of which have been patented, to induce fiber alignment in composites. Some of these processes are novel, while others are simply modifications of processes that have existed for decades. Yet, despite the advantages of these composites and the variety of processing options available for making them, questions remain: Why have applications of aligned-fiber composites not grown more rapidly? Why are they not present in more commonly used items? Can a greater presence in commercial markets be established in the near future? This chapter closes by briefly considering these questions.

In general, a number of factors can account for the slow growth of aligned-fiber composites in commercial markets. The processing issues noted in Section 1.1.8 certainly contribute to this slow growth. The high cost of manufacturing and raw materials also act to reduce their widespread use. Widespread use of composites must also overcome hesitation from designers in using these materials. Beyond this, technological innovations meet resistance merely because they are new and because they often must displace existing, familiar technology. Thus, technological innovations are not always adopted immediately.

More specifically, one reason for the slow growth of aligned-fiber composites is that carbon fiber is relatively expensive. Carbon fiber is a popular reinforcement, especially for polymer-matrix composites, because it offers excellent reinforcing properties. The cost of carbon fiber is not likely to come down until the demand increases significantly. Yet, the growth in applications of aligned-fiber composites that would spur this demand will not occur until the cost of carbon fiber is lower. Barring the intervention of other forces, this catch-22 would appear to block aligned-carbon fiber composites from rapidly moving into low cost applications for the foreseeable future.

Aligned-fiber composites have not significantly penetrated structural applications, partly because of the (necessary) conservatism of the structural design discipline. Where airplanes, bridges and automobiles are involved, safety and reliability are paramount concerns because lives are at stake. Replacing materials wholesale without knowledge of their reliability in those applications would be unwise because materials can fail in complex and unanticipated ways. Interaction with other materials, unexpected environments, slight imperfections in processing or raw materials, failure of other components and ageing can all lead to material failure. Thus, new materials must offer compelling advantages over materials whose reliability has been proven over decades before they will displace existing materials. Where the advantages of aligned-fiber composites are not compelling, growth in these applications has been (and will be) slow.

Beyond these factors, the expectation that advanced technologies will be immediately adopted simply because they are more advanced is not realistic. This general fact is well recognized among those who study the spread of technological advances. One influential author notes that 'getting a new idea adopted, even when it has obvious advantages, is often very difficult. Many innovations require a lengthy period, often of many years, from the time they become available to the time they are widely adopted'.[92] As technologists, it is easy to believe that advanced technologies sell themselves; however, this often is not the case. For example, in securing venture capital to commercialize a new technology, the business plan that is developed to commercialize that technology is generally more important than the technology itself.

Numerous examples are available of technological advances that were not immediately adopted. Even simple technologies such as eating citrus fruit to combat scurvy was not immediately adopted when first discovered.[92] In fact, the effectiveness of lemon juice to prevent scurvy was demonstrated by the British Navy in 1601, then demonstrated by them again 150 years later before they adopted it as standard practice 50 years after that. Even then, another 70 years elapsed before it was adopted by the British merchant marines. Certainly, it can be argued that we are better able today to evaluate and rapidly adopt new technologies than we were centuries ago. Today, such a significant discovery would not languish for two centuries before gaining widespread acceptance. Nevertheless, many human and non-technical factors that slow the implementation of new technologies still exist. One such factor is that the adoption of new technologies involves social change (at least among those who implement the technology), as well as technological change.[92] Such social change is particularly important where an advance must displace an existing technology. For example, the common QWERTY keyboard layout was specifically designed to impede touch typists, yet the DVORAK layout, which was designed to speed touch typing, has not been widely adopted because of the significant impact it would have on computer users.[92] Likewise, aligned-fiber composites must, for the most part, displace existing materials in current applications.

Predicting a coming revolutionary expansion in the use of aligned-fiber composites would be satisfying, but such a revolution does not seem likely. Such revolutions have been anticipated in composite materials before without coming to fruition. The ingredients for a revolution in aligned-fiber composites do not seem to be in place at this time. In particular, previous revolutionary technologies have offered entirely new capabilities or order of magnitude improvements in speed, cost or other important factors. For example, the automobile brought a new age of personal mobility. Digital computers solve problems that once were intractable. Solid state electronic components make today's digital computing age possible, and they make electronic devices significantly smaller and more reliable. Personal computers allow documents and presentations to be continuously refined and improved with little effort (although whether this constitutes an advance in productivity is perhaps open to debate). Cellular telephones provide communication at any time on a scale which was not previously possible. The World Wide Web provides unprecedented access to information, allowing the smallest businesses and individuals to establish a global presence, even if only to present the most trivial information. These technologies have all experienced rapid growth because they alone solve important problems. While aligned-fiber composites do offer significant advantages in many instances, examples are limited in which aligned-fiber composites provide a revolutionary advance in capability. The most active areas of composites research are in such applications and aligned-fiber composites may provide the improvements in material properties that are needed for successful application.

Rather than a revolution in aligned-fiber composites, an evolutionary growth seems more likely, a growth in which these materials find specific applications in which their unique advantages are put to the most good. The aerospace industry offers applications that are well suited to these advantages, but this has been true for years. Military applications are another arena in which high strength and light weight are important and in which cost restrictions are not as great as in commercial markets. Gradual growth spurred by such applications is one available route toward more widespread use. However, the full scope of possible applications cannot be completely predicted. The hip implant noted earlier is an example of this.[42] Here, the ability to match material properties between the implant and living bone provides an important improvement in the likelihood of successful use.

The automotive industry is another potential market for aligned-fiber composites. In order to achieve the mass reduction needed to increase fuel efficiency significantly, extensive use of lightweight composites is necessary. Aligned-fiber composites offer one promising route to obtaining these weight reductions. The weight savings that materials such as aligned-fiber composites offer can contribute significantly to the greatly increased fuel efficiency envisioned for the automobiles of the future.[61] Such cooperative endeavors between industry and government offer additional impetus to overcome immediate economic factors that impede the long term development of aligned-fiber composites. (Of course, such initiatives are also subject to the vagaries of national elections and changing political winds and as such, cannot be counted on.) Before short-fiber composites can be used on such a large scale, numerous other problems must be solved. Most notably, processes must be developed that can produce more than 250 000 parts per year at a rate of one part every 4 min.[61] Such high volume production at low cost is a necessity for automotive applications. Also, fiber production must grow from the current level of 20 Mlb (9 Mkg) to 200 Mlb (90 Mkg) annually, and the cost must come down significantly (to the $3–5/lb, or $6–11/kg, range) in order to be economically feasible. Beyond this, other issues must also be addressed. These include crash worthiness, the health effects of small diameter fibers, flammability, repair of composite parts, recycling, non-destructive testing, resistance to sulfur, mold design, total life cycle costs and durability.[61]

The ingredients are in place for an evolutionary spread of aligned short-fiber composites into wider use. Numerous processes are available for their production and the materials have already been used in a number of applications. Beyond this, the elements that have been identified as contributing to the adoption of new technologies in general also are in place. These elements are a relative advantage, compatibility, complexity, trializability (the ability to try the technology before fully adopting it) and observability.[92] Briefly, aligned-fiber composites offer in most cases a moderate advantage, and in some cases, a significant advantage over existing materials. Compatibility with other materials in use today is not generally a problem. While some complexity is involved in processing, the overall level of complexity in adopting aligned-fiber composites is not great. Finally, these materials can be readily tried and their advantages easily observed. Thus, from the broad perspective of how technologies become widely adopted, aligned-fiber composites are well positioned to do so. Beyond this, however, aligned-fiber composites have an important ingredient that aids in their adoption, that is, a small group of workers to champion them. This ingredient is significant because the spread of most technologies is spurred by such a group of people who believe in and work with a technology without regard to whether or not others have already adopted it.[92]

In the end, predicting what the future holds for aligned-fiber composites, beyond the likely evolutionary growth, is inherently hazardous. As the pace of technological change quickens, seeing the technology horizon becomes even more difficult. Nevertheless, a core level of interest in aligned-fiber composites seems to have remained over the years and that interest does not seem to be waning. The obstacles that remain to be overcome are relatively clear and they are as much economic as technological. As materials and processing technologies advance, these obstacles will diminish and disappear and as they do a broad base of knowledge for aligned-fiber composites will be available to spur more widespread adoption.

# References

1. M. Schwartz. *Composite Materials Handbook*. McGraw Hill, second edition, 1992.
2. R. Meredith. Fibrous polymers. *Contemp. Phys.*, **11**:43, 1970.
3. B.Z. Jang. *Advanced Polymer Composites*. ASM International, 1994.
4. J. Delmonte. *Plastics in Engineering*. Penton, second edition, 1942.
5. E.P. Scala. A brief history of composites in the U.S. – the dream and the success. *J. Materials*, **48**:45, 1996.
6. J.E. Ashton, J.C. Halpin, and P.H. Petit. *Primer on Composite Materials: Analysis*. Technomic, 1969.
7. R.L. McCullough. *Concepts of Fiber-Resin Composites*. Marcel Dekker, 1971.

8. M.J. Bozarth, J.W. Gillespie and R.L. McCullough. Fiber orientation and its effect upon thermoelastic properties of short carbon fiber reinforced polyetheretherketone (PEEK). In *ANTEC '86*, held Boston, USA, page 568. Society of Plastics Engineers, Connecticut, USA, 1986.

9. J.M. Charrier, R.V. Ciplijauskas, S.R. Doshi and F.A. Hamel. Fabrication and thermoforming of coextruded short fiber-reinforced thermoplastic sheets. In *Antec '86*, page 939. Society for Plastics Engineering, 1986.

10. S.R. Doshi, J.M. Charrier, J.M. Dealy and F.A. Hamel. Coextrusion of short fiber-reinforced plastic pipes. In *Antec '86*, page 944. Society for Plastics Engineering, 1986.

11. W.C. Jackson, S.G. Advani and C.L. Tucker. Predicting the orientation of short fibers in thin compression moldings. *J. Compos. Mater.*, **20**:539, 1986.

12. L. Kacir, M. Narkis and O. Ishai. Oriented short glass-fiber composites. I. Preparation and statistical analysis of aligned fiber mats. *Polym. Eng. Sci.*, **15**:525, 1975.

13. T. Itoh, S. Masuda and F. Gomi. Electrostatic orientation of ceramic short fibers in liquid. *J. Electrostatics*, **32**:71, 1994.

14. M.N. Vyakarnam and L.T. Drzal. A new process for aligning fibers in composites. Plastics Engineering, **53**:1, pages 35–37, Jan 1997.

15. P.C. Sturman and R.L. McCullough. Static field electrical conductivity of dilute fiber suspensions. *J. Appl. Phys.*, **72**:2883, 1992.

16. H. Richter. Single fibre and hybrid composites with aligned discontinuous fibres in polymer matrix. In *Advances in composite materials: Proceedings of the Third International Conference on Composite Materials*, held in Paris, volume 1, page 387, 1980. Pergamon Press, Oxford.

17. D.J. Hannat and N. Spring. Steel-fiber-reinforced mortar – technique for producing composites with uniaxial fiber alignment. *Concr. Res.*, **26**:47, 1974.

18. H. Edwards and N.P. Evans. A method for the production of high quality aligned short fibre mats and their composites. In *Advances in composite materials: Proceedings of the Third International Conference on Composite Materials*, page 1620. Pergamon Press, Oxford, 1980.

19. L.E. Dingle. Aligned, discontinuous carbon-fibre composites. In *International Conference on Carbon Fibres; their place in modern technology*, London, page 78. Plastics Institute, London, 1974.

20. P.K. Mallick. *Fiber-reinforced Composites*. Marcel Dekker, second edition, 1993.

21. G.E.G. Bagg, M.E.N. Evans and A.W.H. Pryde. The glycerine process for the alignment of fibres and whiskers. *Composites*, **1**:97, 1969.

22. G.A. Cooper. The structure and mechanical properties of composite materials. *Rev. Phys. Tech.*, **2**:49, 1971.

23. R.E. Lavengood. Strength of short-fiber reinforced composites. *Polym. Eng. Sci.*, **12**:48, 1972.

24. L. Kacir, M. Narkis and O. Ishai. Aligned short glass fibre/epoxy composites. *Composites*, Vol 10, page 89, April 1978.

25. J.C. Malzahn and J.M. Schultz. Tension–tension and compression–compression fatigue behavior of an injection-molded short-glass-fiber/poly(ethylene terephthalate) composite. *Compos. Sci. Tech.*, **27**:253, 1986.

26. D.C. Guell and A.L. Graham. Improved mechanical properties in hydrodynamically aligned, short-fiber composite materials. *J. Compos. Mater.* **30**:2, 1996.

27. J.A. Cornie, R. Guerriero, L. Meregalli and I. Tangerini. Microstructures and properties of zinc-alloy matrix composite materials. In *International Symposium on Advances in Cast Reinforced Metal Composites*, page 155, 1988.

28. F. Truckenmuller and H.G. Fritz. Injection molding of long fiber-reinforced thermoplastics: a comparison of extruded and pultruded materials with direct addition of roving strands. *Polym. Eng. Sci.*, **31**:1316, 1991.

29. E.D. Kragness, M.F. Amateau and G.L. Messing. Processing and characterization of laminated SiC whisker reinforced $Al_2O_3$. *J. Compos. Mater.*, **25**:416, 1991.

30. R.A.J. Sambell, D.H. Bowen and D.C. Phillips. Carbon fibre composites with ceramic and glass matrices. *J. Mater. Sci.*, **7**:663, 1972.

31. J. Dinwoodie, M.D. Taylor and M.H. Stacey. Fiber-reinforced metal matrix composites. US Patent, Mar. 26 1991. No. 5,002,836.

32. G.B. Jeffery. The motion of ellipsoidal particles immersed in a viscous liquid. *Proc. R. Soc. London Ser A*, **102**:161, 1972.

33. G.K. Batchelor. Slender-body theory for particles of arbitrary cross-section in stokes flow. *J. Fluid Mech.*, **44**:419, 1970.

34. G.K. Batchelor. The stress generated in a nondilute suspension of elongated particles by pure straining motion. *J. Fluid Mech.*, **46**:813, 1971.

35. D.M. Bigg. Mechanical property enhancement of semicrystalline polymers – a review. *Polym. Eng. Sci.*, **28**:830, 1988.

36. J. Karger-Kocsis and K. Friedrich. Microstructure and fracture toughness of short fibre rein-forced injection-moulded peek composites. *Plastics Rubber Proc. Appl.*, **8**:91, 1987.

37. A.L. Graham, L. Mondy and D.C. Guell. Anisotropic fiber alignment in composite structures. US Patent, Nov. 16 1993. No. 5,262,106.

38. P.L. Berneburg and R.W. Rice. Ceramic composite valve for internal combustion engines and the like. US Patent, May 29 1990. No. 4,928,645.

39. J.M. Yang, T.J. Chen, S.M. Jeng, R.B. Thayer and J.F. LeCoustaouec. Processing and mechanical behavior of SiC fiber-reinforced $Si_3N_4$ composites. *J. Mater. Res.*, **6**:1926, 1991.

40. A.P. Penton and F.L. Freeman. Processing of graphite epoxy isogrid structure. In *Bicentenial of Materials Progress: Proceedings of the 21st National SAMPE Symposium and Exhibition*, page 934, 1976. Held 6–8 April 1976, Los Angeles. SAMPE, California, USA.

41. D.L. Anton. High temperature intermetallic composites. *Mater. Res. Soc. Symp. Proc.*, **120**:57, 1988.

42. S.K. Gupte and S.G. Advani. Process modeling for manufacture of orthopaedic implants from short fiber composites. *Polym. Compos.*, **15**:7, 1994.

43. M. Jaffe. Applications of liquid crystal polymers. *J. Stat. Phys.*, **62**:985, 1991.

44. M.H. Stacey. Production and characterisation of fibres for metal matrix composites. *Mater. Sci. Tech.*, **4**:227, 1988.

45. A. Kelly and N.H. MacMillan. *Strong Solids*. Clarendon Press, third edition, 1987.

46. A. Kelly and W.R. Tyson. Tensile properties of fibre-reinforced metals: copper/tungsten and copper/molybdenum. *J. Mech. Phys. Solids*, **13**:329, 1965.

47. B. Fisa. Injection molding of thermoplastic composites. In *Composite Materials Technology: Process and Properties*, eds P.K. Mallick and S. Newmann, page 267. Hanser, Munich, 1990.

48. L. Dilandro, A.T. Dibendetto and J.H. Groeger. The effect of fiber-matrix stress transfer on the strength of fiber reinforced composite materials. *Polym. Compos.*, **9**:209, 1988.

49. M.G. Bader and W.H. Bowyer. An improved method of production for high strength fiber-reinforced composites. *Composites*, **4**:150, 1973.

50. F. Chen. Short fiber reinforced thermosets. In *Handbook of Polymer–Fibre Composites*, ed. F.R. Jones, Polymer Science and Technology Series, page 278, Harlow, England, 1994. Longman Scientific & Technical.

51. L. Nguyen Thanh and M. Suéry. Microstructure and compression behaviour in the semisolid state of short-fibre-reinforced A256 aluminium alloys. *Mater. Sci. Eng. A*, **196**:33, 1995.

52. R. Blanc, J.F. Agassant and M. Vincent. Injection molding of unsaturated polyester compounds. *Polym. Eng. Sci.*, **32**:1440, 1992.

53. R.F. Eduljee, J.W. Gillespie and R.B. Pipes. Design methodology for the molding of short fiber thermoset composites. In *Proceedings of the Second American Society of Composites*, Sept 23–25, 1987, Newark, Delaware, ASM, page 199. Technomic, Pennsylvania, 1987.

54. R. Tolbert. Controlled orientation of discontinuous fibers in composites. *US Government Report* AD 879156, 1970.

55. M.R. Piggott. The interface – an overview. In *36th International SAMPE Symposium*, page 1773, 1991. Held in San Diego, California, USA, Apr 15–18 1991, SAMPE, California, USA.

56. J. Karger-Kocis. *Polypropylene: Structure, Blends and Composites*, volume 1. London, Chapman & Hall, 1995.

57. J.F. Mason, C.M. Warwick, P.J. Smith, J.A. Charles and T.W. Clyne. Magnesium–lithium alloys in metal matrix composites – A preliminary report. *J. Mater. Sci.*, **24**:3934, 1989.

58. A. Karnis and S.G. Mason. Particle motions in sheared suspensions. XIX. Viscoelastic media. *Trans. Soc. Rheol.*, **10**:571, 1966.

59. G. Segre and A. Silberberg. Radial particle displacements in poiseuille flow of suspensions. *Nature (London)*, **189**:209, 1961.

60. N. Dontula, N.S. Ramesh, G.A. Campbell, J.D. Small and A.L. Fricke. An experimental study of polymer-filler redistribution in injection molded parts. *J. Reinforced Plastics Compos.*, **13**:98, 1994.

61. Report of workshop on composite vehicle structures. PNGV report of work sponsored by the US Dept. of Energy and USCAR, Sept. 28 1995.

62. G. Cuff. Long fibre reinforced thermoplastics – extending the technology. In *Proceedings of the 16th Reinforced Plastics Congress*, page 15, Nov. 1988. Held in Blackpool, England. The Federation, England.

63. H.D. Keith and F.J. Padden. A phenomenological theory of spherulitic crystallization. *J. Appl. Phys.*, **34**:2409, 1963.

64. D.V. Rosato and D.V. Rosato. *Injection Molding Handbook*. London, Chapman & Hall, 1995.

65. M.J. Folkes. Injection molding – thermoplastics. In *Handbook of Polymer–Fibre Composites*, ed. F.R. Jones, Polymer Science and Technology series, page 165, Harlow, England, 1994. Longman Scientific & Technical.

66. R.S. Bailey. Conventional thermoplastics. In *Handbook of Polymer–Fibre Composites*, ed. F.R. Jones, Polymer Science and Technology series, page 74, Harlow, England, 1994. Longman Scientific & Technical.

67. D.A. Cianelli, J.E. Travis and R.S. Bailey. Processing of new long fiber reinforced thermoplastics. In *43rd Annual SPI Technical Conference*, Cincinnati, Ohio. Society of the Plastics Industry, New York, 1988.

68. S. Toll and P.O. Andersson. Microstructure of long- and short-fiber reinforced injection molded polyamide. *Polym. Compos.*, **14**:116, 1993.

69. M.J. Owen and K. Whybrew. Fibre orientation and mechanical properties in polyester dough moulding compound DMC. *Plastics Rubber*, 1, 6, 231–8, 1976.

70. M.J. Owen, D.H. Thomas and M.S. Found. Flow, fibre orientation, and mechanical property relationships in polyester DMC. In *Proceedings of the 33rd SPI Conference*, held Feb 7–10 1978, Washington DC. SPI, New York.

71. L.A. Goettler. Controlling flow orientation in molding of short-fiber compounds. *Modern Plastics*, **47**:140, 1970.

72. P.S. Allan and M.J. Bevis. Injection moulding – fibre management by shear controlled orientation. In *Handbook of Polymer–Fibre Composites*, ed. F.R. Jones, Polymer Science and Technology Series, page 171, Harlow, England, 1994. Longman Scientific & Technical.

73. L.A. Goettler and A.J. Lambright. Process for controlling orientation of discontinuous fiber in a fiber-reinforced product. US Patent, Nov. 1 1977. 4,056,591.

74. L.A. Goettler and A.J. Lambright. Hose reinforced with discontinuous fibers oriented in the radial direction formed by extrusion. US Patent, Nov. 8 1977. 4,057,610.

75. L.A. Goettler, R.I. Leib and A.J. Lambright. Short fiber reinforced hose – a new concept in production and performance. *Rubber Chem. Tech.*, **52**:838, 1979.

76. S.R. Doshi, J.M. Dealy and J.M. Charrier. Flow induced fiber orientation in an expanding channel tubing die. *Polym. Eng. Sci.*, **26**:468, 1986.

77. W.-Y. Chiu, H.-C. Lu and C.-L. Chang. Relationship between extrusion condition and mechanical properties of FRPP. *J. Appl. Polym. Sci.*, **43**:1335, 1991.

78. R.K. Okine, D.H. Edison and N.K. Little. Properties and formability of an aligned discontinuous fiber thermoplastic composite sheet. *J. Reinforced Plastics Compos.*, **8**:70, 1990.

79. R.K. Okine. Analysis of forming parts from advanced thermoplastic composite sheet materials. *SAMPE J.*, **25**:9, 1989.

80. S.J. Medwin. Long discontinuous ordered fiber structural parts. In *Proceedings of the 34th SAMPE Symposium*, held in Reno, Nevada, volume 34, page 171. Society for the Advancement of Material and Process Engineering, California, USA, May 1989.

81. X. Jin and W. Li. Correlation of mechanical properties with morphology, rheology, and processing parameters for thermotropic liquid crystalline polymer-containing blends. *J. Macro. Sci.-Rev. Macromol. Chem. Phys.*, **C35**:1, 1995.

82. A.A. Handlos and D.G. Baird. Processing and associated properties of *in situ* composites based on thermotropic liquid crystalline polymers and thermoplastics. *Rev. Macromol. Chem. Phys.*, **C35**:183, 1995.

83. D.G. Baird and A.M. Sukhadia. Mixing process for generating *in situ* reinforced thermoplastics. US Patent, May 1 1993. 5,225,488.

84. D.E. Alaman and N.S. Stoloff. Powder fabrication of monoliths and composite NiAl. *Internat. J. Powder Met*, **27**:29, 1991.

85. W.C. Williams and G.C. Stangle. Fabrication of near-net-shape $Al_2O_3$-fiber-reinforced $Ni_3Al$ composites by combustion synthesis. *J. Mater. Res.*, **10**:1736, 1995.

86. K. Suganuma, T. Fujita, K. Niihara, T. Okamoto, M. Koizumi and N. Suzuki. Hot extrusion of AA7178 reinforced with alumina short fibre. *Mat. Sci. Tech.*, **5**:249, 1989.

87. D.L. McDanels. Analysis of stress–strain, fracture, and ductility behavior of aluminum matrix composites containing discontinuous silicon carbide reinforcement. *Met. Trans. A*, **16**:1105, 1985.

88. J.H. Ter Haar and J. Duszczyk. Mechanical properties and microstructure of a P/M aluminum matrix composite with δ-alumina fibres and their relation to extrusion. *J. Mater. Sci.*, **29**:1011, 1994.

89. D.J. Towle and C.M. Friend. Effect of reinforcement architecture on mechanical properties of a short fibre/magnesium RZ5 MMC manufactured by preform infiltration. *Mater. Sci. Eng. A*, **188**:153, 1994.

90. A.G. Evans. The mechanical performance of fibre-reinforced ceramic matrix composites. *Mater. Sci. Eng. A*, **107**:227, 1989.

91. R.M. German and A. Bose. Fabrication of intermetallic matrix composites. *Mater. Sci. Eng. A*, **107**:107, 1989.

92. E.M. Rogers. *Diffusion of Innovations*. The Free Press, 1995.

# Fiber–fiber and fiber–wall interactions during the flow of non-dilute suspensions

S R I D H A R   R A N G A N A T H A N
A N D   S U R E S H   G   A D V A N I

## 2.1   Introduction

Short-fiber composites are reinforced by particles that are slender, but whose length is small compared to the overall dimensions of the part. They include fiber-reinforced thermoplastics for injection molding and extrusion, sheet molding compounds and short-fiber GMTs (glass-mat reinforced thermoplastics). Short-fiber composites can be processed by fast, highly-automated methods such as injection or compression molding. In these processes, the fibers and the resin are transported as a suspension into the mold cavity. The properties of a short-fiber composite part are highly dependent on the way the part is manufactured. As the resin or molding compound deforms to achieve the desired shape, the orientation of the fibers is changing. Fiber orientation changes stop when the matrix solidifies and the orientation pattern becomes part of the microstructure of the finished article. To design short-fiber composite parts effectively, the way that processing-induced fiber orientation influences the properties of the finished part must be anticipated. Hence, the flow of fiber suspensions needs to be understood in order to predict the orientation distribution of the fibers.

The best way to learn about fiber motion is first to analyse how a single fiber behaves in a flowing suspension. Hence, in this chapter, a brief overview is provided first of the background to the motion of single fibers in an infinite sea of liquid and basic orientation characterization methods as a foundation for the later topics. A review of the early work on fiber–fiber interactions, most notably those of S. G. Mason and co-workers is presented. This is followed by a description of more recent work on fiber–fiber interactions. The current state of the art as far as fiber–wall interactions are concerned is examined in the following section. Finally, a summary and future outlook on the subject of fiber–fiber and fiber–wall interactions is discussed. The reader is referred to Chapters 3 and 6 in Advani[1] for additional discussion on short-fiber suspensions.

During the flow of suspensions, the suspended fibers, which are typically non-spherical, tend to orient themselves in certain preferred directions and translate with the fluid. The particles are normally assumed to translate affinely with the fluid, that is, the velocity of the center of each particle is equal to the value of the unperturbed velocity of the fluid at that spatial location. However, the rotation of the particles may lead to a distribution of orientations at different locations, due to the flow field and hence is referred to as flow-induced orientation.

A fiber, which is immersed in a flowing fluid, is subjected to the local velocity gradients in the fluid. Due to these gradients a fluid element will experience deformation which includes rotation and stretching. A fiber will also rotate as the fluid deforms but will not change in length. The disturbance in the flow field caused by an immersed fiber decays as the distance from it increases. If there is another fiber or a wall present within the area where the disturbance is felt, the motion of the fiber is affected. The hydrodynamic force and torque felt by the fiber depend on its shape and the local rheological properties of the suspending fluid. This force and torque determine the rotational and translational velocities of the fiber.

The problem of motion of particles suspended in a flowing liquid has been studied in detail.[2–4] However, most of the existing theoretical models pertain only to the motion of a single particle. Analytical results are not available to predict the motion of particles in flows in which they are allowed to interact with other particles. Several researchers have observed that the qualitative behavior of rigid rod-like fibers varies as the concentration of the fibers in the suspension increases.[5–7] This is mainly because the fibers interact with each other. It is necessary to understand the mechanics of fiber–fiber interactions to be able to quantify its effect and judge its impact on the structure of a flowing suspension. The orientation state of a fiber suspension plays an important role in deciding the material properties in both the processing phase when it is in a fluid form and in the post-manufacturing phase when it is in a solid form. It is well known that physical properties such as conductivity and mechanical properties such as elastic modulus are greater in the direction parallel to a fiber than in a direction perpendicular to the fiber. Similarly, the viscosity of a suspension is also a strong function of the orientation state. Therefore the orientation state of a short fiber suspension/composite is a topic of immense interest.

Usually, fiber suspensions are classified into three concentration regimes according to the fiber volume fraction, $\phi_v$, and fiber aspect ratio, $r_f$ defined as the length $l$/diameter $d$ (Fig. 2.1):

$$\phi_v < \frac{1}{r_f^2} \quad \text{Dilute} \tag{1a}$$

$$\frac{1}{r_f^2} < \phi_v < \frac{1}{r_f} \quad \text{semi-concentrated} \tag{1b}$$

$$\frac{1}{r_f} < \phi_v \quad \text{concentrated} \tag{1c}$$

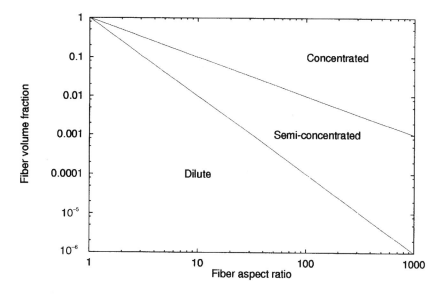

2.1 Classification of fiber suspensions.

The semi-concentrated and the concentrated regimes constitute what was previously referred to in this chapter as the non-dilute regime. For the case of cylindrical fibers

$$nl^3 = \frac{4}{\pi} r_f^2 \phi_v \qquad\qquad [2]$$

where $n$ is the number of fibers per unit volume.

The effect of the fiber concentration on the fiber–fiber interactions may be summarized as follows. In the dilute region, the spacing between the fibers is greater than their length and the fibers are free to move and rotate and hydrodynamic interactions between the fibers are rare. Hence, one may use Jeffery's theory to model the fiber rotations in dilute suspensions. In semi-concentrated suspensions, the spacing between the fibers is less than $l$ but greater than $d$ and hydrodynamic interactions are frequent. In the concentrated region, the spacing between fibers is of the order of $d$ and non-hydrodynamic effects, such as 'physical' collisions between fibers may become important. The physics of the angular motion of particles in dilute and non-dilute suspensions is vastly different.

## 2.2    Single fiber motion

In the absence of Brownian motion, the translation and rotation of a neutrally buoyant ellipsoid suspended in a Newtonian fluid undergoing a homogeneous flow was solved by Jeffery.[8] Consider an ellipsoid of aspect ratio $r_e$ whose centroid is located at the origin and whose major axis is in the direction $\mathbf{p}$ shown in Fig. 2.2.

Jeffery's solution for the angular velocity of this ellipsoid immersed in a simple shear flow $v_x = \dot{\gamma}y$ is given by

$$\dot{\theta} = \left(\frac{r_e^2 - 1}{r_e^2 + 1}\right)\frac{\dot{\gamma}}{4}\sin 2\theta \sin 2\phi \qquad \text{[3a]}$$

$$\dot{\phi} = \frac{-\dot{\gamma}}{(r_e^2 + 1)}(r_e^2 \sin^2 \phi + \cos^2 \phi). \qquad \text{[3b]}$$

From the above, it can be seen that the angular velocities depend linearly on the shear rate. This is as expected because the problem being solved is linear (Stokes flow). Therefore, the angular change in the ellipsoids over a period of time is dependent only on the total strain ($\dot{\gamma}t$) and not on the time $t$ and strain rate $\dot{\gamma}$ individually. It can also be seen that $\dot{\phi}$ is maximum at $\phi = \pi/2$ when the

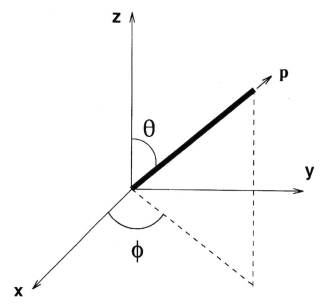

*2.2*  Coordinate system used. **p** denotes the direction of the major axis of the fiber.

ellipsoid is perpendicular to the flow and minimum at $\phi = 0$ when it is aligned with the flow. Equations [3a] and [3b] may be integrated and rewritten as,

$$\tan \phi = \frac{1}{r_e} \tan \left[ \frac{-\dot{\gamma} t r_e}{r_e^2 + 1} + \tan^{-1} (r_e \tan \phi_o) \right]$$  [4a]

$$\tan \theta = \frac{C r_e}{(r_e^2 \sin^2 \phi + \cos^2 \phi)}$$  [4b]

where $\phi_o$ is the value of $\phi$ at $t = 0$ and $C$ is a constant of integration known as the orbit constant. The value of $C$ characterizes the elliptical orbit executed by the particle. For example, $C = 0$ implies that the value of $\theta$ is always zero and hence the particle is always aligned along the vorticity axis. On the other hand if $C = \infty$, $\theta = \pi/2$ and the fiber always lies in the shear plane. All other values of $C$ correspond to orbits between the two described above. The orientation of the particle is seen to be periodic and the value of the period, $T$, may be obtained from Eqn. [4a] as

$$T = \frac{2\pi}{\dot{\gamma}} \left( \frac{r_e^2 + 1}{r_e} \right).$$  [5]

For particles of large aspect ratio, (characteristic length : characteristic diameter), it is possible to determine asymptotic solutions in terms of a small parameter that is a function of the aspect ratio. This is known as slender body theory and was originated by Burgers.[9] Slender body theory uses a line of force singularities, known as stokeslets, along the particle centerline to model the flow disturbance caused by the particle. The strength of the singularities are calculated such that the no-slip condition on the surface of the particle and the far-field boundary conditions are satisfied. The terms in the asymptotic solution decay as $1/\ln(r_f)$, where $r_f$ is the aspect ratio of the slender body. Therefore the solution converges slowly and many terms may be required to obtain an accurate solution for even particles of moderately large aspect ratio. Tuck[10] used this to estimate the drag on a slender body of arbitrary shape immersed in a viscous fluid undergoing Stokes flow and Tillet[11] used it to estimate the drag felt by a stationary axisymmetric slender body in a uniform flow, when the body is aligned parallel or perpendicular to the flow. Cox[12] derived a general relationship for the force per unit length acting on a long slender body with a circular cross-section, but of otherwise arbitrary shape. Batchelor[13] showed that for long slender particles with a straight centerline and arbitrary cross-section, the longitudinal and transverse relative motion of the particle and the suspending fluid may be approximated by that of an equivalent circle and an equivalent ellipse, respectively.

Bretherton[14] showed that in simple shear flow the angular motion of any axisymmetric body with fore-aft symmetry could be modeled as that of an equivalent ellipsoid. Therefore, the rotation of any such body can be described by Jeffery's theory. The approach most commonly taken to describe the rotation of particles is to determine the equivalent ellipsoid from theoretical considerations or experimental results and to use Jeffery's equations. The equivalent ellipsoidal axis ratio, $r_e$, for cylinders has been experimentally obtained by several researchers.[15–20] Anczurowski and Mason[20] found that the $r_e$ for a cylinder of aspect ratio, $l/d = r_f = 0.86$ was 1.0, which marks the transition of a cylinder from a rod to a disk. They also determined that for $r_f > 1.68$, $r_e$ was less than $r_f$ and for $r_f < 1.68$, $r_e$ was greater than $r_f$.

The motion of a single fiber in an unbounded fluid is well understood. However, in the presence of another fiber or a wall near by, the motion is more complicated. To model realistic suspension flow situations, both these effects need to be understood. However, before the role of such interactions can be studied, it should be possible to characterize fiber orientation quantitatively when a description of the orientation state of a suspension that has more than one fiber is required.

## 2.3    Orientation characterization

A single fiber such as the one shown in Fig. 2.2 may be described by the angles $\theta$ and $\phi$. However, when a suspension of several fibers needs to be described that approach becomes too cumbersome to be practical. Moreover, such precise information about the orientation of each and every fiber is typically unnecessary. The distribution function is the most general way to represent the orientation state of a suspension. This function, $\psi(\theta, \phi)$, is defined such that the probability of finding a fiber between angles $\theta$ and $\theta + d\theta$ and $\phi$ and $\phi + d\phi$ is given by $\psi(\theta, \phi) \sin \theta \; d\theta \; d\phi$. The probability distribution function must satisfy two physical conditions. First, one end of the fiber is indistinguishable from the other, so $\psi$ must be periodic,

$$\psi(\theta, \phi) = \psi(\pi - \theta, \phi + \pi) \qquad [6]$$

and second, every fiber must have a direction, so the integral over all possible directions of the orientation space must equal unity:

$$\int_0^{2\pi} \int_0^{\pi} \psi(\theta, \phi) \sin \theta \, d\theta d\phi = 1.0. \qquad [7]$$

This is known as the normalization requirement. If the orientation changes in space, $\psi$ is a function of position in addition to $\theta$ and $\phi$. While the distribution function is a complete and unambiguous description of the fiber orientation state, the disadvantage of using it to represent orientation is that it makes calculations to predict orientation in flowing suspensions very cumbersome and does not have a convenient interpretation.[21] To furnish easily interpreted measures of orientation, a number of orientation parameters have been defined.

One such measure is the Hermans orientation parameter.[22] This is a scalar measure of the strength of orientation for the special class of axisymmetric distribution functions, which implies that the distribution is a function of only one angle. It is not possible to extend this definition to other types of distribution functions. Also, it is not possible directly to calculate the evolution of the value of this parameter from the governing equation for fiber orientation.[23]

These disadvantages are overcome by the use of orientation parameters known as orientation tensors which are the moments of the distribution function.[21] The orientation tensors of odd order are zero as $\psi(\mathbf{p}) = \psi(-\mathbf{p})$. There is however an infinite number of even order tensors that can be obtained from the distribution function. Orientation tensors retain the generality of the distribution function and provide a compact, efficient and convenient representation. The components of the second and fourth order tensor, $a_{ij}$ and $a_{ijkl}$, are defined as

$$a_{ij} = \langle p_i p_j \rangle \qquad [8a]$$

$$a_{ijkl} = \langle p_i p_j p_k p_l \rangle \qquad [8b]$$

where $\langle \ \rangle$ is defined as,

$$\langle g \rangle = \int_0^{2\pi} \int_0^{\pi} g\, \psi(\theta, \phi) \sin\theta \mathrm{d}\theta \mathrm{d}\phi. \qquad [9]$$

From the definition of orientation tensors, it is clear that they possess complete symmetry with respect to the indices. Also since the distribution function is normalized and $\mathbf{p}$ is a unit vector, the trace of the second order tensor is unity

$$a_{ij} = a_{ji} \qquad [10a]$$

$$a_{ii} = 1 \qquad [10b]$$

$$a_{ijkl} = a_{ijlk} = a_{jikl} = a_{ikjl} = \cdots \qquad [10c]$$

$$a_{ij} = a_{ijkk}. \qquad [10d]$$

Therefore, the second and fourth order orientation tensor have only 5 and 14 independent components. The diagonal components of the second order orientation tensor represent the strength of the orientation in the respective direction. The off-diagonal terms indicate the deviation of the principal axes of orientation from the geometric axes with respect to which the components are calculated. In a two-dimensional case the second order orientation tensor has only two independent components and Fig. 2.3 shows some typical fiber distributions and corresponding values of the second order orientation tensor components.

It has been shown[21] that the prediction of an $n$th order tensor property of a composite requires only the knowledge of the $n$th order orientation tensor. Thus, if it is necessary to calculate the thermal conductivity tensor of a short fiber composite, then the second order orientation tensor is all that is needed. This

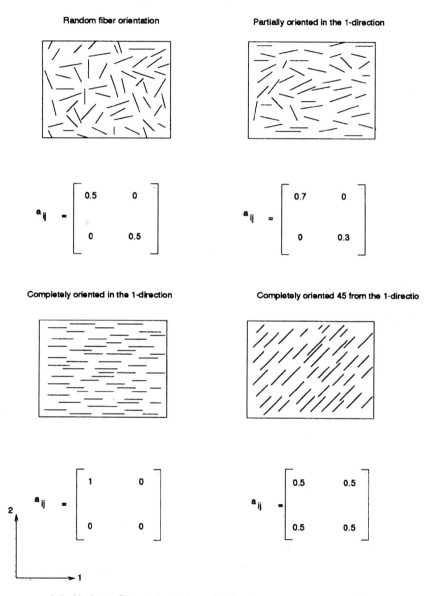

**2.3** Various fiber orientation distributions and corresponding second order orientation tensor components.

is true even though the second order tensor does not completely describe the orientation state. Similarly to predict the fourth order elastic stiffness, the fourth order orientation tensor is sufficient. The description of the orientation state may be further simplified and defined by a single scalar value, $f$, which is the strength of the orientation defined in terms of the invariants of the second order orientation tensor.[24]

$$f = 1 - \alpha \det|a_{ij}| \qquad\qquad [11]$$

where $\alpha$ takes on the value 4 for two-dimensional orientations and 27 for three-dimensional orientations. The constant $\alpha$ is chosen such that the value of $f$ is zero at the random orientation state and unity at an aligned orientation state. Even though $f$ provides a convenient representation of the strength of the orientation, it does not indicate the direction of the orientation. Also, it is not possible to perform orientation evolution calculations in terms $f$. However, it is a useful parameter to depict the results of orientation calculations.

As the representation goes from the distribution function to the orientation tensors to the scalar strength of orientation, the degree of complexity decreases and the amount of information contained in the representations also decreases. Therefore, depending on the required accuracy and the ease of calculations, one of the above descriptions for the fiber orientation state may be chosen. The significance of the values of the orientation tensor components is clear. Therefore, the orientation tensors provide a good compromise between the simplicity and the accuracy of the representation of the fiber orientation state in a suspension. Also they can be directly calculated by solving the equations of flow and fiber orientation.[21]

## 2.4    Fiber–fiber interactions

### 2.4.1    Early work

Equal sized non-interacting particles in a dilute suspension undergoing simple shear flow exhibit an undamped periodic motion, leading to a periodic change in the orientation state of the suspension. Anczurowski and Mason[25] derived expressions for the equilibrium orientation of dilute suspensions of rods and discs. The distribution of orbit constants at steady state, however, had to be measured experimentally. Previously, Jeffery[8] had proposed that the orbit constants should settle down to a distribution that would minimize the suspension viscosity. In other words, $C=0$ for $r_e > 1$ and $C=\infty$ for $r_e < 1$. Eisenschitz[26] had postulated that all the particles would retain their initial orbits. Therefore, if a suspension had an isotropic distribution at the onset, it would be the same at steady state. Anczurowski and Mason[19] measured the orientation state of suspensions of rods and discs undergoing Couette flow. The equilibrium orbit constant distribution was found to be between Jeffery's and Eisenschitz's estimates. They concluded that even in a dilute suspension, given sufficient time, particles interact with each other which leads to a change in their orbits. Earlier, Saffman[27] had suggested that collisions (interactions) between particles would tend to change their orbit such that the energy dissipation is minimized or the volume swept by the particle is minimized. This hypothesis predicts $C=0$ for

$r_e > 1$ at steady state which does not agree with the experimental results. It is clear that none of the simple models for orbit constant distribution provided good experimental agreement.

Anczurowski and Mason[19] showed that the steady orientation state of a dilute suspension is not a function of the concentration. Experimentally, they noted that this inference was valid for $nl^3 < 0.2$. For $nl^3 > 0.2$, the steady orientation was found to be dependent on the concentration. They also clearly demonstrated that the orientation in a dilute suspension is different from that in a suspension without collisions. They argue, therefore, that even in a dilute suspension fiber–fiber interactions play a role in determining the steady orientation state.

Anczurowski et al.[28] also observed the transient rotation of rods in Couette flow. They noted that the distribution of the phase angle $\phi$ reached steady state quickly while the orbit constant distribution took a longer time. They also noticed that orbit changes and phase angle changes resulted from particle collisions. These particle collisions were categorized as close if the two rods passed within one particle diameter ($d$) of each other or distant if they passed within one particle length ($l$) of each other. The frequency of close collisions in a simple shear flow field (shear rate $G$) is of the order $nld^2G$ and that of distant collisions is of the order $nl^3G$. Even though the change in orientation is greater due to a close collision than a distant collision, they concluded that the distant collisions have a more significant effect on the orientation state. The distant collisions primarily damp out the otherwise periodic motion of cylinders in simple shear flow.

The motion of particles in a dilute suspension is reversible, that is, by reversing the flow field, particle orientations at previous times can be recovered exactly. Arp and Mason[6] in a shear flow experiment showed that on reversing the flow, a pair of hydrodynamically interacting fibers which do not collide or touch one other, reverse their translational and rotational motion and retrace their original paths. However, when they appear to make contact, they change their trajectories and on flow reversal the rods do not retrace their original trajectories but acquire new orbits.

The effect of particle–particle long range interactions was analysed by Cox[29] using slender body theory. This theory is capable of predicting the force and torque on a test fiber in a suspension due to the imposed flow field and due to particles farther away than its length. Even though the theory is general enough to sum multiparticle interactions, it does not take into account reflections or indirect interactions. The disturbance in the velocity field caused by one fiber is reflected by the fiber it interacts with, which also affects the motion of the other fibers in its neighborhood. Therefore, this theory is directly applicable only to model two-particle interactions. Okagawa et al.[30] have modeled fiber–fiber interactions in suspensions where two-particle interactions are the only ones present, that is, $nl^3 \ll 1$, where $n$ is the number of fibers per unit volume and $l$ is the length of the fiber. The particles were categorized into four types based on the combination of the orbit constant $C$ being $O(1)$ or $O(1/r_e)$ and whether the particle was aligned with the flow or away from the flow. The effect of pair interactions between all possible combinations of such particles on their rotation was studied. The most dominant interaction was between two particles, one with $C = O(1)$ and aligned with the flow and the other particle with $C = O(1)$ and not aligned with the flow. The distance of separation between the particles when such an interaction becomes important was given to be the order of $l(r_e/\log r_e)^{1/3}$. The combined probability of finding two particles which satisfy the orientation criterion and whose centers are within the distance of interaction defined above is $\dot{\gamma} n l^3/\log r_e$. Therefore, the characteristic time for this type of interaction, denoted by $\tau$, is $\log r_e/(\dot{\gamma} n l^3)$. If $\tau$ is much greater than the time period of rotation of the particles, then the orientation distribution would exhibit damped oscillations since the probability of an interaction in one time period of rotation is small. The value of $\tau$ decreases as the concentration increases (interaction frequency increases). If the concentration increases to such a level that $\tau$ is much less than the period of rotation of the particle, the change in the distribution function is expected to be monotonic since the particles undergo multiple interactions in each time period. This work clearly demonstrates that the effect of interactions start to become significant at reasonably small concentrations.

Early work showed that the physics of fiber–fiber interactions was crucial to the understanding of the orientation and motion of fibers. Also, the deviation from Jeffery's theory was shown to increase as the concentration increased. More recent work has focused on deriving quantitative estimates of such deviations. Most of the recent work on fiber–fiber interactions in suspensions may be broadly classified into one of the following categories: (1) diffusion models which account for interactions between fibers in terms of a diffusivity term in the governing equation for the fiber orientation state, (2) slender body theory, based solutions that model the interactions between the fibers explicitly, (3) Stokesian dynamics solutions that computationally account for the long and short range interactions between fibers, and (4) full numerical simulations that solve Stokes equation 'exactly' given the configuration of particles. These methods are described below and their relative advantages and disadvantages are discussed.

## 2.4.2  Diffusion models

The evolution of the orientation state of non-dilute fiber suspensions may be modeled as a diffusive process based on the following assumptions.[31] All the fibers in the suspension are rigid cylinders of equal length and diameter. The particle centroids are distributed homogeneously in a Newtonian suspending liquid. The only forces acting on the particles are due to the suspending liquid and due to fiber–fiber interactions. If the change in particle orientation due to interactions is treated as a random variable and superposed with the hydrodynamic effect of the flow field, the following equation is obtained for the fiber angular velocity

$$\dot{\mathbf{p}} = -\tfrac{1}{2}(\omega \cdot \mathbf{p}) + \tfrac{1}{2}\lambda(\dot{\gamma} \cdot \mathbf{p} - \dot{\gamma} : \mathbf{p}\,\mathbf{p}\,\mathbf{p}) - D\frac{1}{\psi}\frac{\partial\psi}{\partial\mathbf{p}} \qquad [12]$$

where $\omega$ is the vorticity tensor, $\gamma$ is the rate of strain tensor, $\lambda = (r_e^2 - 1)/(r_e^2 + 1)$ and $D$ is a diffusion coefficient that models the effect of the fiber–fiber interactions on their angular velocity. This equation is identical to that which describes Brownian motion. This topic has been studied in detail by several researchers. Consequently, there have been several solutions proposed for various special cases of this problem. Even though the fibers have been modeled as non-Brownian, the interactions introduce an effect similar to the rotary diffusivity introduced by Brownian motion.

If $D$ is set to zero, or the effect of fiber–fiber interactions is neglected, Jeffery's equation for the angular velocity is recovered. It is seen that the expression for the angular velocity is not only a function of the orientation of the fiber, but also a function of the local orientation state of the suspension. The effect of the diffusion term is twofold. One is that it randomizes the orientation state, that is, as the value of $D$ increases the orientation state becomes more random. Second, it also makes the distribution function in simple shear flow asymmetric. The magnitude of randomization and asymmetry depend on the value of the diffusivity. The parameter $D$ contains the effect of the flow field and the suspension parameters. The flow field affects the interaction rate as defined in the assumptions. Folgar[32] showed that the interaction rate for many simple flows such as simple shear flow and uniaxial extensional flow was proportional to the strain rate. Therefore, in the Folgar–Tucker model[31] the flow field effect was characterized by the magnitude of the strain rate tensor, $[\dot{\gamma}]$, and $D$ was expressed as

$$D = C_{\mathrm{I}}[\dot{\gamma}] \qquad [13]$$

where $C_{\mathrm{I}}$ is a dimensionless number that models the effect of the suspension parameters on the strength of the interactions. From the definition of $D$ it is seen that once the flow ceases, the interactions also cease because $[\dot{\gamma}]$ goes to zero. The value of $C_{\mathrm{I}}$ for a given suspension was assumed to be a constant independent of the orientation state, as a first approximation. Subsequently, by fitting experimental steady orientation results to numerical results, values of $C_{\mathrm{I}}$ were calculated for various suspensions. Advani and Tucker[21] recast the Folgar–Tucker model governing equation in terms of orientation tensors to describe the orientation state of fiber suspensions as follows:

$$\frac{Da_{ijkl}}{Dt} = v_{im}a_{jklm} + v_{jm}a_{iklm} + v_{km}a_{ijlm} + v_{lm}a_{ijkm} - 4v_{mn}a_{ijklmn}$$
$$+ C_{\mathrm{I}}[\dot{\gamma}][-20a_{ijkl} + 2(a_{ij}\delta_{kl} + a_{ik}\delta_{jl} + a_{il}\delta_{jk}$$
$$+ a_{jk}\delta_{il} + a_{jl}\delta_{ik} + a_{kl}\delta_{ij})] \qquad [14]$$

where, $v_{ij}$ is the modified velocity gradient tensor defined as,

$$v_{ij} = \left(\frac{\lambda + 1}{2}\right)\frac{\partial u_i}{\partial x_j} + \left(\frac{\lambda - 1}{2}\right)\frac{\partial u_j}{\partial x_i} \qquad [15]$$

$$\frac{Da_{ij}}{Dt} = v_{im}a_{mj} + v_{jm}a_{mi} - 2v_{mn}a_{ijnm}$$
$$+ 2C_I[\dot{\gamma}](\delta_{ij} - 3a_{ij}) \qquad [16]$$

Note that this creates a closure problem and has been addressed elsewhere.[24]

Folgar and Tucker[5] observed the fiber orientation of semi-concentrated and concentrated suspensions. The fibers they used were nylon and the suspending fluid was silicone oil. The densities of the fibers and the fluid were matched to ensure no appreciable settling occurred during the course of the experiments. The fibers were cut to a specified nominal length and the distribution in length was measured. Still photographs of the suspension were taken along the vorticity axis in a Couette flow between concentric cylinders. The planar orientations of the suspensions were computed by digitizing the fiber images from these photographs. These results were fitted to the numerically computed orientation distribution and the value of $C_I$ that best fit the experimental results was obtained. The $C_I$ values they obtained for various suspensions ranged between $O(10^{-1})$ and $O(10^{-3})$. The agreement between experimentally obtained two-dimensional steady distribution functions and those calculated numerically was found to be good. However, the transient orientation evolution calculated using the $C_I$ value computed at steady state and the experimentally observed transients were not in agreement. The numerical solution predicted faster transients than those observed experimentally. Thus, the suspension behavior indicated that $C_I$ was a function of the orientation state and had a higher value at a random orientation state than the one at a more aligned steady state. Jackson et al.[7] and Advani and Tucker[33] have used this diffusive model to predict the flow-induced fiber orientation state in compression molded parts. They found that the experimental results and the theoretical predictions matched well for values of the interaction coefficient of the order of $10^{-2}$.

Ranganathan and Advani[34] proposed an approach to predict the interaction coefficient using the average interfiber spacing as a parameter. It was assumed that the angular motion of the fibers is affected only by the flow field and interactions with other fibers due to flow. The interaction coefficient is a measure of the intensity of interactions in a suspension. Therefore, it would be expected to change with the orientation state of the suspension. Qualitatively, they argued that the intensity of fiber-fiber interactions should vary inversely with the interfiber spacing, $a_c$ as the effect of the interactions will be pronounced when the fibers are in close proximity. It has also been shown[35,36] for certain flow configurations, that the stresses around fibers in semi-concentrated suspensions increase as the spacing between them reduces, which implies hydrodynamic interactions will increase. Also reducing the average spacing will increase the probability of direct interactions. Therefore, the interfiber spacing would include the effects of both the hydrodynamic interactions and the physical space available for fiber rotation. They combined these two effects and conducted phenomenological studies to determine the exact functional relationship between $C_I$ and $a_c/d$.

Ranganathan[37] conducted fiber orientation measurement experiments in a Couette flow between concentric cylinders. The outer cylinder was rotated at a prescribed speed and the inner one was held at rest. The fibers were nylon and the suspending fluid was a highly viscous polyalkylene glycol which was water soluble. The densities of the fibers and the fluid were matched by dissolving the appropriate amount of water in the polyalkylene glycol. The nylon fibers were cut in-house in order to have good control over the length distribution. Several razor blades separated by spacers of accurate dimension were used to chop the fibers to the prescribed length. Two different types of nylon fibers were used in the experiments. The bulk of the fibers were natural nylon that has a refractive index very close to that of the polyalkylene glycol–water mixture and therefore are almost transparent as a suspension. A small number of black nylon fibers essentially of the same aspect ratio and density as the natural ones were also added to the suspension to aid in visualizing the orientation. Experiments were performed with fibers of two different aspect ratios at various volume fractions that were mostly in the semi-concentrated regime.

Still photographs were taken along the vorticity axis. The ends of the tracer fibers were digitized to obtain the orientation state of the suspension. Since the length distribution was narrow it was assumed that the out-of-plane orientation could also be estimated from the two-dimensional picture. Based on these experiments a linear inverse relation between $C_I$ and the interfiber spacing was found to give acceptable agreement between the observed and the computed values of fiber orientation.

$$C_I = \frac{K}{a_c/d} \qquad (12)$$

where $K$ is a constant of proportionality. Experimentally, $K$ was determined to be $10^{-2}$.

The effect of having an interaction coefficient that is dependent on orientation on the predicted evolution of orientation is shown in Fig. 2.4.

It is seen that a slower transient is predicted in the case where the interaction coefficient is allowed to change with orientation as opposed to the case where it is held constant even though they end up at about the same steady state value.

Kamal and Mutel[38] also used the concept of rotational diffusivity to model the transient variation of the orientation distribution function. However, they do not isolate the shear rate dependence of the rotational diffusivity. They presented the results in terms of a Peclet number ($\dot{\gamma}/D$) which is equivalent to the reciprocal of the interaction coefficient. They modeled the diffusivity, $D$, as being

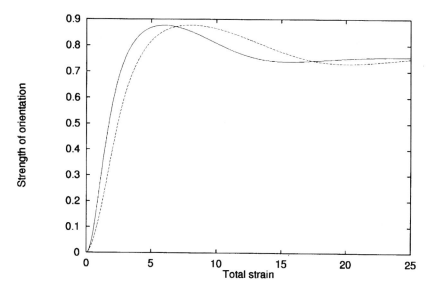

*2.4* Effect of constant and varying interaction coefficient on time evolution of strength of orientation. Non-dimensionalized number of fibers per unit volume $= 20$, aspect ratio $= 26$. Case 1 (—), orientation independent interaction coefficient. Case 2 (---), orientation dependent interaction coefficient. $C_I$ varies from 0.186 at random orientation to 0.0069 at steady state.[35]

proportional to the shear viscosity of the suspension, which depends on the orientation distribution function. The argument used to substantiate this approach was that the viscous dissipation in a suspension is an indication of the disturbance in the flow and hence the interaction intensity. In this way they incorporated the effect of orientation into the diffusivity. Their expression for the viscous dissipation is $\langle \sin^4 \theta \sin^2 2\phi \rangle$ where $\langle \ \rangle$ represents an averaged quantity in the orientation space weighted by the orientation distribution function. The dependence of $D$ on the fiber volume fraction and the fiber aspect ratio were not modeled. Their approach predicts different steady state orientations for different shear rates. This is in agreement with their experimental results.[39] On the other hand, experimental results reported by Bibbo and Armstrong[40] indicate that the strain rate does not affect the steady state orientation for suspensions of moderate volume fractions. These results are contradictory. Further studies are required to resolve this issue.

Recently, Bay[41] injection molded end-gated rectangular plaques and center-gated circular disks at various volume fractions using a variety of thermoplastic matrices. Based on the observed experimental values of the orientation state at various locations he estimated a value for the interaction coefficient that best fit the data. In the concentrated regime he found that $C_I$ decreased with increasing volume fraction of fibers. The following empirical relationship described his data best

$$C_I = 0.0184 \exp(-0.7148 \phi r_f). \tag{18}$$

This is opposed to the trend observed by Folgar[32] and Ranganathan.[37] However, their experiments were in the semi-concentrated regime while Bay's experiments were in the concentrated regime. The tendency of the orientation to become more random as concentration increases in the semi-concentrated regime while it becomes more aligned with increasing concentration in the concentrated regime may be due to the increase in physical constraints for the fiber motion with increasing volume fraction (see the chapter by Milliken and Powell in Advani[1] for a discussion of this topic).

Also in the category of diffusive models for fiber orientation, Stover[42] has recently suggested using a weak anisotropic rotary diffusivity with non-zero components $D_{\theta\theta}$ and $D_{\phi\phi}$, as opposed to the isotropic rotary diffusivity suggested by Folgar and Tucker[31] to model the steady orientation state of semi-concentrated short-fiber suspensions. Due to the weak diffusion assumption, an asymptotic solution for the steady orientation as a function only of the ratio of the diffusivities, $D_{\theta\theta}$ and $D_{\phi\phi}$, was obtained. Therefore, the number of empirical parameters in the orientation model was still restricted to one.

Stover *et al.*[43] observed the orientation of fiber suspensions in a Couette flow. They used a refractive index and density-matched suspension and introduced a tracer fiber to monitor the orientation state of the suspension. The orientation of the tracer fiber was time-averaged to obtain the orientation distribution function of the suspension. Earlier, Anczurowski and Mason[19] showed that in monodisperse fiber suspensions the ergodic hypothesis holds, that is, time-averaged distribution and particle-averaged distribution are equivalent. The tracer fiber center was two fiber lengths away from the walls. They noted that the tracer fiber rotated for the most part according to Jeffery's equation. However, occasionally, it would rotate in a direction opposite to that caused by the flow field due to interactions with the other fibers.

They expressed the experimentally observed orbit distribution in terms of a normalized orbit constant $C_b$ defined as $C/(1 + C)$. The $C_b$ distribution in dilute suspensions was shifted more towards lower orbits compared with the isotropic distribution, which corresponds to the orbit constant distribution in an isotropic suspension.[19] For semi-dilute suspensions they found that the distribution was shifted towards higher orbits compared to that for the dilute case. However, the distribution was still shifted towards lower orbits compared with the isotropic distribution. The ratio $D_{\theta\theta}/D_{\phi\phi}$ was calculated to be 1.5 by matching experimental results and numerical predictions for semi-dilute suspensions.

## 2.4.3  Slender body theory-based solution

Shaqfeh and Koch[44] have used slender body theory and diagrammatic resummation of the interactions to calculate the steady orientation state of suspensions in extensional flow. They assumed that the steady orientation state of dilute and semi-dilute suspensions could be modeled as a small perturbation from the completely aligned state predicted by dilute suspension theory. This perturbation or dispersion in orientation is caused by the hydrodynamic interactions between the fibers in a suspension. They consider a uniaxial extensional flow where 1 is the direction of principal extension and 2 and 3 are the orthogonal directions. For the dilute case, only two-particle interactions need to be considered. The dispersion in orientation is expressed in terms of $\langle p_2^2 \rangle$ and $\langle p_3^2 \rangle$, where $p_2$ and $p_3$ are the projections of the fiber orientation vector in the 2 and 3 directions, respectively and $\langle \ \rangle$ signifies a suspension average. The quantities $\langle p_2^2 \rangle$ and $\langle p_3^2 \rangle$ are equivalent to the orientation tensor components $a_{22}$ and $a_{33}$. If there was no orientation dispersion, $\langle a_{11} \rangle = 1$, $\langle a_{22} \rangle = \langle a_{33} \rangle = 0$. However, taking into account two-particle interactions they predict the orientation dispersion in the dilute regime to be

$$\langle a_{22} \rangle = \langle a_{33} \rangle = 0.01496 \frac{nl^3}{(\ln(r))^2} \qquad [19]$$

In a semi-dilute suspension ($nl^3 \gg 1$, $\phi \ll 1$), the hydrodynamic disturbance created by a fiber is screened at lengths of the order of the screening length. Shaqfeh and Koch argue that for a fiber to be oriented away from the principal axis of extension, there have to be significant velocity disturbances in the perpendicular direction to its axis. However, the short range screening attenuates the disturbances and keeps the orientation state mostly aligned in the 1 direction. They obtain the following expression for the dispersion

$$\langle a_{22} \rangle = \langle a_{33} \rangle = 0.020 \frac{\ln(nl^3)}{nl^3}. \qquad [20]$$

Therefore, the predicted dispersion in orientation increases with increasing concentration in the dilute regime but decreases with increasing concentration in the semi-dilute regime.

Slender body theory does not account for lubrication forces that are generated when two surfaces get very close to each other. Therefore, if used in a dynamic simulation it would not prevent particles from passing through one another. It has been shown however that the probability of close interactions in dilute suspensions is small.[45] In the semi-dilute case there will be more close particle interactions. Such interactions are expected to reduce the dispersion in orientation. Therefore, Shaqfeh and Koch recommend their estimate as an upper bound for orientation dispersion. Shaqfeh and Koch[44] also derived estimates for the orientation dispersion in planar extensional flows. In the dilute case they obtained

$$a_{22} = 0.0091 \frac{nl^3}{(\ln r)^2} \tag{21}$$

$$a_{33} = 0.0274 \frac{nl^3}{(\ln r)^2} \tag{22}$$

and in the semi-dilute case,

$$a_{22} = 0.0116 \frac{\ln(nl^3)}{nl^3} \tag{23}$$

$$a_{33} = 0.0367 \frac{\ln(nl^3)}{nl^3} \tag{24}$$

A similar trend to that observed in uniaxial extension is seen above. However, $a_{22}$ and $a_{33}$ are not equal because the restoring torques experienced by particles are greater in the 2-1 plane than in the 3-1 plane. This leads to lower steady state values of $a_{22}$ compared with $a_{33}$.

Koch and Shaqfeh[46] provide an estimate for the angular velocity of an individual fiber in a semi-dilute suspension. This work assumes that all the fibers except the test fiber are aligned in an equilibrium orientation and uses slender body theory. In other words, the mechanics of a single fiber perturbed from its equilibrium state is studied. In simple shear flow the effect of the surrounding medium of aligned particles is always to reduce the angular velocity of the test fiber as it gets back to its equilibrium orientation. In contrast, in uniaxial extension, the test fiber rotates faster in some orientations and slower in others compared with Jeffery's solution. In planar extension, the effect of the surrounding medium is shown to be much more complex.

Rahnama et al.[47] have recently used slender body theory to compute the orientation of a fiber suspension undergoing simple shear flow. The fibers in such a flow are aligned in the flow direction for the majority of the time. Alternatively, most of the fibers are aligned with the flow at any given time. In their calculations, they neglect higher order interactions than two-particle interactions in both the dilute and semi-dilute regimes since the probability of a fiber interacting with two or more misaligned fibers simultaneously is small. Based on scaling arguments they show that the dominant interactions are those between an aligned and a misaligned fiber. The effect of fiber–fiber interactions on the orientation state is expressed in terms of a rotary diffusivity and a drift velocity. The governing equation for orientation is written as,

$$\frac{\partial \psi}{\partial t} + \nabla_p \cdot (\dot{p}^h \psi) = \nabla_p \cdot (D \cdot \nabla_p \psi - \psi \dot{p}^h). \tag{25}$$

Expressions for the diffusivity $D$ and the drift velocity $\dot{p}^h$ are derived for the dilute and semi-dilute cases. The diffusivity tensor is assumed to have only $D_{\theta\theta}$ and $D_{\phi\phi}$ components. The values of these parameters are functions of the

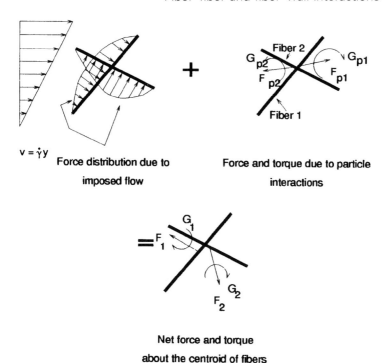

Force distribution due to
imposed flow

$v = \dot{\gamma} y$

Force and torque due to particle
interactions

Net force and torque
about the centroid of fibers

*2.5* Superposition of the force and torque exerted on the two particles
by each other and the imposed flow field.

orientation state of the suspension. Therefore the equation needs to be solved iteratively.

The effect of the drift velocity in the dilute regime was to increase the degree of alignment along the flow direction. In the semi-dilute regime, however, the drift velocity was seen to cause more fibers to have higher orbit constants. The diffusivity ratio predicted in this study is a function of the fiber aspect ratio and fiber volume fraction. For a given aspect ratio, with increasing volume fraction, the diffusivity ratio was predicted to decrease. The quantitative predictions from this study and the experimental results of Stover[42] were found to agree within the statistical errors associated with the experiments.

Ranganathan[37] superposed the lubrication solution for the force and torque exerted on two surfaces separated by very small gaps[48] and the slender body theory solution for the force and torque exerted by a fluid on fibers[29] to determine the motion of two fibers interacting in close proximity (Fig. 2.5).

The net force and torque on the fibers are set to be zero and their linear and angular velocity computed. The parametric studies conducted revealed that the angular and linear velocity of the fibers were different from those predicted by dilute suspension theory. The change in angular velocity affects the orientation state of the suspension. The change in linear velocity causes non-affine fiber motion which leads to spatial non-homogeneity of fiber concentration in a suspension. The deviation in the angular velocity from that predicted by Jeffery's theory and the non-affine motion were found to increase as the distance of separation between the particles decreased or as the particle aspect ratio decreased. Aside from these qualitative observations, no quantitative correlations for the velocity and angular velocity of the particles with respect to the suspension parameters were obtained.

## 2.4.4 Stokesian dynamics

There have been efforts to use a computer simulation approach to tackle the problem of fiber–fiber interactions in suspensions. J. F. Brady and co-workers have made a significant contribution to this approach and have coined the term Stokesian dynamics to describe their methodology.

Brady and Bossis[49] have reviewed the Stokesian dynamics approach and also discuss the various types of problems that can be studied using this method. This approach takes into account the effect of the long range interactions between the particles in addition to the short range lubrication forces. The force acting on the particles is regarded as a sum of the hydrodynamic forces due to the fluid, external forces and a stochastic Brownian component. The exact relation between the forces and the velocity is dependent on the particle size and shape.

The hydrodynamic interactions between particles in a suspension is modeled in two parts. The long range interactions between the particles are accounted for in the mobility matrix (which relates velocity to force). This is then inverted to obtain the resistance matrix (which relates force to velocity) to which the lubrication interactions are added in a pairwise manner. Care is taken not to account for any interaction that occurs many times. The number of times these matrices need to be computed during a simulation can be optimized by examining how they change with respect to the suspension structure. The matrix characterizing the long range interactions changes when particle separation changes by the order of the particle size. The lubrication component, however, changes when the distance of separation changes by the order of inter-particle spacing. Therefore, the mobility matrix needs to be inverted less often than the number of times the resistance matrix needs to be evaluated. This feature optimizes the calculation time. Some examples of published literature illustrating the Stokesian dynamics technique are given below.

Durlofsky et al.[50] computed the drag coefficients and trajectories of a system of settling spheres of equal diameters. A lubrication solution for the resistance and mobility functions, which relates the force and torque acting on the particles to the linear and angular velocity of the particles, for two spheres at small separation distances[51–53] was used to model the near field interactions. The simulation predictions and the analytical solutions for a small number of spheres were found to be in good agreement. Durlofsky and Brady[54] used the Stokesian dynamics approach to simulate the flow of a bounded suspension of spheres between two infinite parallel plates. The suspension viscosity was found to exhibit large fluctuations at high solid volume fractions. These were correlated to the formation and break up of clusters of spheres. At even higher volume fractions the spheres formed a single aggregate and translated as such.

Claeys and Brady[55] have developed an extension of this technique that can handle elongated particles. Using this technique, they show that the period of rotation of an ellipsoid is reduced when another particle is nearby. The fiber suspension rheology they predict was found to be in good agreement with previous experimental and theoretical computations. The extension to elongated particles, however, increases the number of degrees of freedom that need be dealt with since the orientation of the particles must now be tracked.

The Stokesian dynamics approach is very computer intensive and the number of particles that can be used in each computation is limited. However, it is an extremely valuable tool that can be used to evaluate theories about particle interactions. It can also provide insight into behavior of suspensions at least in a qualitative sense. For example, the importance of aggregate formation in concentrated suspensions of spheres.

## 2.4.5  Computer simulations/full numerical solutions

The problem of hydrodynamic interactions between the particles in a suspension can be solved from a purely computational point of view. The creeping flow equations, that is, Stokes' equations are solved in the flow domain around each fiber in the suspensions and the force and torque on each fiber are calculated to determine the motion of the fiber. No-slip boundary conditions at fluid–solid interfaces are applied rigorously in addition to other flow field conditions. These equations can be solved by the finite element or the boundary element method. The boundary element method reduces the dimensionality of the problem by one due to the fact that the equations are transformed from a three-dimensional form to surface integrals. The disadvantage of the boundary element method is that it involves inverting a non-sparse matrix. These two effects offset each other to a certain extent.

Improving the efficiency of the matrix inversion calculation, therefore, has been a topic of interest. Karrila et al.[56] proposed an efficient iterative solution for the system of equations obtained when using such a technique to solve multiparticle problems. Improvement in the inverse calculation time is critical since any dynamic simulation requires several inversions of the system. Some recent works address this issue[56–58] by optimizing the matrix inversion strategy.

Chan[59] addressed the problem of hydrodynamic interactions between ellipsoidal particles suspended in a highly viscous fluid using the boundary element method. Chan et al.[60] have shown that a Galerkin boundary element technique can be used to predict the motion of particles. They used this technique to solve the motion of fibers in close proximity. The solution was found to be computationally intensive.

Yoon and Kim[61] use a boundary collocation method to study the multi-particle problem. They state that this technique is superior to the boundary integral technique both from stability and computational efficiency points of view.

The boundary element method is the most promising numerical technique to solve Stokes' equations exactly in a suspension of particles. There have been some recent advances that use parallel computing strategies to speed up the computations. Kim and Amann[62] have used the completed double–layer boundary integral equation method (CDL-BIEM) to compute the microstructure evolution in suspensions. The main computational effort required is in solving a large linear system of equations. They have implemented an iterative solution well suited for parallel computers. Based on some test cases, they conclude that their solution will scale well for use in massively parallel machines. Pakdel and Kim[63,64] have analyzed the motion of two particles in close proximity using CDL-BIEM. They found that the results converge fast for particle separations that are greater than 10% of the particle radius with uniform refinement. However, for smaller gaps, they concluded that something more sophisticated such as a Chebyshev refinement is required. Seeling and Phan-Thien[65] have also used a completed double layer boundary element algorithm to solve suspension problems in parallel computers.

Ultimately, the goal is to have a predictive model for the orientation state of a fiber suspension, rather than having to perform a simulation for every specific configuration of fiber centroid locations and orientations. However, to build a predictive model, the numerical simulation approach could prove to be a valuable tool. Such simulations may be used to gain insight into the behavior of particles in the presence of a solid boundary and/or other particles in its neighborhood. The information thus acquired may then be used to build a predictive model. The last step is non-trivial, however. More details about these techniques are covered in Chapter 3 by Phan-Thien and Zheng.

## 2.5    Concentrated suspensions

When suspensions are in the concentrated regime, non-hydrodynamic effects start to become significant. Recently, Toll and Manson[66] have proposed a model for fiber–fiber interactions in such suspensions. They modeled the interactions between straight fibers in planar orientation as completely mechanical (physical). Any hydrodynamic interactions are ignored. The number of contact points on a typical fiber in a suspension can be obtained from the volume fraction, fiber aspect ratio and orientation state.[67] The number of contact points is expected to be high enough for such concentrations. Hence discrete point contacts are modeled effectively to be a continuous distribution. The results of this approach were found to be sensitive to the model chosen for the friction between the particle surfaces as they undergo these non-hydrodynamic interactions. After deriving a general formulation for the rheology of a concentrated suspension, some specific functional forms of the friction were used to compute the viscosity–volume fraction relation in suspensions.

When a linear friction function is assumed, that is, friction is proportional to sliding velocity and independent of the contact force, then the shear viscosity is shown to be proportional to $\phi^2$. Compare this to dilute suspension theory which predicts a linear dependence of shear viscosity on $\phi$ and experiments in semi-concentrated suspensions[68] which exhibit a $\phi^3$ type dependence. If Coulomb friction is assumed the stresses would be proportional to $\phi^5$. Thus, it is clear that the assumed functional form of the friction function plays an important role in the predicted behavior of the suspension.

Finally, this theory is strictly applicable only when non-hydrodynamic effects dominate over hydrodynamic effects. This is expected to be the case in concentrated suspensions. The applicability of this theory is therefore limited to such suspensions.

The ultimate goal of modeling fiber–fiber interactions is to predict the orientation state of suspensions in complex flows such as those encountered during the manufacture of short-fiber composites. In order for this to be feasible, a model for the orientation state is required in terms of the suspension parameters and the flow dynamics. The diffusion models described previously fit this requirement well and are currently being used in simulations such as the C-MOLD program for flow and fiber orientation in injection molding. Other classes of models such as Stokesian dynamics or full numerical simulations can help develop a correct physical understanding of the effect of interactions on the fiber orientation state. The insight gained from such approaches must then be used to formulate models that can be used in mold filling simulations.

## 2.6   Fiber–wall interactions

The influence of a flow boundary (wall) on the motion of a fiber arises due first to the mechanical constraint that a fiber cannot penetrate a wall and second to the wall being present within a distance from the fiber such that the hydrodynamic disturbance caused by the fiber has not decayed at the wall.

There have been several studies recently that focus on this effect. There is no consensus yet as to exactly how far away from the wall a particle needs to be before the wall effect may be ignored. Li and Ingber[69] used a two-dimensional boundary element simulation and calculated the period of rotation for an ellipse in plane Couette flow and concluded that the effect of the wall is felt if the channel width is less than about five times the particle length. Ingber and Mondy[70] used a three-dimensional boundary element formulation to compute fiber motion near a wall. Their solution for the angular motion of a single fiber was within 5% of Jeffery's analytical solution. In the presence of a wall near the fiber they found that the angular velocity increases when the fiber is transverse to the flow and decreases away from that orientation. They also show that a two-dimensional formulation does not predict the correct physical behavior of three-dimensional systems. For example, the predicted angular motion from a two-dimensional formulation for an ellipse rotating near a wall is very different from a three-dimensional formulation prediction for an ellipsoid, of the same aspect ratio as the ellipse, rotating near a wall.

Yang and Leal[71,72] used slender body theory to estimate the motion of an elongated particle suspended in a simple shear flow near a flat fluid–fluid interface. Their result may be used to study the motion of a fiber near a solid–fluid interface too. They concluded that due to the presence of an interface the motion of the end of an elongated fiber near the interface is retarded. This leads to an increase in the torque exerted on the fiber which leads to an increase in the angular velocity. Also, the center of the particle in the neighborhood of the wall is expected to oscillate periodically perpendicular to the interface since there can be no interfacial deformation.

Wakiya[73] studied the motion of a spheroid in the presence of a wall. The spheroid was found to experience both a drag and a lift depending on its orientation with respect to the wall. Happel and Brenner[74] have used the method of reflections to solve for the force experienced by an arbitrary shaped particle translating, without rotating, in the vicinity of a wall. They found that for a sphere translating inside a circular tube along the axial direction, the drag initially decreases and then increases as the distance from the wall is decreased.

The qualitative behavior of a sedimenting rod in the vicinity of a vertical wall is dependent on its orientation. Russel et al.[75] used slender body theory to derive an expression for the angular velocity of sedimenting rods.

Barta and Liron[76,77] studied the slow motion of a slender body in the vicinity of a wall using a singularity method with asymptotic expansions. The resulting equations were valid for separations from the wall of the order of the particle length. They found that the wall effect was more pronounced with increasing aspect ratio of the particle. They also found that when the particle is not oriented parallel or perpendicular to the wall, there is a component of force that is not in the direction of motion. This should lead to particle migration.

Ascoli et al.[78] used the boundary integral method to estimate the normalized drag force on a particle in the vicinity of a wall. The results from this study were compared to asymptotic analytical solutions for simple shapes and the agreement between them was found to be good.

Using slender body theory, De Mestre and Russel[79] developed expressions for the force and torque exerted on a circular cylinder moving near a plane wall. Their analysis is valid when the particle–wall distance is of the order of the length of the particle. In general, they found that as the particle gets closer to the wall, the drag on the particle increases.

Bibbo *et al.*[80] found that a gap greater than 1.2 times the particle length was sufficient to obtain the rheological properties of suspensions experimentally without significant wall effects. This could be because in semi-concentrated suspensions the interactions between the fibers dominate over any effects the wall may have on the overall rheological behavior of the suspension.

Stover and Cohen[81] observed the motion of elongated particles in the vicinity of a wall in a Poiseuille flow. They used a Hele–Shaw cell with transparent walls so that the fiber motion could be observed. Two cameras from orthogonal directions were used to track the angular motion of the fiber. The Newtonian fluid they used was a mixture of corn syrup and water. They matched the fluid and fiber density so that settling effects were negligible. When the fiber centroid was more than one fiber length from the wall, the period of rotation was unaffected by the wall. Fibers very close to the wall underwent an irreversible non-hydrodynamic interaction that looked like 'pole vaulting'. This interaction moved the centroid to about half a fiber length from the wall. The motion after this was reversible and periodic with the fiber passing very close to the wall when it was transverse to the flow. They concluded that the motion of a fiber is described well by Jeffery's equation even in the presence of a wall, if an effective aspect ratio is used for the fiber based on the experimentally observed period of rotation. Thus the effect of the wall seems to leave the qualitative motion of a fiber unaltered.

Burget[82] has recently conducted fiber motion studies in simple shear flow. The flow field was generated between two belts moving at equal velocity in opposite directions (Fig. 2.6).

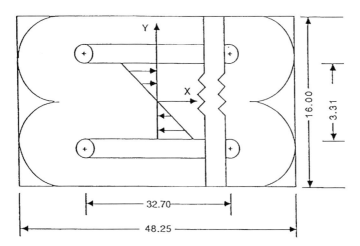

*2.6* Schematic representation of the shear flow apparatus used to study fiber–wall interactions along with the velocity profile and dimensions (in inches).[80]

The fluid used in these experiments was a mixture of corn syrup and water. The mixture was found to be Newtonian with a viscosity of about 6000 cP. The velocity field in the gap between the belts was characterized by injecting bubbles at various points in the flow and monitoring their motion. The measured velocity field agreed well with the expected linear profile. Fibers of aspect ratio 27.5, 43.3 and 50.5 were used in their study. The speed of the belts was also varied in order to obtain four different shear rates.

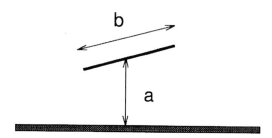

*2.7*  Fiber motion near a planar wall. *a* is the distance from the centroid of the fiber to the wall and *b* is the length of the fiber.[80]

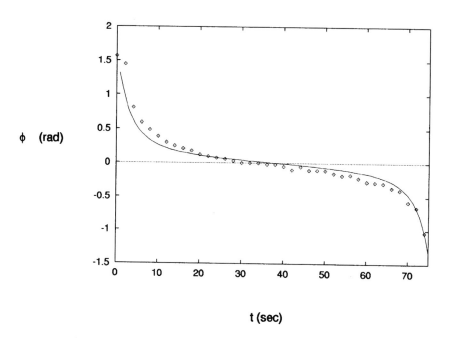

*2.8*  Comparison of experimental and Jeffrey's equation for rotation of a single fiber in simple shear flow for $a/b=5.41$ and $\dot{\gamma}=0.26$ s$^{-1}$.[80] —, Jeffery; ◇, experimental.

Figure 2.7 gives the explanation of symbols used in Burget's work. Fibers far away from the belt rotated according to Jeffery's equation over the entire period (Fig. 2.8).

However, as the fiber center moved closer to the wall, the fibers seemed to rotate faster. They modeled the increase in fiber angular velocity by defining an effective shear rate which was higher than the actual shear rate of the flow field. This effective shear rate was found to rise exponentially as the fiber moved closer to the wall (Fig. 2.9).

The results for fibers that were less than one fiber length from the wall may be summarized as follows. When the fiber is not aligned with the flow, Jeffery's equation with an effective shear rate (higher than the actual shear rate) describes the motion of the particle well. However, once the fiber is aligned with the wall, it tends to stay that way without executing any more rotations. The motion of the fiber could not be observed for very long since the field of view was limited. Therefore, this stabilizing effect of the wall was not studied in detail. Particles of smaller aspect ratio could be used so that the particle can be observed for a longer fraction of its period of rotation. However, if the particle aspect ratio becomes too small, slender body theory may no longer be applicable.

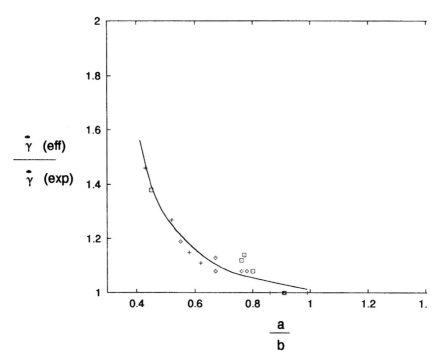

2.9 Normalized effective shear rate versus a/b. A curve fit suggests that the shear rate increases exponentially with decreasing distance from the wall of the apparatus.[80] ◇, exp ($r_e = 50.5$); +, exp($r_e = 43.3$); □, exp ($r_e = 27.5$), where $r_e$ is the effective fiber aspect ratio.

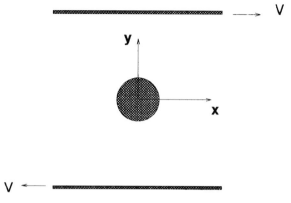

*2.10*  Placement of a circular cylinder in simple shear flow.[80]

Burget[82] superimposed the results of De Mestre and Russel[79] for the effect of a plane wall on the motion of a fiber with the flow field effects as given by Cox[29] in order to estimate the effective shear rate theoretically. The predictions from this superposition compared well with the effective shear rates obtained by curve fitting Jeffery's equation to experimentally observed fiber motion. Burget also studied the effect of obstacles such as a cylinder in the flow field on the angular velocity of a fiber in simple shear flow. Using the schematic shown in Fig. 2.10, it was found that as the fiber moves closer to the wall of the obstacle, it experiences a higher effective shear rate as in the case of a plane wall.

The effect of fiber–wall interactions is still not well understood in a quantitative sense. However, there has been some significant progress in the recent past. The experimental evidence indicates that the qualitative nature of the fiber motion is unaltered due to the presence of a wall except for the stabilizing effect found by Burget.[82] This needs to be studied further. If the fibers near the wall align quickly and remain in that orientation, the rheology of the suspension in the vicinity of the wall could be very different from that in the bulk. This would affect the flow field and the orientation evolution significantly. Therefore, the boundary conditions imposed at the wall need to account for this. Theoretical approaches such as the boundary element method may be used to develop a quantitative understanding of the fiber–wall interactions.

## 2.7    Summary and outlook

There are several ways to represent the orientation state of fiber suspensions. These include orientation distribution functions that provide the most descriptive representation, orientation tensors that provide sufficient information and can be directly computed from flow and scalar orientation parameters that are best suited for representing results and cannot typically be used to compute the evolution of the orientation state of suspensions. The advantage that orientation tensors offer is that they provide a direct link to the rheology and other properties of suspensions.

Fiber–fiber interactions have been the focus of this chapter. There have been various steps taken towards improving the understanding of these interactions in non-dilute suspensions. In semi-concentrated suspensions, diffusive models and slender body theory-based models offer the least computationally intensive solutions. The limitations are that if one were to use the diffusive models, an estimate is required for the interaction coefficient (diffusivity). A few relationships have been suggested to obtain this value. Slender body theory is strictly applicable only for long slender rigid bodies. This should be kept in mind while applying these theories to short-fiber suspensions. Stokesian dynamics provides a good hybrid in the sense that it uses fundamental solutions along with computations to make the calculations efficient. However, it is restricted to those cases where fundamental solutions are available. The purely numerical methods are computationally intensive and cannot be used to solve formidable problems yet. However, the solution is 'exact' in a sense and may be used to gain insight into the behavior of suspensions. As more powerful computers emerge and computational techniques become more refined, this approach is bound to be more feasible. However, from a practical viewpoint, such an approach could never be used in a mold filling simulation. Diffusive models are best suited to those applications with orientation tensors as primary variables. Many commercial codes of injection molding such as Moldflow, C-Flow and Polyflow[83,84] are now available that compute the orientation tensors in a complex geometry with diffusive models. The future goal should be to improve the predictive capability of such models to represent the fiber–fiber interactions and fiber–wall interactions more accurately.

Compared to fiber–fiber interactions, fiber–wall interactions have not attracted much attention. There is still no consensus on the effect a wall has on the motion of fibers. While some studies claim that the presence of a wall more than one fiber length away has no effect, others predict the effect to be significant even at a few fiber lengths away. Some recent numerical and experimental studies are trying to address this controversy. However, more studies in this area will also shed light on how the boundary conditions for fiber suspension flows in mold filling simulations should be addressed and also provide predictive capability of orientation distribution near the mold walls and inserts. This will prove useful in designing the tool to obtain the desired properties of fiber-reinforced molded structures.

Most of the work directed towards understanding fiber–fiber and fiber–wall interactions has dealt with only Newtonian suspending liquids. While it is important to understand the mechanics of these interactions in simple liquids, it should also be borne in mind that most of the applications such as injection molding of short fiber composites involves non-Newtonian liquids. There have been some studies[85–89] in the past that have addressed the motion of a single particle suspended in a non-Newtonian liquid both experimentally and theoretically. These studies have noted some similarities and some differences between the Newtonian and the non-Newtonian cases. Fiber–fiber interactions are influenced strongly by the hydrodynamic decay of the disturbance caused by fibers. The characteristics of this decay should change with the rheology of the suspending liquid. Further work is required in this area to attain an understanding of the effect of suspending fluid rheology on fiber–fiber interactions.

## References

1. C.L. Tucker III and S. G. Advani. Processing of short fiber systems. In *Flow and Rheology in Polymer Composites Manufacturing*, ed. S. G. Advani volume 10 of Composites Materials Series, Chapter 6, pp.147–202. Elsevier, Amsterdam, 1994.

2. H. L. Goldsmith and S. G Mason. *Rheology: Theory and Applications*. Academic Press, 1967.

3. R. G. Cox and S. G. Mason. Suspended particles in fluid flow through tubes. *Ann. Rev. Fluid Mech.*, 3:291, 1971.

4. L. G. Leal. Particle motions in viscous fluids. *Anni Rev. Fluid Mech.*, 12:453, 1976.

5. F. P. Folgar and C. L. Tucker. Orientation behavior of fibers in concentrated suspensions. *J. Reinforced Plastics Comp.*, 3:98, 1984.

6. P. A. Arp and S. G. Mason. Interactions between two rods in shear flow. *J. Colloid Interface Sci.*, 59:378, 1977.

7. W. C. Jackson, S. G. Advani, and C. L. Tucker. Predicting the orientation of short fibers in thin compression moldings. *J. Comp. Mater.*, 26:539, 1986.

8. G. B. Jeffery. The motion of ellipsoidal particles immersed in a viscous fluid. *Proc. R. Soc. London, Ser A.*, 102:161, 1923.

9. J. M. Burgers. Second report on viscosity and plasticity. *Technical Report*, North Holland, Amsterdam, 1938.

10. E. O. Tuck. Some methods for flows past blunt slender bodies. *J. Fluid Mech.*, 18:619, 1964.

11. J. P. K. Tillett. Axial and transverse stokes flow past slender axisymmetric bodies. *J. Fluid Mech.*, 44:401, 1970.

12. R. G. Cox. The motion of long slender bodies in a viscous fluid. Part I. General theory. *J. Fluid Mech.*, 44:791, 1970.

13. G. K. Batchelor. Slender-body theory for particles of arbitary cross-section in Stokes flow. *J. Fluid Mech.*, 44(3):419, 1970.

14. F. P. Bretherton. The motion of rigid particles in a shear flow at low Reynolds number. *J. Fluid Mech.*, 14:284, 1962.

15. B. J. Trevelyan and S. G. Mason. Particle motions in sheared suspensions. I. Rotations. *J. Colloid Sci.*, 6:354, 1951.

16. S. G. Mason and R. St J. Manley. Particle motions in sheared suspensions: orientations and interactions of rigid rods. *Proc. R. Soc., Ser A*, 238:117, 1957.

17. H. L. Goldsmith and S. G. Mason. Particle motions in sheared suspensions. xiii. The spin and rotation of disks. *J. Fluid Mech.*, 12:88, 1962.

18. C. E. Chaffey, M. Takano and S. G. Mason. Particle motions in sheared suspensions. xvi. Orientation of rods and disks in hyperbolic and other flows. *Can. J. Phys.*, **43**:1269, 1965.

19. E. Anczurowski and S. G. Mason. The kinetics of flowing dispersions. iii. Equilibrium orientations of rods and discs (experimental). *J. Colloid Interface Sci.*, **23**:533, 1967.

20. E. Anczurowski and S. G. Mason. Particle motions in sheared suspensions. xxiv. Rotation of rigid spheroids and cylinders. *Trans. Soc. Rheol.*, **12**(2):209, 1968.

21. S. G. Advani and C. L. Tucker. The use of tensors to describe and predict fiber orientation in short fiber composites. *J. Rheol.*, **31**:751, 1987.

22. P. H. Hermans. *Contributions to the Physics of Cellulose Fibers.* Elsevier, 1946.

23. C. L. Tucker. Predicting fiber orientation in short fiber composites. In *Manufacturing International '88, The Manufacturing Science of Composites, Atlanta*, ed. T. G. Gutowski, volume IV, page 95. American Society of Mechanical Engineers, 1988.

24. S. G. Advani and C. L. Tucker. Closure approximations for three-dimensional structure tensors. *J. Rheol.*, **34**(3):367, 1990.

25. E. Anczurowski and S. G. Mason. The kinetics of flowing dispersions. ii. Equilibrium orientations of rods and discs (theoretical). *J. Colloid Interface Sci.*, **23**:522, 1967.

26. R. Eisenschitz. *Z. Physik. Chem. A*, **158**:85, 1932.

27. P. G. Saffman. *J. Fluid Mech.*, **1**:540, 1956.

28. E. Anczurowski, R. G. Cox and S. G. Mason. The kinetics of flowing dispersions. iv. Transient orientation of cylinders. *J. Colloid Interface Sci.*, **23**:547, 1967.

29. R. G. Cox. The motion of long slender bodies in a viscous fluid. Part ii. Shear flow. *J. Fluid Mech.*, **45**:625, 1971.

30. A. Okagawa, R. G. Cox and S. G. Mason. The kinetics of flowing dispersions. vi. Transient orientation and rheological phenomena of rods and discs in shear flow. *J. Colloid Interface Sci.*, **45**(2):303, 1973.

31. F. P. Folgar and C. L. Tucker. Orientation behaviour of fibers in concentrated suspensions. *J. Reinforced Plastics Compos.*, **3**:98, 1984.

32. F. P. Folgar. Fiber *Orientation Distribution in Concentrated Suspensions: A Predictive Model.* PhD thesis, University of Illinois at Urbana-Champaign, 1983.

33. S. G. Advani and C. L. Tucker. A numerical simulation of short fiber orientation in compression molding. *Polym. Compos.*, **11**(3):164, 1990.

34. S. Ranganathan and S. G. Advani. Fiber–fiber interactions in homogeneous flows of nondilute suspensions. *J. Rheol.*, **35**(8):1499, 1991.

35. R. Shanker. *The Effect of Nonhomogeneous Flow Fields and Hydrodynamic Interactions on the Rheology of Fiber Suspensions.* PhD thesis, University of Delaware, December 1991.

36. S. M. Dinh and R. C. Armstrong. A rheological equation of state for semiconcentrated suspensions. *J. Rheol.*, **28**(3):207, 1984.

37. S. Ranganathan. *Mechanics of Fiber–fiber Interactions during the Flow of Non-dilute Short Fiber Suspensions.* PhD thesis, University of Delaware, 1993.

38. M. R. Kamal and A. T. Mutel. The prediction of flow and orientation behavior of short fiber reinforced melts in simple flow systems. *Poly. Compos.*, **10**(5):337, 1989.

39. M. R. Kamal and A. T. Mutel. Rheological properties of suspensions in Newtonian and non-Newtonian fluids. *J. Polym. Eng.*, **5**(4):293, 1985.

40. M. A. Bibbo and R. C. Armstrong. Rheology of semi-concentrated fiber suspensions in Newtonian and non-Newtonian fluids. In *Manufacturing International '88, The Manufacturing Science of Composites*, ed. T. G. Gutowski, volume IV, page 105. American Society of Mechanical Engineers, 1988.

41. R. S. Bay. *Fiber Orientation in Injection Molded Composites: A Comparison of Theory and Experiment*. PhD thesis, University of Illinois at Urbana-Champaign, August 1991.

42. C. A. Stover. *The Dynamics of Fibers Suspended in Shear Flows*. PhD thesis, Cornell University, 1991.

43. C. A. Stover, D. Koch and C. Cohen. Observations of fiber orientation in simple shear flow of semi-dilute suspensions. *J. Fluid Mech.*, **238**:277, 1992.

44. E. S. G. Shaqfeh and D. L. Koch. Orientation dispersion of fibers in extensional flows. *Phys. Fluids A*, **4**, 1990.

45. J. G. Evans. *The Flow of a Suspension of Force-free Rigid Rods in a Newtonian Fluid*. PhD thesis, University of Cambridge, July 1975.

46. D. L. Koch and E. S. G. Shaqfeh. The average rotation rate of a fiber in the linear flow of a semi-dilute suspension. *Phy. Fluids A*, **2**:2093, 1990.

47. M. Rahnama, D. L. Koch and E. S. G. Shaqfeh. The effect of hydrodynamic interactions on the orientation distribution in a fiber suspension subject to simple shear flow. *Phys. Fluids*, **7**(3):487, 1995.

48. R. G. Cox. The motion of suspended particles almost in contact. *Internat. J. Multiphase Flow* **1**:343, 1974.

49. J. F. Brady and G. Bossis. Stokesian dynamics. *Ann. Rev. Fluid Mech.*, **20**:111, 1988.

50. L. Durlofsky, J. F. Brady and G. Bossis. Dynamic simulation of hydrodynamically interacting particles. *J. Fluid Mech.*, **180**:21, 1987.

51. P. A. Arp and S. G. Mason. The kinetics of flowing dispersions. viii. Doublets of rigid spheres (theoretical). *J. Colloid Interface Sci.*, **61**:21, 1977.

52. D. J. Jeffrey and Y. Onishi. Calculation of the resistance and mobility functions for two unequal rigid spheres in low Reynolds number flow. *J. Fluid Mech.*, **139**:261, 1984.

53. S. Kim and R. T. Mifflin. The resistance and mobility functions for two equal spheres in low Reynolds number flow. *Phys. Fluids*, **28**:2033, 1985.

54. L. J. Durlofsky and J. F. Brady. Dynamic simulation of bounded suspensions of hydrodynamically interacting particles. *J. Fluid Mech.*, **200**:39, 1989.

55. I. L. Claeys and J. F. Brady. Suspensions spheroids in Stokes flow. 1. Dynamics of a finite number of particles in an unbounded fluid. *J. Fluid Mech.*, **251**:411, 1993.

56. S. J. Karilla, Y. O. Feuntes and S. Kim. Parallel computational strategies for hydrodynamic interactions between rigid particles of arbitrary shape in a viscous fluid. *J. Rheol.*, **33**:913, 1989.

57. S. J. Karilla and S. Kim. Integral equations of the second kind for stokes flow: direct solution for physical variables and removal of inherent accuracy limitations. *Chem. Eng. Commun.*, **82**:123, 1989.

58. M. S. Ingber. Dynamic simulation of the hydrodynamic interactions among immersed particles in Stokes flow. *Internat. J. Numer. Method Fluids*, **10**:791, 1990.

59. C. Chan. *Hydrodynamic Interactions in Large Aspect Ratio Fiber Suspensions*. PhD thesis, University of Delaware, 1991.

60. C. Y. Chan, A. N. Beris and S. G. Advani. Second order boundary element method calculation of hydrodynamic interactions between particles in close proximity. *Internat. J. Numer. Method Fluids*, **14**:1063, 1992.

61. B. J. Yoon and S. Kim. Boundary collocation method for the motion of two spheroids in Stokes flow. Hydrodynamic and colloidal interactions. *Internat. J. Multiphase Flow*, **16**(4):639, 1990.

62. S. Kim and M. Amann. Simulation of microstructure evolution on high performace parallel compter architectures: Communications scheduling strategies for CDL-BIEM. *International Conference on Boundary Element Technology*, Madison, Wisconsin, Computational Mechanics Publications: Southampton, page 863, 1992.

63. P. Pakdel and S. Kim. On the capabilities of the double layer representation for Stokes flow. Part i. Analytical solutions. *Eng. Analysis Boundary Elements*, **13**(4):339, 1994.

64.  P. Pakdel and S. Kim. On the capabilities of the double layer representation for Stokes flow. Part ii. Iterative solutions. *Eng. Analysis Boundary Elements*, **13**(4):349, 1994.

65. C. Seeling and N. Phan Thien. Completed double layer boundary element algorithm in many body problems for a multi-processor: An implementation on the CM-5. *Computational Mech.*, **15**(1):31, 1994.

66. S. Toll and J. A. E. Manson. Dynamics of a planar concentrated fiber suspension with non-hydrodynamic interaction. *J. Rheol.*, **38**(4):985, 1994.

67. S. Toll. On the 'tube' model for fiber suspensions. *J. Rheol.*, **37**:123, 1993.

68. R. L. Powell. Rheology of suspensions of rod-like particles. *J. Statistical Phys.*, **62**:1073, 1991.

69. J. Li and M. S. Ingber. A boundary element study of the motion of rigid particles in internal Stokes flow. In *Boundary Elements XIII*, eds. C. A. Brebbia and G. S. Gibson, page 149. Elsevier Applied Science, 1991.

70. M. S. Ingber and L. A. Mondy. A numerical study of three dimensional jeffery orbits in shear flow. *J. Rheol.* **38**(6):1829, 1994.

71. S. M. Yang and L. G. Leal. Particle motion in stokes flow near a plane fluid-fluid interface. Part 1. Slender body in a quiescent fluid. *J. Fluid Mech.*, **136**:393, 1983.

72. S. M. Yang and L. G. Leal. Particle motion in stokes flow near a plane fluid-fluid interface. Part 2. Linear shear and axisymmetric straining flows. *J. Fluid Mech.*, **149**:275, 1984.

73. S. Wakiya. Viscous flows past a spheroid. *J. Phys. Soc. Jpn.*, **12**(10):1130, 1957.

74. J. Happel and H. Brenner. *Low Reynolds Number Hydrodynamics.* Prentice-Hall, 1965.

75. W. B. Russel, E. J. Hinch, L. G. Leal and G. Tieffenbruck. Rods falling near a vertical wall. *J. Fluid Mech.*, **83**:273, 1977.

76. E. Barta and N. Liron. Slender body interactions for low Reynold's numbers. Part 1. Body–wall interactions. *SIAM J. Appl. Math.*, **48**(5):992, 1988.

77. E. Barta and N. Liron. Slender body interactions for low Reynold's numbers. Part 2. Body–body interactions. *SIAM J. Appl. Math.*, **48**(6):1262, 1988.

78. E. P. Ascoli, D. S. Dandy and L. G. Leal. Low Reynolds number hydrodynamic interaction of a solid particle with a planar wall. *Internat. J. Numer. Method. Fluids*, **9**:651, 1989.

79. N. J. DeMestre and W. B. Russel. Low Reynolds number translation of a slender cylinder near a plane wall. *J. Eng. Math.*, **9**:81, 1975.

80. M. A. Bibbo, S. M. Dinh and R. C. Armstrong. Shear flow properties of semiconcentrated fiber suspensions. *J. Rheol.*, **29**(6):905, 1985.
81. C. A. Stover and C. Cohen. The motion of rod-like particles in the pressure-driven flow between two flat plates. *Rheol. Acta*, **29**(3):199, 1990.
82. K. M. Burget. *An Investigation of Fiber–Wall Interactions in Simple Shear Flow.* Master's thesis, University of Delaware, 1994.
83. M. J. Crochet, B. Debbaut, R. Keunings and J. M. Marchal. A multi-purpose finite element program for continuous polymer flows. *Computer Modeling for Extrusion and Other Continuous Polymer Processes*, eds. K. T. O'Brien and E. C. Bernhardt. page 25. Hanser, Munich, 1992.
84. G. Suh, C. K. Yoon and S. McCarthy. Verification study on fiber orientation software. In *Proceedings of ANTEC*, Detroit. Society of Plastics Engineers, page 679, 1995.
85. F. Gauthier, H. L. Goldsmith and and S. G. Mason. Particle motions in non-Newtonian media. ii. Poiseuille flow. *Trans. Soc. Rheol.*, **15**(2):297, 1971.
86. F. Gauthier, H. L. Goldsmith and S. G. Mason. Particle motions in a non-Newtonian media. i. Couette flow. *Rheol. Acta*, **10**:344, 1971.
87. L. G. Leal. The slow motion of slender rod-like particles in a second-order fluid. *J. Fluid Mech.*, **69**:305, 1975.
88. J. Brunn. The slow motion of a rigid particle in a second order fluid. *J. Fluid Mech.*, **82**:529, 1977.
89. P. J. Brunn. The motion of a slightly deformed sphere in a viscoelastic fluid. *Rheol. Acta*, **18**:229, 1979.

# 3

# Macroscopic modelling of the evolution of fibre orientation during flow

NHAN PHAN-THIEN AND RONG ZHENG

All current constitutive theories of fibre suspensions consist of an evolution equation for the microstructure, which is idealized either by a unit vector or a second order tensor field, and a stress rule, allowing the calculation of the stress tensor from the unit vector field. These major theories are reviewed in this chapter. We also describe two numerical implementations of the theories for the simulation of the fibres orientation in flow processes. Practical applications including the flow past a sphere and a complex three-dimensional injection moulding problem are described.

## 3.1 Introduction

Suspensions abound in natural and man-made materials: blood, paint, slurries, mineral concentrates, mine tailings, clay, cement, fibre-reinforced polymeric materials, etc. When the suspended particles are slender (usually referred to as fibres), the orientation of these particles in suspension strongly affects the rheological behaviour of the suspension, as well as the properties of the composite material produced from such a suspension. Because of its importance in industries, the flow of fibre suspensions has been of interest in fluid mechanics during the last few decades.

The concentration of fibre suspensions is usually classified into three regimes: dilute, semi-concentrated (or semi-dilute) and concentrated. The suspension is called dilute if there is only one fibre in a volume of $V = l^3$, where $l$ is the length of the fibres; the volume fraction therefore satisfies $\phi < d^2 l / V$, or, $\phi a_R^2 < 1$, where $d$ is the diameter of the fibre and $a_R = l/d$ is its aspect ratio. In dilute suspensions, each fibre can therefore freely rotate. The concentration region $1 < \phi a_R^2 < a_R$ is called semi-concentrated, where each fibre is confined in the volume $d^2 l < V < dl^2$. The spacing between the fibres is greater than the fibre diameter but less than the fibre length. In this regime the fibres have only two rotating degrees of freedom. Finally, the suspension with $\phi a_R > 1$ is called concentrated, where the average distance between fibres is less than a fibre diameter, and therefore fibres cannot rotate independently except around their symmetry axes. Any motion of the fibre must necessarily involve a cooperative motion of surrounding fibres.

The central problem of theoretical suspension rheology is to develop a suitable constitutive relation by which the macroscopic rheology properties of a suspension can be predicted from a knowledge of the properties of the particles (e.g. geometry, volume fraction and orientation) and the suspending fluid. There are two different approaches to the constitutive quest: the continuum and the microstructure modelling approaches. The microstructural models provide a direct link between the microstructural properties and the macroscopic rheological behaviour of suspensions. Several investigations on anisotropic constitutive models of fibre suspensions have been reported. The microstructural models developed by Batchelor (1970), Doi and Edward (1978a, b), Hinch and Leal (1972, 1976), Dinh and Armstrong (1984), Lipscomb et al. (1988) and Phan-Thien (1995) have similar functional forms to those derived from continuum mechanics in the early work of Ericksen (1960) and Hand (1962). It is acknowledged now that the behaviour of dilute suspensions in a simple shear flow is well understood, following Leal and Hinch's work (1971, 1972, 1973). Research on suspension rheology has been directed toward complex flows and non-dilute regimes.

The primary interest in constitutive modelling of suspensions is the determination of the reduced viscosity $\eta_r = \eta/\eta_s$ (ratio of the suspension viscosity to the solvent viscosity). Viscosities of fibre suspensions depend not only on the volume fraction of the fibres, but also on the orientation distribution of the fibres. Therefore, an evolution equation in the fibre orientations is required; this then can be used in the anisotropic constitutive model to describe the flow behaviour of the suspension. The starting point of almost all theoretical work has been the evolution equation by Jeffery (1922) for the motion of an isolate rigid spheroid in a Newtonian fluid. The significance of Jeffery's theory is that the presence of non-spherical particles in a Newtonian solvent can give rise to non-Newtonian flow behaviour. For dilute suspensions, it is reasonable to neglect the interactions between particles. For non-dilute suspensions, however, particle–particle interactions have to be considered, which can affect the flow behaviour. Folgar and Tucker (1984) developed an evolution equation for concentrated fibre suspensions, where the fibre–fibre interactions are taken into account by adding a diffusion term to Jeffery's equation. Dinh and Armstrong (1984) discussed the dynamics of non-Brownian particles and derived a constitutive equation for semi-dilute suspensions; the model takes into account the fibre–fibre interaction and uses a distribution function to describe the orientation state.

It has been observed that particles in concentrated suspensions subjected to an inhomogeneous shear could migrate across the streamlines, from regions of high strain rates to regions of low strain rates (Karnis et al., 1966; Arp and Mason, 1977; Gadala-Maria and Acrivos, 1980; Leighton, 1985; Hookham, 1986; Graham et al., 1991; Abbott et al., 1991; Koh et al., 1994). Some attempts to understand the basic mechanisms behind the shear-induced migration have been made. A recent review was given by Acrivos (1995). A notable example is the diffusion equation derived by Phillips *et al.* (1992) based on the scaling arguments of Leighton and Acrivos (1987), where the particle flux is considered to be a balance between a contribution due to a spatially varying interaction frequency and a contribution due to a spatially varying viscosity. The model yields excellent predictions in circular Couette flow and reasonable results for in eccentric cylinder flow geometries for suspensions of spheres (Abbott *et al.*, 1991; Phan-Thien *et al.* 1995). The implications of migration effects in suspensions of fibres have also been observed (Mondy *et al.* 1994).

Numerical methods developed for the solution of Jeffery's equations or extensions of these equations have been successful, to some extent, in predicting the orientation state of fibre suspension in complex flows. Most of these numerical analyses assumed that the (Newtonian) velocity field is unaffected by the orientation of the fibres (Goettler *et al.*, 1981; Givler *et al.*, 1983; Advani and Tucker, 1987; Gupta and Wang, 1993; Altan and Rao, 1995; Zheng *et al.*, 1996). The coupled problem, where the orientation of the fibres and the flow kinematics are solved simultaneously to provide the solution, has also been attempted numerically. Papanastasiou and Alexandrou (1987), Lipscomb *et al.* (1988), Zheng *et al.* (1990b), Phan-Thien *et al.* (1991b), Phan-Thien and Graham (1991), Rosenberg *et al.* (1990) and Altan *et al.*, (1992) are among those workers who have solved the velocity and orientation fields simultaneously. It was found (e.g. by Lipscomb *et al.*, 1988) that the uncoupled calculation may be grossly in error in a complex flow containing strong elongational components, since in such a flow the streamlines can be drastically altered due to the fibres.

Available suspension theories almost invariably assume that the suspending medium is a Newtonian fluid. However, of greatest interest in practical problems are particles dispersed in polymeric solvents, which in most cases are viscoelastic (a comprehensive description of the behaviour of viscoelastic fluids, with mathematical models, can be found in Bird *et al.*, 1987a and Tanner, 1988). Joseph and Liu (1993) have shown that a viscoelastic solvent dramatically changes the nature of a fibre orbit by examining the motion of a sedimenting needle in Newtonian and viscoelastic fluids. This suggests that the assumption of a Newtonian base flow may be inadequate for describing the behaviour of suspensions in viscoelastic fluids. Despite the inadequacy of this assumption, rapid progress has been seen in the application of the existing fibre suspension theories to practical problems, such as the flow past a sphere (Phan-Thien *et al.*, 1991b; Phan-Thien and Graham, 1991; Rosenberg *et al.*, 1990) and the fibre-filled polymeric materials in injection moulding (Altan *et al.*, 1990; Zheng *et al.*, 1990b; Friedl and Brouwer, 1991; Bay and Tucker, 1992; Frahan *et al.*, 1992; Gupta and Wang 1993; Crochet *et al.*, 1994; Henry *et al.*, 1994; Zheng *et al.*, 1996).

Much of the interest in the flow past a sphere arises from the possibility of measuring the effective viscosity by observing the fall of a precision sphere through the test fluid. The apparatus required is relatively simple to construct and the underlying theory for Newtonian fluids is well understood. Walters and

Tanner (1992) have also reviewed studies of the motion of a sphere through viscoelastic fluids. When the test fluid is a fibre suspension, the problem is considerably more complex. First, as the initial orientation of the fibres cannot be controlled to an arbitrary accuracy, the effect of the initial orientation must be considered. Second, the fluid viscosity could be influenced by a local rearrangement of the microstructure due to the motion of the sphere. Third, the time dependent feature of the drag force on the sphere must be considered. Furthermore, complexity also arises from the particle migration.

Another important application area is injection moulding, which is the most widely used process for forming plastic products. The numerical simulation of injection moulding has been part of computer aided engineering (CAE) for many years. Several injection-moulding-specific computational programs are now commercially available (e.g. Moldflow, C-Flow, Cadmould, Timon, Simuflow, CAE-Mold, Caplas, McKam, etc.). The prediction of fibre orientations in injection moulding has recently been a subject of interest owing to the increasing use of short fibre-filled composite materials. Flow of a fibre suspension during injection moulding results in a flow-induced fibre orientation. After the composite solidifies, the anisotropic orientation of the fibres is frozen into the matrix and significantly affects the properties (e.g. dimensional accuracy, elastic moduli and thermal expansion coefficients) of the final product. Therefore an accurate prediction of fibre orientation is highly relevant for producing injection-moulded parts with good quality.

The objective of this chapter is to describe some mathematical models and numerical methods for the macroscopic modelling of the evolution of fibre orientation during flow. We shall begin in Section 3.2 with a review of some models of fibre suspensions. In view of the large number of contributions that have been made in this field, it is not possible to review them all here; the selection somewhat reflects the authors' interest. In Section 3.3, we introduce some numerical methods for the simulation of fibre orientations. Again, we make no attempt to discuss the variety of available numerical techniques developed for this purpose; we will rather focus on a boundary element method for which we have had some success in our early work. Finally, in Section 3.4, we present examples of practical applications including the flow past a sphere and a complex injection moulding problem.

## 3.2    Theory

### 3.2.1    Evolution equations

It is customary to use a unit vector to specify the orientation of a small, rigid, axially symmetric particle. The orientation vector directed along the axis may be denoted by $\mathbf{p}$. Then the motion of the particle with the bulk fluid deformation can be described by some functional $\dot{\mathbf{p}}(\mathbf{p}, t)$, where $\dot{\mathbf{p}}$ is the material derivative

$D\mathbf{p}/Dt$. The difference between various models to describe the particle motion is the expression used for $\dot{\mathbf{p}}$. Any fibre orientation model must satisfy conditions of continuity, periodicity and normalization. In addition, in suspension mechanics, the principle of material objectivity may only be applied when particle inertia is negligible (Ryskin and Rallison, 1980). All microstructural models implicitly assume negligible microinertia, the effect of which at the macroscale level has not been investigated in sufficient detail. In addition, the fluid inertia is ignored in most cases, since the particle Reynolds number is negligible ($\sim 10^{-3}$). However, the weak inertial secondary flow may have accumulated effects on the re-ordering of the microstructure (Petit and Noetinger, 1988).

### 3.2.1.1  Jeffery's Orbit

The best known model for a single fibre orientation is Jeffery's equation, derived by modelling the fibre as an inertialess rigid ellipsoid suspended in a Newtonian fluid. The microstructure of the fluid is characterized by a unit vector field $\mathbf{p}$, which evolves in time according to, in the absence of Brownian motion,

$$\dot{\mathbf{p}} = \mathbf{W} \cdot \mathbf{p} + \lambda(\mathbf{D} \cdot \mathbf{p} - \mathbf{D} : \mathbf{ppp}), \qquad [1]$$

in which $\lambda$ is given by

$$\lambda = \frac{a_R^2 - 1}{a_R^2 + 1}$$

and $\mathbf{W} = (\nabla \mathbf{u}^T - \nabla \mathbf{u})/2$ is the vorticity tensor, $\mathbf{D} = (\nabla \mathbf{u}^T + \nabla \mathbf{u})/2$ is the strain rate tensor with superscript $T$ denoting the transpose operation and $a_R$ is the aspect ratio of the microstructure. Note that as $\dot{\mathbf{p}} \cdot \mathbf{p} = 0$, the magnitude of $\mathbf{p}$ is preserved in this time evolution. If $\mathbf{p}$ is initially a unit vector, then it remains a unit vector at all times. There are two physical interpretations for $\mathbf{p}$. First, it can be regarded as the local orientation of an individual fibre. Second, in the case where there is some Brownian motion present, then it represents the averaged configuration. The term $\mathbf{W} \cdot \mathbf{p}$ indicates that $\mathbf{p}$ rotates with the fluid, while the term $\mathbf{D} \cdot \mathbf{p}$ represents the component of the strain with the fluid. Since $\mathbf{p}$ is of unit length, the stretching component $\mathbf{D} : \mathbf{ppp}$ must be subtracted, producing the last term. In shear flows of non-interacting fibres, the fibres exhibit a closed periodic rotation known as a Jeffery orbit. Equation [1] can be rewritten as

$$\dot{\mathbf{p}} = \mathscr{L} \cdot \mathbf{p} - \mathscr{L} : \mathbf{ppp} \qquad [2]$$

where the 'effective' velocity gradient tensor is $\mathscr{L} = \mathbf{L} - \zeta \mathbf{D}$, with $\zeta = 1 - \lambda = 2/(a_R^2 + 1)$. This is reminiscent of the effective velocity gradient tensor that has been used in a number of non-affine network theories.

With Brownian motion, there is a random excitation in Eqn [1], represented by some white noise on the space orthogonal to $\mathbf{p}$, with a strength determined from the fluctuation–dissipation theorem. In this case, we write

$$\dot{\mathbf{p}} = \mathscr{L} \cdot \mathbf{p} - \mathscr{L} : \mathbf{ppp} + (1 - \mathbf{pp}) \cdot \mathbf{F}^{(b)}(t), \qquad [3]$$

where the Brownian motion $\mathbf{F}^{(b)}(t)$ has zero mean and delta correlation function:

$$\langle \mathbf{F}^{(b)}(t + s)\mathbf{F}^{(b)}(t) \rangle = 2D_r \delta(s)\mathbf{1} \qquad [4]$$

and $D_r$ is the rotational diffusivity of the process $\delta(s)$ is the delta Dirac function and the angular brackets denote the ensemble average with respect to the probability density function of the process concerned. The factor $(1 - \mathbf{pp})$ in front of $\mathbf{F}^{(b)}$ is the statement that only rotational Brownian motion is allowed. To complete the description of the micromechanics, the probability density function $\psi(\mathbf{p})$ must be specified. This quantity satisfies the Liouville equation (Chandrasekhar, 1943)

$$\frac{\partial \psi}{\partial t} = \frac{\partial}{\partial \mathbf{p}} \cdot \left\{ \left\langle \frac{\Delta \mathbf{p} \Delta \mathbf{p}}{2\Delta t} \right\rangle \frac{\partial \psi}{\partial \mathbf{p}} - \left\langle \frac{\Delta \mathbf{p}}{\Delta t} \right\rangle \psi \right\}. \qquad [5]$$

With

$$\left\langle \frac{\Delta \mathbf{p}}{\Delta t} \right\rangle = \mathscr{L} \cdot \mathbf{p} - \mathscr{L} : \mathbf{ppp}$$

and

$$\left\langle \frac{\Delta \mathbf{p} \Delta \mathbf{p}}{2\Delta t} \right\rangle = D_r(1 - \mathbf{pp})$$

the diffusion or Fokker–Planck equation for the probability density becomes

$$\frac{\partial \psi}{\partial t} = \frac{\partial}{\partial \mathbf{p}} \cdot \left\{ D_r(1 - \mathbf{pp})\frac{\partial \psi}{\partial \mathbf{p}} - (\mathscr{L} \cdot \mathbf{p} - \mathscr{L} : \mathbf{ppp})\psi \right\}. \qquad [6]$$

In most published work, the diffusivity is usually written as a scalar $D_r$. However, the operator $\partial/\partial \mathbf{p}$ is interpreted as the two-dimensional gradient operator on a unit sphere surface, which is essentially equivalent to Eqn [6].

At high concentrations, Jeffery's evolution equation is no longer valid; in addition, the pairwise and higher distributions must also be specified to account for the multiparticle interactions. However, the dilute theory has been used to approximate the behaviour of suspensions beyond the dilute region. Ingber and Mondy (1994) have reported numerical simulations of three-dimensional Jeffery orbits in shear flows. They examined wall effects, particle interactions and non-linear shear flows and found that Jeffery theory provides a good approximation of the orientation trajectory for the particle in both linear and non-linear shear flows, even in proximity to other particles or walls.

### 3.2.1.2 Equations of change

The evolution equation for a dynamical quantity $Q(\mathbf{p})$ can be found by averaging it with respect to $\psi$. It can be derived without having to go through the diffusion equation by noting that

$$\frac{dQ}{dt} = \frac{\partial Q}{\partial p_k}\dot{p}_k = \frac{\partial Q}{\partial p_k}[\mathscr{L}_{km}p_m - \mathscr{L}_{mn}p_m p_n p_k + (\delta_{km} - p_k p_m)F_m^{(b)}(t)]. \quad [7]$$

Thus,

$$\frac{d}{dt}\langle Q \rangle = \mathscr{L}_{km}\left\langle p_m \frac{\partial Q}{\partial p_k}\right\rangle - \mathscr{L}_{mn}\left\langle p_m p_n p_k \frac{\partial Q}{\partial p_k}\right\rangle$$

$$+\left\langle\left(\frac{\partial Q}{\partial p_k}F_k^{(b)} - \frac{\partial Q}{\partial p_k}p_k p_m F_m^{(b)}\right)\right\rangle. \quad [8]$$

To evaluate the last two terms in the previous equation we should keep in mind the fast timescale of the Brownian force and use it to our advantage. First, we define

$$C_k = \frac{\partial Q}{\partial p_k}, \quad D_k = \frac{\partial Q}{\partial p_m}p_m p_k.$$

These vectors evolve in time according to, from Eqn [3],

$$\dot{C}_k = \frac{\partial^2 Q}{\partial p_k \partial p_l}[\mathscr{L}_{lm}p_m - \mathscr{L}_{mn}p_m p_n p_l + (\delta_{lm} - p_l p_m)F_m^{(b)}(t)],$$

and

$$\dot{D}_k = \frac{\partial^2 Q}{\partial p_m \partial p_n}p_m p_k[\mathscr{L}_{nr}p_r - \mathscr{L}_{rs}p_r p_s p_n + (\delta_{nr} - p_n p_r)F_r^{(b)}(t)]$$

$$+ \frac{\partial Q}{\partial p_m}p_m[\mathscr{L}_{kr}p_r - \mathscr{L}_{rs}p_r p_s p_k + (\delta_{kr} - p_k p_r)F_r^{(b)}(t)]$$

$$+ \frac{\partial Q}{\partial p_m}p_k[\mathscr{L}_{mr}p_r - \mathscr{L}_{rs}p_r p_s p_m + (\delta_{mr} - p_m p_r)F_r^{(b)}(t)].$$

Our plan is to write down formal solutions to the previous two equations at time $t + \Delta t$ in terms of those at time $t$, where $\Delta t$ is much larger than the fluctuation timescale $\tau_c$ of the Brownian force, but small compared to the relaxation times of the process, so that the two previous equations can be regarded as linear. Thus,

$$C_k(t + \Delta t) = C_k(t) + \frac{\partial^2 Q}{\partial p_k \partial p_l}(\delta_{lm} - p_l p_m)B_m^{(b)}(\Delta t) + E_k \Delta t$$

$$D_k(t + \Delta t) = D_k(t) + \frac{\partial^2 Q}{\partial p_m \partial p_n}p_m p_k(\delta_{nr} - p_n p_r)B_r^{(b)}(\Delta t)$$

$$+ \frac{\partial Q}{\partial p_m}p_m(\delta_{kr} - p_k p_r)B_r^{(b)}(\Delta t)$$

$$+ \frac{\partial Q}{\partial p_m}p_k(\delta_{mr} - p_m p_r)B_r^{(b)}(\Delta t) + F_k \Delta t \qquad [10]$$

where $E_k$ and $F_k$ are the 'drift' terms that do not involve the Brownian force, and

$$\mathbf{B}^{(b)}(\Delta t) = \int_t^{t+\Delta t} \mathbf{F}^{(b)}(t')dt'. \qquad [11]$$

The average of the last two terms in Eqn [8] is

$$\left\langle \frac{\partial Q}{\partial p_k}F_k^{(b)} - \frac{\partial Q}{\partial p_k}p_k p_m F_m^{(b)} \right\rangle = \langle C_k(t + \Delta t)F_k^{(b)}(t + \Delta t)\rangle$$

$$- \langle D_k(t + \Delta t)F_k^{(b)}(t + \Delta t)\rangle.$$

This average contains terms linear in the Brownian force ($C_k(t)$, $D_k(t)$ and the drift terms $E_k$ and $F_k$) which will not survive the averaging process, and terms that are quadratic in the Brownian force. These latter terms will contribute to the average, since*

$$\langle \mathbf{B}^{(b)}(t + \Delta t)\mathbf{F}^{(b)}(t + \Delta t)\rangle = \int_t^{t+\Delta t} \langle \mathbf{F}^{(b)}(t')\mathbf{F}^{(b)}(t + \Delta t)\rangle dt'$$

$$= D_r(\mathbf{1} - \mathbf{pp}).$$

Thus

$$\langle C_k F_k^{(b)} - D_k F_k^{(b)} \rangle = D_r \left\langle \left[ (\delta_{kl} - p_k p_l)\frac{\partial^2 Q}{\partial p_k \partial p_l} - 2p_k \frac{\partial Q}{\partial p_k} \right] \right\rangle$$

and the equation of change for $Q(\mathbf{p})$ is

$$\frac{d}{dt}\langle Q\rangle = \mathcal{L}_{km}\left\langle p_m \frac{\partial Q}{\partial p_k} \right\rangle - \mathcal{L}_{mn}\left\langle p_m p_n p_k \frac{\partial Q}{\partial p_k} \right\rangle$$

$$+ D_r \left\langle (\delta_{kl} - p_k p_l)\frac{\partial^2 Q}{\partial p_k \partial p_l} - 2p_k \frac{\partial Q}{\partial p_k} \right\rangle. \qquad [12]$$

*The factor of 1/2 arises because the delta correlation function overlaps the upper limit of the integral.

In particular, with $B = p_i p_j$,

$$\frac{\partial B}{\partial p_k} = \delta_{ik} p_j + \delta_{jk} p_i, \quad \frac{\partial^2 B}{\partial p_k \partial p_l} = \delta_{ik} \delta_{jl} + \delta_{jk} \delta_{il}$$

and

$$\frac{d}{dt} \langle \mathbf{pp} \rangle = \langle \mathcal{L} \cdot \mathbf{pp} + \mathbf{pp} \cdot \mathcal{L}^T - 2\mathcal{L} : \mathbf{pppp} \rangle + 2D_r \langle \mathbf{1} - 3\mathbf{pp} \rangle$$

leading to

$$\Lambda \left( \frac{\Delta}{\Delta t} \langle \mathbf{pp} \rangle + 2\mathcal{L} : \langle \mathbf{pppp} \rangle \right) + \langle \mathbf{pp} \rangle = \frac{1}{3} \mathbf{1} \qquad [13]$$

where $\Lambda = 1/6D_r$ is the relaxation time of the tumbling motion, and $\Delta/\Delta t$ is the upper convective derivative, defined with the effective velocity gradient tensor $\mathcal{L}$. This is exactly the expression given by Bird $et~al.$ (1987b), when $\mathcal{L}$ is identified with $\mathbf{L}$, for a rigid dumbbell in a homogeneous flow field. The advantages of this derivation are that surface integrations on the unit sphere are avoided and the Brownian motion need not be represented by white noise. Indeed, for arbitrary large amplitude fluctuations, the method outlined in Phan-Thien (1995) can be applied, leading to the same results but with $D_r$ being related to the integral of the correlation function and the fluctuation timescale of the fluctuations.

### 3.2.1.3 Folgar–Tucker Model

In Folgar and Tucker's model (1984), the diffusivity $D_r$ is assumed of the form $C_I \dot{\gamma}$, where $\dot{\gamma} = (2\mathrm{tr}\mathbf{D}^2)^{1/2}$ is the generalized strain rate and the aspect ratio is assumed infinite. The parameter $C_I$ is known as the interaction coefficient, which has been experimentally in the range of $10^{-2}$–$10^{-3}$ (Bay and Tucker 1992). Yamane $et~al.$ (1994) have recently predicted the interaction coefficient in a numerical simulation of semi-dilute suspensions of rod-like particles in a shear flow. However, the predicted values of $C_I$ are about two orders of magnitude smaller than those suggested by Bay and Tucker. It seems that this phenomenological constant must be a function of the volume fraction of the fibres and its aspect ratio; it may even be a tensorial quantity, reflecting the anisotropy of the fluid.

### 3.2.1.4 Closure approximations

In general, the flow problem of fibre orientation is intrinsically unsteady, since the vector $\mathbf{p}$ is time dependent. The orientation distribution function $\psi$ completely describes the fibre orientation state in the orientation space. Alternatively, one can solve for the moments of $\mathbf{p}$, starting with the second order moments (prescribing the probability function is equivalent to specifying the full set of moments of **p**).

However, since the evolution equation for the orientation tensor $\langle \mathbf{pp} \rangle$ contains the fourth order tensor $\langle \mathbf{pppp} \rangle$, closure approximations have to be made in order to close the set of evolution equations. These approximations are simply formulae that approximate a higher order tensor in terms of lower order tensors. The simplest such formula is the quadratic closure, employed by Doi (1981) and others,

$$\langle \mathbf{pppp} \rangle = \langle \mathbf{pp} \rangle \langle \mathbf{pp} \rangle. \tag{14}$$

This closure is exact in the limit of perfectly aligned fibres. When the Peclet number (for example, $Pe = O(\eta_s \dot{\gamma} a^3 / kT)$, where $\dot{\gamma}$ is a typical strain rate, $\eta_s$ is the solvent viscosity, $a$ is the size of the particles and $kT$ is the Boltzmann temperature) is small, the fibre orientation tends to be randomized and the approximation is not recommended.

A more sophisticated approximation, called the composite closure, was derived by Hinch and Leal (1976) as

$$\langle \mathbf{pppp} \rangle : \mathbf{D} = \frac{1}{5}[6\langle \mathbf{pp} \rangle \cdot \mathbf{D} \cdot \langle \mathbf{pp} \rangle - \langle \mathbf{pp} \rangle \langle \mathbf{pp} \rangle : \mathbf{D} - 2\mathbf{I}\langle \mathbf{pp} \rangle^2 : \mathbf{D}$$
$$+ 2\mathbf{I}\langle \mathbf{pp} \rangle : \mathbf{D}]. \tag{15}$$

This closure was designed to have the correct limits in both strong (perfectly aligned fibres) and weak flows (perfectly random orientation).

Cintra and Tucker (1995) developed a new family of closure approximations, called orthotropic fitted closure, by transforming the fourth order tensor in the principal axis system of the second order tensor and expressing its three independent components in terms of the second order principal values. The formula of the closure approximation was fitted to numerical solutions of the probability density function in a few well defined flow fields.

A variety of other closure approximations have been proposed. Further details may be found in Advani and Tucker (1987, 1990), Szeri and Leal (1992, 1994) and Verleye and Dupret (1993). It is known that the validity of the closure schemes depends on the type of flow and the degree of alignment of the fibres.

## 3.2.2 Anisotropic fluid models

The orientations of the fibres need to be able to be related to the stress tensor, given the kinematics (the constitutive equation). In general, all the constitutive equations that describe the rheology of suspensions have two parts contributing to the total stress:

$$\sigma_{ij} = \sigma_{ij}^{(s)} + \sigma_{ij}^{(p)} \tag{16}$$

Table 3.1. Asymptotic values of $A_i$, $i = 1$ to 4

| Cases | $A_1$ | $A_2$ | $A_3$ | $A_4$ |
|---|---|---|---|---|
| $a_R \to \infty$ | $\dfrac{a_R^2}{2(\ln 2a_R - 1.5)}$ | $\dfrac{6\ln 2a_R - 11}{a_R^2}$ | 2 | $\dfrac{3a_R^2}{\ln 2a_R - 0.5}$ |
| Rod-like | | | | |
| $a_R = 1+\delta$ | $\dfrac{395}{147}\delta^2$ | $\dfrac{15}{14}\delta - \dfrac{395}{588}\delta^2$ | $\dfrac{5}{2}\left(1 - \dfrac{2}{7}\delta + \dfrac{1}{3}\delta^3\right)$ | $9\delta$ |
| Sphere-like $\delta \ll 1$ | | | | |
| $a_R \to 0$ | $\dfrac{10}{3\pi a_R} + \dfrac{208}{9\pi^2} - 2$ | $-\dfrac{8}{3\pi a_R} + 1 - \dfrac{128}{9\pi^2}$ | $\dfrac{8}{3\pi_R}$ | $-\dfrac{12}{\pi a_R}$ |
| Disk-like | | | | |

where $\sigma_{ij}^{(s)}$ is the viscous contribution of the suspending fluid and $\sigma_{ij}^{(p)}$ is the particles-contributed stress. The difference between various constitutive models is the expression of the particles-contributed stress.

### 3.2.2.1 Dilute suspensions: transversely isotropic fluid (TIF)

The transversely isotropic fluid (TIF), first proposed by Ericksen (1960) from a continuum mechanic point of view, is an appropriate description of the micro-mechanics of a dilute suspension. In this model, the particle-contributed stress generated by the microstructure is given by

$$\sigma^{(p)} = 2\eta_s\phi\{A_1\mathbf{D}:\mathbf{pppp} + A_2(\mathbf{D}\cdot\mathbf{pp} + \mathbf{pp}\cdot\mathbf{D}) + A_3\mathbf{D} + D_rA_4\mathbf{pp}\},$$
[17]

where, $\eta_s$ is the viscosity of the solvent, $\phi$ is the volume fraction, $D_r$ is the rotational diffusivity of the spheroids and the constants $A_i$ ($i = 1$ to 4) are given in terms of the aspect ratio of the microstructure, as tabulated in Table 3.1.

The TIF model is clearly not very different from that for rigid dumbbells (see Bird et al., 1987b).

### 3.2.2.2 Semi-concentrated suspensions: Dinh–Armstrong model

The theoretical background in suspension mechanics has been well established, following the classical work of Batchelor (1970). In brief, a volume $V$ which is large enough to contain many particles but small enough so that the field variables hardly change on the scale $V^{1/3}$, i.e. $l \ll V^{1/3} \ll L$, is considered, where $l$ is a typical dimension of a suspended particle and $L$ is a typical size of the apparatus. The effective stress tensor seen from a macroscopic level is simply the average stress over this representative volume. If the solvent is

Newtonian and the particles are rigid, then the particle-contributed stress is given, from the divergence-free nature of the stress and the divergence theorem (Batchelor, 1970), by

$$\sigma_{ij}^{(p)} = \frac{1}{V}\sum_{p}(\mathscr{S}_{ij}^{(p)} + \tfrac{1}{2}\epsilon_{ijk}T_k^{(p)}) \tag{18}$$

where $T_k^{(p)}$ is the torque and $\mathscr{S}_{ij}^{(p)}$ is the 'stresslets', both exerted on particle $p$, $\epsilon_{ijk}$ is the alternating tensor and the summation is over all particles in the volume $V$. The stresslet on particle $p$ is given by

$$\mathscr{S}_{ij}^{(p)} = \frac{1}{2}\int\int\{x_i t_j + x_j t_i\}\psi(\mathbf{r}, \mathbf{p}, t)\mathrm{d}\mathbf{p}\mathrm{d}S, \tag{19}$$

in which $t_i = \sigma_{ij}n_j$ is the traction on the surface ($n_i$ denotes the outward-normal unit vector), and the integral is taken over the surface of the particle and all the configurations that it can assume. A distribution function $\psi$ has been introduced to account for the probability of having the particle at point $\mathbf{r}$ with an orientation $\mathbf{p}$ at time $t$.

Dinh and Armstrong (1984) applied Batchelor's theory and developed a constitutive model for semi-concentrated suspensions, which includes Jeffery's equations for a particle with infinite aspect ratio. They assume that there are no mechanical contacts between the fibres, therefore all interactions between fibres are hydrodynamic. In this model, the bulk stress due to the presence of the fibres in a homogeneous flow field is given by

$$\sigma_{ij}^{(p)} = \phi\eta_{\mathrm{s}}\frac{\pi l^3 N_{\mathrm{p}}^3}{6\ln(2H/d)}\int L_{kl}p_k p_l p_i p_j\psi(\mathbf{p})\mathrm{d}\mathbf{p} \tag{20}$$

where $L_{kl} = \partial u_k/\partial x_l$ ($\mathbf{L} = \nabla\mathbf{u}^{\mathrm{T}}$) is the velocity gradient tensor; $N_{\mathrm{p}}$ is the number of particles per unit volume; $l$ and $d$ are the fibre length and diameter, respectively and $H$ is the average distance between a fibre to its nearest neighbour, given by

$$H = (N_{\mathrm{p}}l^2)^{-1}$$

for random orientation and

$$H = (N_{\mathrm{p}}l)^{-1/2}$$

for fully aligned orientation. The orientation distribution function $\psi$ satisfies

$$\frac{\partial\psi}{\partial t} + \nabla_{\mathbf{p}}\cdot[(\mathbf{L}\cdot\mathbf{p} - \mathbf{L}:\mathbf{p}\mathbf{p}\mathbf{p})\psi] = 0 \tag{21}$$

where $\nabla_{\mathbf{p}}$ is the two-dimensional gradient operator on the surface of a unit sphere. The form of the constitutive equation is very similar to that of the rigid dumbbell model (e.g. Bird et al. 1987b).

### 3.2.2.3  Concentrated suspensions: Phan-Thien–Graham and Phan-Thien models

In Eqn [17] the dominant term at large aspect ratios is the term involving **pppp**, because it is of $O(a_R)$ as compared to $O(a_R^{-1})$ of the $\mathbf{D}\cdot\mathbf{pp} + \mathbf{pp}\cdot\mathbf{D}$ term. At large Peclet number the Brownian motion is also negligible. Thus the TIF model can be simplified to include only the leading terms. The model, however, can only predict that the specific viscosity is proportional to the volume fraction, while in reality, as the volume fraction increases, the specific viscosity is observed to increase rapidly with the volume fraction (Milliken *et al.*, 1989; Powell, 1991). An analogous constitutive form was constructed by Phan-Thien and Graham (1991) to predict the transition behaviour of non-dilute suspensions. In this model, the particle-contributed stress is given by

$$\sigma^{(p)} = 2\eta_s \phi f(\phi, a_R)\mathbf{D} : \langle \mathbf{pppp}\rangle, \tag{22}$$

where $f$ is a function of the volume fraction and the aspect ratio. The form of the constitutive equation is precisely that proposed by Dinh and Armstrong (1984), despite a different starting point.

The function $f$ is assumed to take the form

$$f(\phi, a_R) = \frac{a_R^2(2 - \phi/A)}{4(\ln 2a_R - 1.5)(1 - \phi/A)^2} \tag{23}$$

where the parameter $A$ is determined from experimental shear data using an empirical equation proposed by Kitano *et al.* (1981):

$$\eta_r = \frac{1}{(1 - \phi/A)^2} \tag{24}$$

where $\eta_r$ is the reduced viscosity. The physical interpretation of the parameter $A$ is the maximum allowable volume fraction of the suspension. A linear regression through the data of Kitano *et al.* (1981) leads to

$$A = 0.53 - 0.013a_R, \quad 5 < a_R < 30. \tag{25}$$

At the aspect ratio $a_R = 20$, the maximum allowable volume fraction is about 0.27.

A new microstructural model for concentrated suspensions of spheres has been derived by Phan-Thien (1995), based on the average affine motion of the particles and the lubrication forces between them. The model has several attractive features, including an instantaneous and linear response in the strain rate, a relaxation in the start up flow, a full recovery in the restart of the flow in the same direction, a partial recovery in the restart of the flow in the opposite direction and a universal response in the strain.

## 3.3    Numerical methods

There are two approaches towards numerical modelling of the fibre motion. In the first approach, the Stokes equations are solved for each particle and the forces and torques on a particle are used to calculate its motion (e.g. Tran-Cong and Phan-Thien, 1989; Phan-Thien *et al.*, 1991a; Ilic *et al.*, 1992; Ingber and Mondy, 1994). For non-dilute suspensions, the mobility of a multiparticle system needs to be solved. The best tool to date for doing this may be the Stokesian dynamic simulations (Brady and Bossis, 1988) and the completed double layer boundary element integral method (Kim and Karrila, 1991; Phan-Thien and Kim, 1994), which is based on iterative solutions and is particularly well suited to parallel computers (Phan-Thien and Tullock, 1994). The second approach uses the constitutive relation for stress–strain behaviour of the suspension and solves the momentum equations (which are no longer the Stokes equations and may be non-linear) for the stress and velocity fields. The motion of the microstructure is obtained by solving an auxiliary differential equation (i.e. the evolution equation). In the present chapter we limit ourselves to the second approach. This approach can be implemented by a variety of techniques: finite differences, finite volumes, finite elements, boundary elements and others. Here we review the use of boundary element methods.

### 3.3.1    The boundary element method

Traditional boundary element methods are particularly suitable for solving linear problems. However, the methods have also been extended to solve non-linear problems by iterative or incremental procedures. The variations of the boundary element method for a wide range of non-linear problems including non-Newtonian fluid flows and finite deformation of rubber-like solids have been described in detail elsewhere (Bush, 1984; Zheng, 1991) and here we will only briefly describe the technique for handling the non-linearity brought about by the particle-contributed stress. We begin with rewriting the total stress tensor in the form

$$\boldsymbol{\sigma} = -p\mathbf{1} + 2\eta_N\mathbf{D} + \boldsymbol{\epsilon} \qquad [26]$$

where $\eta_N$ is a positive, viscosity-like constant and $\boldsymbol{\epsilon}$ represents the non-linear component of the extra-stress tensor. The constant $\eta_N$ is arbitrary. In the case of the TIF model (Eqn [17]), one can set $\eta_N = \eta_s$ and $\boldsymbol{\epsilon} = \boldsymbol{\sigma}^{(p)}$.

When Eqn [26] is substituted into the equations of motion $\nabla\cdot\boldsymbol{\sigma} = \mathbf{0}$, the result can be written as

$$-\nabla p + \eta_N\nabla^2\mathbf{u} = -\nabla\cdot\boldsymbol{\epsilon}. \qquad [27]$$

Thus the stress component $\boldsymbol{\epsilon}$ gives rise to a pseudobody force term. The development of an integral equation then follows well-documented methods previously described in elasticity or fluid mechanics (Bush and Tanner, 1983; Brebbia *et al.*, 1984; Banerjee, 1994). The resulting equation can be written as

$$C_{ij}(\mathbf{x})u_j(\mathbf{x}) = \int_\Gamma [t_j(\mathbf{y})u_{ij}^*(\mathbf{x}, \mathbf{y}) - u_j(\mathbf{y})t_{ij}^*(\mathbf{x}, \mathbf{y})]\mathrm{d}\Gamma$$

$$- \int_\Omega \epsilon_{jk}(\mathbf{y}) \frac{\partial u_{ij}^*(\mathbf{x}, \mathbf{y})}{\partial x_k} \mathrm{d}\Omega, \qquad [28]$$

where $\Omega$ is the region within which a solution is sought, $\Gamma$ is the smooth boundary of this region, $\mathbf{x}$ is the field point, $u_{ij}^*$ and $t_{ij}^*$ are fundamental solutions (Brebbia *et al.*, 1984), $t_j$ is the traction vector (defined as $\mathbf{t} = \boldsymbol{\sigma} \cdot \mathbf{n}$, and $\mathbf{n}$ is the outward directed unit normal vector at the surface), $u_j$ is the velocity vector, $C_{ij} = \delta_{ij}$ if $\mathbf{x}$ lies in the domain $\Omega$, $C_{ij} = (1/2)\delta_{ij}$ for a point on the boundary which is smooth at the point $\mathbf{x}$.

The general iterative procedure can be described by the following steps:

1  Solve Eqn [28] for the unknown values of boundary velocity and/or traction using a given body force field (an initial guess or the results calculated at the previous iteration).
2  Evaluate the velocity field in the domain using Eqn [28].
3  Solve the evolution equation for the fibre orientation (the vector $\mathbf{p}$ or the tensor $\langle \mathbf{pp} \rangle$) using the kinematics computed in steps 1 and 2.
4  Calculate the stresses using the constitutive relation.
5  Check if convergence has been achieved, if not, evaluate the domain integrals in Eqn [28], then return to step (1).

The domain integral can be calculated numerically using Gaussian integration by dividing the domain $\Omega$ into a suitable number of triangular or quadrilateral cells within which the pseudobody force is approximated as a linear function of the spatial position. Although such a discretization has the appearance of that used in finite element methods, the cells are for the sole purpose of integrating the domain integral numerically; they do not add new unknowns to the problem. Efforts have also been made to find alternative formulations which transform the domain integrals into boundary integrals. This transformation has already been used successfully in several problems (e.g. Henry and Banerjee, 1988, Azevedo and Brebbia, 1988, Zheng *et al.*, 1991a,b; Zheng and Phan-Thien, 1992a). Step 3 will be discussed in more detail in the next subsection.

## 3.3.2  Solution of the evolution equations

Let us consider Jeffery's equation. Either Eqn [13] can be solved for the orientation tension $\langle \mathbf{pp} \rangle$ (a closure approximation is needed), or Eqn [1] can be solved for the vector $\mathbf{p}$, when Brownian forces are not present.

To solve Eqn [13] it is recognized that it is similar in form to the Phan-Thien–Tanner (PTT) constitutive equation (Phan-Thien and Tanner, 1977; Phan-Thien, 1978), which is hyperbolic in nature. In the past the PTT equations have been solved either by a streamline integration scheme (e.g. Bush, 1984; Sun, 1995), or by a pseudotime stepping scheme (e.g. Zheng, 1991). The same methods can be borrowed. The major shortcoming of the procedure is the need for closure approximations.

In the absence of Brownian forces, we need only to solve for **p**. To eliminate the third order term **ppp** in Eqn [1], an alternative formulation is constructed by first introducing a vector **q** which satisfies

$$\mathbf{p} = \frac{\mathbf{q}}{q},$$  [29]

where $q = (\mathbf{q} \cdot \mathbf{q})^{1/2}$. That is, **p** is the unit vector along **q**. It then can be shown that Eqn [1] is equivalent to

$$\frac{\partial \mathbf{q}}{\partial t} + \mathbf{u} \cdot \nabla \mathbf{q} - \mathscr{L}\mathbf{q} = \mathbf{0}.$$  [30]

The fourth order Runge–Kutta method is adopted to solve the equation for **q**, using the velocity **u** field obtained from the equations of motion at the previous time step. Then **p** can be easily obtained from Eqn [29]. The scheme using **q** as the unknowns was found to be more robust than the scheme directly handling **p** or **pp**. Both Eqns [1] and [13] are non-linear, while Eqn [30] is linear (given a velocity field) and therefore it may be easier to solve. It is also interesting to note the markedly different behaviour between the vectors **p** and **q** as functions of time in a simple shear flow. The components of **p** can vary dramatically within a very short time period. This kind of behaviour is difficult for any numerical scheme to capture accurately, unless a sufficiently small time step is used. By contrast, the components of **q** vary in time much more smoothly, just behaving like sinusoidal functions and hence are easier to handle numerically.

Once $\langle \mathbf{pp} \rangle$ or **p** has been obtained, the extra stresses can be calculated straightforwardly from the constitutive equations, such as Eqn [17].

## 3.4    Applications

In this section we present solutions for the flow of fibre suspensions past a sphere and in injection moulding, which illustrate some applications of the fibre orientation theory to practical problems. For the problem of flow past a sphere, we adopt the above-described boundary element method and solve the full axisymmetric equations of motion, coupled with the fibre orientation process. For the injection moulding problem, the numerical methods used are combinations of boundary elements, finite differences, finite control volumes and finite element methods. However, the injection flow calculation is based on a number of simplifications and is decoupled from the orientation calculation (i.e. without using an anisotropic constitutive model).

## 3.4.1 Flow past a sphere

This example is to simulate the falling-ball rheometry measurements. The problem involves the determination of the drag force on a body. For this class of problem the boundary element method has its advantage, since the total force per unit area (i.e. the traction) on the boundary forms one of the primary solution variables. We consider the flow generated by a sphere falling along the centreline of a cylindrical tube containing a model suspension fluid. The radius of the sphere is $a$ and the radius of the tube is $2a$. The half length of the cylinder is chosen to be $6a$. In a frame of reference that is translated with the sphere, the sphere is at rest and the tube wall is seen moving with a constant velocity (the falling speed $U$ of the sphere, but in the opposite direction of the falling sphere). Henceforth, all length scales are normalized with respect to $a$ and velocities are normalized with respect to $U$; the time is therefore made dimensionless with respect to $a/U$. Furthermore, the fluid far away from the sphere is seen moving rigidly with the tube and all associated stresses are zero there.

The prescribed boundary conditions are:

- At the entry of the flow domain ($z = -6a$) plug flow conditions are applied, where the axial velocity $u = U$, the radial velocity $v = 0$.
- Along the tube wall ($r = a$) $u = U$ and $v = 0$,
- Along the centreline ($r = 0$) symmetry boundary conditions apply, where $v = 0$ and the axial traction $t_x = \sigma_{rz} = 0$,
- On the surface of the sphere, $u = 0 = v$,
- At the outlet of the flow domain ($z = 6a$), the axial traction is set to zero (no net force action on the fluid) and the radial velocity $v = 0$.

In addition, all stress components are set to zero initially. Three different conditions were chosen for the initial unperturbed fibre orientations:

1  The director **p** is initially aligned with the tube axis, or $\theta = 0°$, where $\theta$ is the angle between **p** and the tube axis.
2  The director **p** is initially perpendicular to the tube axis, or $\theta = 90°$.
3  The director **p** is initially randomized; $\theta$ is chosen from a sequence of pseudorandom numbers between $-180°$ and $180°$.

Some typical meshes used in the simulation are shown in Fig. 3.1, the finest mesh contains 132 boundary elements, with the smallest element length of $0.105a$, located near the poles of the sphere.

The suspension was modelled by the TIF and the Phan-Thien–Graham models, incorporating Jeffery's evolution equation. The Phan-Thien–Graham

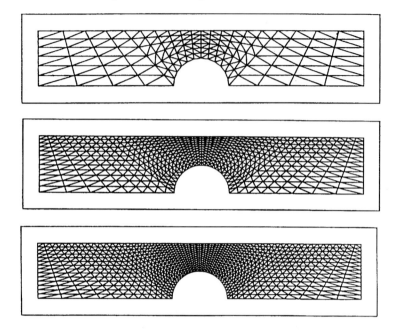

*3.1* Typical meshes used in the boundary element simulations.

model covers a wide range of concentration regimes from dilute to concentrated. It was found that at low volume fractions the results of two models are very close.

### 3.4.1.1  Kinematics

At low volume fractions, the kinematics of the suspension are similar to the Newtonian kinematics. As the volume fraction increases, the kinematics are shifting from the Newtonian case. Strictly speaking, there are no streamlines in this problem, since the kinematics are intrinsically unsteady. However, the instantaneous 'streamlines' serve as a useful device for comparing the results with the Newtonian solution. The non-Newtonian streamlines are slightly non-symmetric in contrast to the Newtonian streamlines. The asymmetry is seen most clearly if the contours of the modified stream function are plotted

$$\psi_1 = \psi - \tfrac{1}{2} r^2$$

as shown in Fig. 3.2; $\psi_1$ is the stream function that corresponds to the moving falling sphere in a stationary cylinder. The fore-and-aft symmetry is no longer present in the non-dilute suspension fluid case. Similarly, the contours of the axial and radial velocities are not symmetric about the plane $x = 0$.

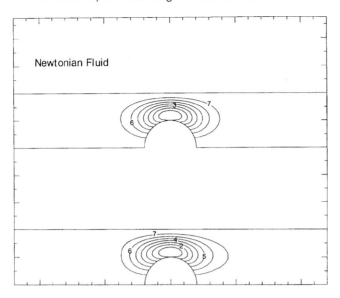

3.2 Streamlines corresponding to a moving sphere and fixed cylinder show the asymmetry in the suspension flow; $a_R = 20$, $\phi = 0.1$. Initially the fibres are aligned perpendicular to the flow direction ($\theta = 90°$).

### 3.4.1.2  Evolution of fibre orientation

The orientation of microstructure along the tube is shown in Fig. 3.3 at time $t = 40$ for $a_R = 20$, $\phi = 0.1$, but with a different initial orientation of fibres. It is interesting to find that the evolution of the microstructure can depend dramatically on the initial conditions of **p**. However, at large time the overall orientations of the fibres are quite similar, despite their different initial states.

It is seen that the major disturbances in the microstructure occur in the layer between the sphere and the tube, extending to about one sphere diameter upstream and downstream of the sphere. These disturbances can be quite severe: **p** can flip its direction across a narrow layer of fluid ($\approx 0.1a$). Along the centreline of the tube, approaching the sphere from upstream, fibres align perpendicular to the flow direction. Away from the sphere, fibres tend to align themselves with the tube axis in the downstream wake region near the tube axis. The alignment of the fibres behind the sphere has been observed experimentally with semi-concentrated systems (R. Morr and A.L. Graham, unpublished data). The falling ball, therefore, may be used as a device to align fibres. A physical explanation of this phenomenon is the existence of large extensional stresses near the rear stagnation pole of sphere (Chicott and Rallison, 1988; Zheng et al., 1990a, Phan-Thien et al., 1991c). Such a flow field plays a dominant role in aligning the fibres.

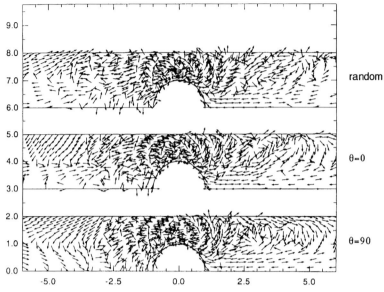

**3.3** Orientation of the fibres at dimensionless time $t = 40$ at $a_R = 20$, $\phi = 0.1$, but at three different initial configurations of the fibres.

### 3.4.1.3 *Reduced viscosity*

The reduced viscosity $\eta_r$ can be determined by the drag on the sphere with respect to the Newtonian value in the same flow geometry. An accurate Newtonian drag value for the given geometry is found to be 5.947 (Zheng *et al.*, 1991c). With both the TIF and the Phan-Thien–Graham model, the flow is intrinsically unsteady, and the traction on the surface of the sphere is a periodic function of time, with a frequency that depends on the local shear rate. The latter is also a function of time containing a spectrum of frequencies. It has been found (Phan-Thien *et al.*, 1991b), however, that the amplitude of the unsteady drag force is only about 1% of the mean value at long time observation; the long time average values of the drag forces are only weakly dependent on the original orientation of the fibres.

To calculate the reduced viscosity, we simply take the time-averaged value of the dimensionless drag, from $t = 4$ to $t = 10$ for suspensions of fibres of aspect ratio less than 10. At higher aspect ratios, it is necessary to take the average over a longer time interval due to the large period of the oscillation in **p**. In the simple shear flow, the period of oscillation in **p** is proportional to the aspect ratio at large aspect ratios. The same behaviour is found in the flow past a sphere: the higher the aspect ratio, the higher is the period of oscillation in **p**. At $a_R = 20$, it is necessary to run the problem up to $t = 40$ in order to calculate the averaged drag force accurately.

Using the TIF model for a dilute suspension of fibres with an aspect ratio 10, the mean value of the drag is 6.26, with a variation of 0.05 (or about 1%) about the mean value for a different initial orientation of the fibres. Thus the reduced viscosity of the suspension can be determined to be

$$\eta_r = 1 + 5.3\phi, \tag{31}$$

which is comparable with the shear viscosity of $\eta_r = 1 + 4.8\phi$ obtained from a simple shear flow (Phan-Thien *et al.*, 1991b). For suspensions of fibres of aspect ratio 20, the reduced viscosity is predicted to be given by

$$\eta_{eff} = \eta(1 + 17.4\phi). \tag{32}$$

This is quite different from the reduced shear viscosity measured from a simple shear flow. The difference has been observed in experiments (Milliken *et al.*, 1989) and is caused by the anisotropy of the fluid.

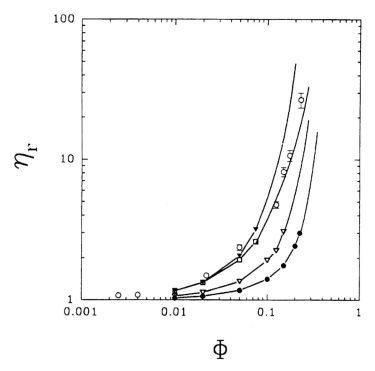

*3.4* The reduced viscosity calculated from the averaged drag force on the sphere, plotted against the volume fraction of the fibres. The data of Milliken *et al.* (1989) are also plotted for comparison. A good agreement was obtained with the maximum volume fraction of $A = 0.46$. The agreement is not as good with $A = 0.27$ (as predicted by the empirical formula of Kitano *et al.*, 1981). ●, $a_R = 5$; ▽, $a_R = 10$; ▼, $a_R = 20$, $A = 0.27$; □, $a_R = 20$, $A = 0.46$; ○, Milliken *et al.* (1989).

The results obtained from the Phan-Thien–Graham model are summarized in Fig. 3.4, where the reduced viscosity is plotted against the volume fraction at three different aspect ratios. Also plotted in the same figure are the experimental falling sphere data on suspensions of blunt-ended rods and nylon fibres at aspect ratio $a_R = 19.8$. It is clear that the theory agrees reasonably well with the experimental data. However, with the maximum volume fraction $\phi_m = 0.27$, as predicted from the empirical formula of Kitano *et al.* (1981), the non-linear behaviour sets in at $\phi \approx 0.075$, which is somewhat early.

With $\phi_m = 0.46$, a much better fit to the experimental results can be achieved. The transition from the linear to cubic behaviour can be better seen by plotting the specific viscosity, $\eta_{sp} = \eta_r - 1$, versus the volume fraction, as was done in Phan-Thien and Graham (1991). Subtracting one from $\eta_r$ has the effect of accentuating the linear regime and with a limited number of data points between $0.1 < \phi < 0.2$, a cubic relation can certainly be fitted to the $\eta_{sp}$ against $\phi$ curve. Whether or not there is a physical transition in the specific viscosity from a linear to the cubic dependence in the volume fraction is a matter of debate.

In summary, the use of a boundary element method has been demonstrated in solving the problem of a sphere falling in a tube containing suspension fluids. The method is capable of predicting the time-dependent fibre orientation in the tube and the drag force on the sphere, through the coupled flow-induced orientation and orientation-dependent constitutive models. The long time behaviour of the drag force depends weakly on the initial configuration of the microstructure. The reduced viscosity determined from the drag force by a time average agrees with the experimental data of Milliken *et al.* (1989). In this study the sphere moving at a constant velocity is considered. It is also possible to obtain a solution for the unsteady motion of a sphere falling under gravity; such a method has been described elsewhere (Zheng and Phan-Thien, 1992b) applied to viscoelastic fluids. The time-dependent fall velocity in concentrated suspensions has been observed experimentally by Chow *et al.* (1993).

With respect to viscosity measurements, the numerical result seems to support the falling ball method as a practical means of measuring the suspension viscosity. The major disadvantage is that the flow around the sphere is non-viscometric (i.e. it is a mixture of shear and elongational flows) and as shown in the numerical results, the suspension microstructure is significantly disturbed by the motion of the sphere. A better alternative is to use a needle as the falling object (Park and Irvine, 1988, Park *et al.*, 1990; Phan-Thien *et al.*, 1993; Zheng *et al.*, 1994), which promotes a dominant shear flow and yet retains the simplicity of the method.

## 3.4.2    Injection moulding

Injection moulding involves two processes: plasticizing polymer (or compound), and shaping the polymer in the mould. Several books have described the various aspects of the processes (Tadmor and Gogos, 1977; Tucker, 1989; Isayev, 1991; Kennedy, 1993; Advani, 1994). We shall only consider the latter process, which starts with filling the cavity with hot polymer melts, followed by a packing stage to force additional material at high pressure into the cavity to compensate for the thermal contraction of the polymer. Due to the heat transfer between the hot polymer and the colder mould wall, the part continues to cool down until it has solidified and is rigid enough for ejection. During solidification, residual stresses develop in the moulded part. The residual stresses can cause the part to shrink and warp after it is ejected. The warpage occurs due to the shrinkage variations throughout the complex part, including (1) asymmetric shrinkage about the mid-surface of the moulded part; (2) non-uniform planar volumetric shrinkage; (3) shrinkage differences in different directions (orientation shrinkage); and (4) differential thermal strain due to geometry effects (such as corners). For short-fibre-filled plastics, anisotropy of the elastic moduli and the thermal expansion coefficients (due to the flow-induced fibre orientation) can significantly affect the shrinkage and warpage pattern of the injection moulded part.

The objective of the present simulation is to predict the fibre orientation and the final deformed shape of a short-fibre-reinforced injection-moulded part. The relationship between the fibre orientation and the warpage will be the main focus of attention, thus causing omission of some important aspects, such as crystallization and the fountain flow (see, for example, Kamal, 1995). The influence of fountain flow on the fibre orientation is discussed elsewhere (Zheng *et al.*, 1990b; Zheng, 1991).

### 3.4.2.1  Problem formulation

With time-dependent flows in geometrically complex domains, it is not computationally efficient to use anisotropic constitutive models. A generalized Newtonian model is therefore used for the flow computation, which is essentially decoupled from the orientation process. According to Tucker (1991), such a simplification is a reasonable assumption, provided the flow channel is sufficiently thin, which is the case considered here.

The modelling of the transient flow during the filling stage uses a generalized quasi-steady state Hele–Shaw flow model that is based on lubrication approximations. The Hele–Shaw approximation reduces the fully three-dimensional flow problem to a two-dimensional formulation for the pressure, as a function of $x$ and $y$,

$$\frac{\partial}{\partial x}\left(S\frac{\partial P}{\partial x}\right) + \frac{\partial}{\partial y}\left(S\frac{\partial P}{\partial y}\right) = 0 \qquad [33]$$

where $S$ defined as

$$S = \int_0^h \frac{z^2 \, dz}{\eta} \qquad [34]$$

in which $z$ is the gapwise direction and $h$ is the half gap width that can be a function of $x, y$. $\eta$ is the temperature-dependent and shear-rate dependent viscosity.

Once Eqn [33] has been solved for the pressure distribution $P(x, y)$, the complete velocity profile can be found from

$$u(x, y, z) = -\frac{\partial P}{\partial x} \int_z^h \frac{z_1 \, dz_1}{\eta}, \qquad [35]$$

$$v(x, y, z) = -\frac{\partial P}{\partial y} \int_z^h \frac{z_1 \, dz_1}{\eta}. \qquad [36]$$

The governing equation [33] is elliptic in nature and requires a boundary condition on each boundary of the flow domain in the $x$–$y$ plane. They are conveniently given as either the inlet flow rate (which may be a function of time) or the pressure boundary condition at the gates; a zero pressure, $P = 0$, on the advanced flow front (a free surface); and a zero normal pressure gradient $\partial P / \partial n = 0$ along the edges of the mould. The no-slip condition is not necessarily satisfied on the edges. Instead, the fluid may 'slip', leading to inaccurate velocity predictions within a thin boundary layer of order $h$. The validity of the assumptions used in the Hele–Shaw model has been justified by Guell and Lovalenti (1995) for the case of homogeneous, inelastic fluids in thin cavities.

Packing begins when the cavity is completely filled and continues until there is no more melt entering the cavity. From the fluid mechanics point of view, packing is a transient, non-isothermal, compressible flow caused by the pressure gradients, the thermal contraction and the compressibility of the polymer. The flow is also accompanied by solidification. An equation of state relating the pressure $P$, the temperature $T$ and the specific volume $V_c$ is employed. The general form of the state equation can be written as

$$V_c = f(T, P). \qquad [37]$$

Although the pressure–volume–temperature ($P$-$V$-$T$) relationships are cooling-rate dependent, the present model is based on a quasi-equilibrium assumption. Eqn [37] leads to the following relation

$$\frac{dV_c}{V_c} = \alpha_v dT - \beta dP \qquad [38]$$

where $\alpha_v$ is the volume thermal expansion coefficient defined as

$$\alpha_v = \frac{1}{V_c}\left(\frac{\partial V_c}{\partial T}\right)_P$$

and $\beta$ is the isothermal compressibility defined as

$$\beta = -\frac{1}{V_c}\left(\frac{\partial V_c}{\partial P}\right)_T.$$

Both the filling and packing analyses require simultaneous solution of the temperature field, which is obtained by solving a three-dimensional energy equation.

Fibre orientation distributions are calculated based on the available velocity field, using the tensorial form of Folgar–Tucker's equation, with a hybrid closure approximation (Advani and Tucker, 1987). The interaction coefficient $C_I$ is assumed to be an exponential function of $\phi a_R$, as suggested by Bay (1991) and was determined empirically using Bay's experimental data over a wide range of concentrations ($\phi a_R \sim 1.0$ to $6.0$).

The fibre orientation prediction is linked to the warpage analysis through available micromechanical theories for anisotropic mechanical properties of composites. It is beyond the scope of the present article to review these theories. The interested reader is referred to Eduljee et al. (1994) for recent progress in this area. In essence, the anisotropic mechanical properties are predicted by estimating the properties for a unidirectional system in which the fibres are assumed to be fully aligned and then calculate the properties of the composite through averaging the unidirectional properties over all directions, weighted by the orientation distribution function. For a unidirectional composite, the Halpin–Tsai equation (Halpin and Kardos, 1976) is used to evaluate Young's modulus ($E_{11}$ and $E_{22}$), shear modulus ($G_{12}$) and Poisson's ratio ($v_{12}$), where we use the number 1 for longitudinal fibre direction and 2 for the transverse direction. The equation can be written in a general form (Krishnamachari, 1993) as

$$\frac{Q_c}{Q_m} = \frac{1 + \zeta\chi\phi}{1 - \chi\phi}, \tag{39}$$

where

$$\chi = \frac{Q_f/Q_m - 1}{Q_f/Q_m + \zeta}.$$

In the above equation, the symbol $Q$ stands for any of the mechanical properties, $E_{11}$, $E_{22}$, $G_{12}$, or $v_{12}$. The subscript c stands for composite, m for matrix and f for fibre. The values of parameter $\zeta$ is given in Table 3.2.

Longitudinal and transverse thermal expansion coefficients, $\alpha_1$ and $\alpha_2$ are evaluated by Chamberlain's model (Bowles and Tompkins, 1989) for the case of

Table 3.2. Summary of parameter $\zeta$ for Haplin–Tsai's equation

| $Q$ | $E_{11}$ | $E_{12}$ | $G_{12}$ | $v_{12}$ |
|-----|----------|----------|----------|----------|
| $\zeta$ | $2a_R$ | 2 | 1 | $\infty$ |

isotropic fibres in an isotropic matrix. Given the thermal expansion coefficients of the fibre, $\alpha_f$, and of the matrix, $\alpha_m$, the Chamberlain model approximates the thermal expansion coefficients of composites as

$$\alpha_1 = \frac{E_f \alpha_f \phi + E_m \alpha_m (1 - \phi)}{E_f \phi + E_m (1 - \phi)} \tag{40}$$

and

$$\alpha_2 = \alpha_m + \frac{2(\alpha_f - \alpha_m)\phi}{v_m(F - \phi) + (F + \phi) + (E_m/E_f)(1 - v_f)(F - \phi)} \tag{41}$$

where $F$ is a factor that accounts for fibre packing type; it equals 0.9069 for parallel hexagonal packing and 0.7854 for parallel square packing.

Knowing the unidirectional properties and the fibre orientation tensors, the orientation averaged stiffness $C_{ijkl}$ and thermal expansion coefficients $\alpha_{ij}$ can be predicted using the procedure proposed by Advani and Tucker (1987).

It was assumed that the solidified polymer behaves as an elastic solid. The constitutive equation is given as follows,

$$\sigma_{ij} = C_{ijkl}(e_{kl} - e_{kl}^0) \tag{42}$$

where $e_{ij}$ is the strain tensor and $e_{ij}^0$ is the thermal strain defined by

$$e_{ij}^0 = \int_{T_0}^{T_r} \alpha_{ij} dT, \tag{43}$$

where $T_r$ is the room temperature and $T_0$ is the material's temperature at the time when the pressure drops to the atmospheric value $P_a$. Note that, at different spatial points in the moulded part, the pressure may drop to $P_a$ at different times and thus $T_0$ also varies from point to point.

The stresses must satisfy the equilibrium equation:

$$\frac{\partial \sigma_{ij}}{\partial x_j} = 0. \tag{44}$$

Here we are concerned only with elastostatics and therefore inertia terms are set to zero. In addition, the body forces are also neglected.

*Table 3.3.* Summary of material data

| Properties | Polymer | Fibres |
|---|---|---|
| Young's modulus | $2.8 \times 10^3$ MPa | $7.2 \times 10^4$ MPa |
| Poisson's ratio | 0.42 | 0.2 |
| Thermal expansion | $6.5 \times 10^{-5} {}^{\circ}\mathrm{C}^{-1}$ | $5.0 \times 10^{-6} {}^{\circ}\mathrm{C}^{-1}$ |
| Volume fraction | 0.82 | 0.18 |
| Average aspect ratio | – | 20 |

*3.4.2.2  Sample results*

The sample calculation considers a complex part shown in Fig. 3.5. Sherman and Calder (1992) analysed the same example, but using a multivariable regression technique to estimate shrinkage; their methodology is totally different from that described in the present chapter. Sherman and Calder also provide a set of experimental data. The material data and processing conditions that we have used here were taken from their paper.

The part was moulded with glass-fibre reinforced plastics. The properties of the polymer and the fibre components are listed in Table 3.3.

The inlet boundary condition for the filling analysis is a constant injection flow rate of $9.773 \times 10^{-5}$ m$^3$ s$^{-1}$. For the packing analysis a constant holding pressure that equals 55 MPa is imposed; the total packing time and the total cooling time are 10 s and 20 s, respectively. The initial conditions of the melt temperature and mould temperature are 283°C and 38°C, respectively. The inlet condition for the fibre orientation is assumed a random state. Different inlet fibre orientation conditions have little impact on the final orientation results.

The fibre orientation calculations were carried out in both filling and packing stages. Although the final fibre orientation state is mainly determined during

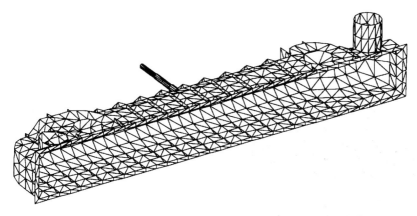

*3.5* Complex injection moulding part considered by Sherman and Calder (1992).

filling, the effect of the packing stage may not be negligible for the final orientation distribution in the core region.

The result of an orientation distribution is displayed graphically by the orientation ellipse. The axes of these ellipses adopt the direction and length of the eigenvectors and eigenvalues of the orientation tensor $\langle\mathbf{pp}\rangle$, respectively. The major axis indicates the direction of preferred fibre orientation, while the roundness of the ellipse is a measure of the degree of alignment. That is, a circle represents a random orientation, whereas a sharp line represents a near-full alignment.

The part has been divided into ten layers in the thickness direction. Fig. 3.6 shows the fibre orientation in the mid-surface of the cavity. At this layer, the flow is shear free and is dominated by elongation. Around the entrance gate is a radial diverging flow, i.e. the flow is compressive in the radial direction and elongational in the direction tangent to the flow front. The fibres align themselves along the elongational direction and show a pattern of concentric circles around the gate. The orientation ellipses show that the fibres are nearly fully aligned, except in the regions far away from the gate.

Figure 3.7 shows the fibre orientation in the layer near the wall (which is $0.8h$ from the mid-surface). The fibre orientation in this layer clearly differs from the orientation in the mid-surface. The flow is a combination of elongational and shear flows. The radial diverging nature of the flow tends to align the fibres transverse to the flow, whereas the shearing flow tends to align the fibres along the flow. Near the entrance gate, there is also a gapwise converging flow component due to the frozen layer, which increases the tendency of flow-aligned fibre orientation. Accordingly, the principal direction of the fibres is basically parallel to the flow, but the degree of alignment is relatively low, as indicated by the roundness of the orientation ellipses.

The warpage of the solid part was calculated separately after the filling, parking and cooling analyses, using a standard finite element method for thin-

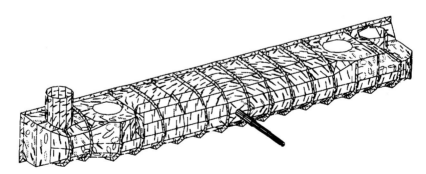

*3.6* Fibre orientation in the mid-surface of the cavity. The fibres align themselves along the elongational direction and show a pattern of concentric circles around the gate.

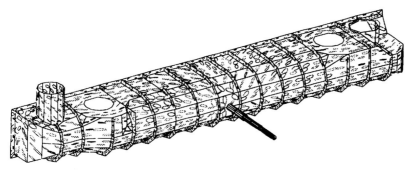

*3.7* Fibre orientation in the layer near the wall (which is 0.8*h* from the mid-surface).

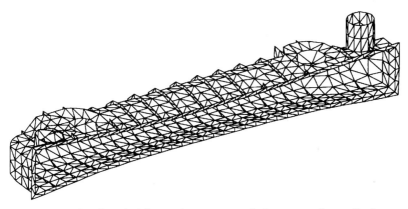

*3.8* Predicted deformed geometry of the part, where displacements have been amplified by a factor of 2 for better visualization.

shell structural analysis. The previous calculated fibre orientation, orthotropic thermomechanical properties and orthotropic thermal strains are used by the finite element analysis to calculate the deformation of the modelled part.

Figure 3.8 shows the predicted deformed geometry, where displacements have been amplified by a factor of 2 for better visualization. The prediction shows inward deflections in both the base's long side walls. Both the direction and magnitude of the warpage are found to compare well with the experimental measurements reported by Sherman and Calder (1992) (see also Zheng *et al.*, 1996).

## 3.5    Conclusions

In this chapter a brief review of constitutive models describing the rheology of fibre suspensions is given, with an emphasis on the microscopic modelling of flow-induced fibre orientation. We have demonstrated the capability of different numerical methods for predicting fibre orientation in complex flows. The simulation of the falling ball rheometer indicates that the boundary element method is competitive with other available numerical techniques in solving fibre-flow interactions, using anisotropic constitutive models. In the numerical analysis of injection moulding, micromechanical theories for unidirectional composites have been combined with the predicted fibre orientation distribution to calculate anisotropic thermal strains, which are then used in the prediction of warpage of the injection moulded part. The simulation should be useful for a better understanding of the relationship among the processing conditions, the fibre orientation distribution and the properties of the final injection moulded part.

## Acknowledgement

This work is supported by an Australian Research Council Collaborative Grant (with Moldflow Pty Ltd). The support is gratefully acknowledged.

## References

Abbott, J.R., Tetlow, N., Graham, A.L., Altobelli, S.A., Fukushima, E., Mondy, L.A., Stephens, T.S. Experimental observations of particle migration in concentrated suspensions: Couette flow, *J. Rheol.*, 1991 **35** 773–795.

Acrivos, A. Shear-induced particle diffusion in concentrated suspension of noncolloidal particles, *J. Rheol.*, 1995 **39** 813–826.

Advani, S.G. (ed.) *Flow and Rheology in Polymer Composites Manufacturing*, Amsterdam, Elsevier Science, 1994.

Advani, S.G. and Tucker, C.L. The use of tensors to describe and predict fiber orientation in short fiber composites, *J. Rheol.*, 1987 **31** 751–784.

Advani, S.G. and Tucker, C.L. Closure approximations for three-dimensional structure tensors, *J. Rheol.*, 1990 **34** 367–386.

Altan, M.C. and Rao, B.N. Closed-form solution for the orientation field in a center-gated disk, *J. Rheol.*, 1995 **39** 581–599.

Altan, M.C., Subbiah, S., Güceri, S.I. and Pipes, R. B. Numerical prediction of three-dimensional fiber orientation in Hele–Shaw Flows, *Polym. Eng. Sci.*, 1990 **30** 848–859.

Altan, M.C., Güceri, S.I. and Pipes, R.B. Anisotropic channel flow of fiber suspensions, *J. Non-Newtonian Fluid Mech.*, 1992 **14** 65–83.

Arp, P.A. and S.G. Mason. Kinetics of flowing dispersions. 9. Doublets of rigid spheres, *J. Colloid Interface Sci.*, 1977 **61** 44–61.

Azevedo, J.P.S. and Brebbia, C.A. An efficient technique for reducing domain integral to the boundary *Proceedings of Xth International Conference on BEM in Engineering*, Computational Mechanics Publications, Southampton, UK, 1988, pp. 347–361.

Banerjee, P.K., *The Boundary Element Methods in Engineering*, London, McGraw-Hill, 1994.

Batchelor, G.K. The stress system on a suspension of force-free and torque-free particles, *J. Fluid Mech.*, 1970 **44** 545–570.

Bay, R.S. *Fiber Orientation in Injection Molded Composites: A Comparison of Theory and Experiment*, PhD Thesis, University of Illinois at Urbana-Champaign, 1991.

Bay, R.S. and Tucker, C.L. Fiber orientation in simple injection moldings, Part 1: Theory and numerical methods, *Polym. Compos.,* 1992 **13** 317–321.

Bird, R.B., Armstrong, R.C. and Hassager, O., *Dynamics of Polymeric Liquids: Vol. 1: Fluid Mechanics*, 2nd edn, New York, John Wiley & Sons, 1987a.

Bird, R.B., Curtiss, C.F., Armstrong, R.C. and Hassager, O., *Dynamics of Polymeric Liquids: Vol. 2: Kinetic Theory*, 2nd edn, New York, John Wiley & Sons, 1987b.

Bowles, D.E. and Tompkins, S.S. Prediction of coefficients of thermal expansion for unidirectional composites, *J. Compos. Mater.*, 1989 **23** 370–388.

Brady, J.F. and Bossis, G. Stokesian dynamics, *Ann. Rev. Fluid Mech.*, 1988 **20** 111–157.

Brebbia, C.A., Telles, J.F.C. and Wrobel, L.C. *Boundary Element Techniques*, Berlin, Springer-Verlag, 1984.

Bush, M.B. *The Application of Boundary Element Method to Some Fluid Mechanics Problems*, PhD Thesis, University of Sydney, 1984.

Bush, M.B. and Tanner, R.I. Numerical solution of viscous flows using integral equation methods, *Internat. J. Numerical Method. Fluids*, 1983 **3** 71–92.

Chandrasekhar, S. Stochastic problems in Physics and Astronomy, *Rev. Mod. Phys.*, 1943 **15** 1–89.

Chicott, M.D. and Rallison, J.M. Creeping flow of dilute polymer solutions past cylinders and spheres, *J. Non-Newtonian Fluid Mech.*, 1988 **29** 381–432.

Chow, A.W., Sinton, S.W. and Iwamiya, J.H. Direct observation of particle microstructure in concentrated suspensions during the falling-ball experiment, *J. Rheol.*, 1993 **37** 1–16.

Cintra Jr., J.S. and Tucker, C.L. Orthotropic closure approximations for flow-induced fiber orientation, *J. Rheol.*, **39** 1995 1095–1122.

Crochet, M.J., Dupert, F. and Verleye, V. in *Flow and Rheology in Polymer Composites Manufacturing,* ed. S.G. Advani, Amsterdam, Elsevier Science, 1994, pp 415–463.

Dinh, S.H. and Armstrong, R.C. A rheological equation of state for semiconcentrated fiber suspensions, *J. Rheol.*, 1984 **28** 207–227.

Doi., M. Molecular dynamics and rheological properties of concentrated solutions of rodlike polymers in isotropic and liquid crystalline phases, *J. Polym. Sci. Polym. Phys. Ed.*, 1981 **19** 229–243.

Doi, M. and Edwards, S.F. Dynamics of rod-like macromolecules in concentrated solution. Part 1, *J. Chem. Soc. Faraday Trans. II,* 1978a **74** 560–570.

Doi, M. and Edwards, S.F. Dynamics of rod-like macromolecules in concentrated solution. Part 2, *J. Chem. Soc, Faraday Trans. II,* 1978b **74** 918–932.

Eduljee, R.F., McCullough, R.L. and Gillespie Jr., J.W. The influence of aggregated and dispersed textures on the elastic properties of discontinuous-fiber composites, *Compos. Sci. Tech.,* 1994 **50** 381–391.

Ericksen, J.L. Anisotropic fluids, *Arch. Rat. Mech. Anal.*, 1960 **4** 231–237.

Folgar, F.P. and Tucker, C.L. Orientation behavior of fibers in concentrated suspensions, *J. Reinforced Plastics Compos.*, 1984 **3** 98–119.

Frahan, H.H., Verleye, V., Dupret, P. and Crochet, M.J. Numerical prediction of fibre orientation in injection moulding, *Polym. Eng. Sci.*, 1992, **32**, 254–266.

Friedl, C. and Brouwer, R. Fiber orientation prediction, *SPE Tech. Papers,* 1991 **37** 326–329.

Gadala-Maria, F. and Acrivos, A. Shear-induced structure in a concentrated suspension of solid spheres, *J. Rheol.,* 1980 **24** 799–811.

Givler, R.C., Crochet, M.J. and Pipes, R.B. Numerical predictions of fiber orientation in dilute suspensions, *J. Compos. Mater.,* 1983 **17** 330–343.

Goettler, L.A., Lambright, A.J., Leib, R.I. and DiMauro, P. J. Extrusion-shaping of curved hose reinforced with short cellulose fibers, *Rubber Chem. Tech.,* 1981 **54** 277–301.

Graham, A.L., Altobelli, S.A., Fukushima, E., Mondy, L.A. and Stephens, T.S. NMR Imaging of shear-induced diffusion and structure in concentrated suspensions undergoing Couette flow, *J. Rheol.,* 1991 **35** 191–201.

Guell, D. and Lovalenti, M. An examination of assumptions underlying the state of the art in injection molding modeling. *SPE ANTEC 95,* 1995 **1** 728–732.

Gupta, M. and Wang K.K. Fiber orientation and mechanical properties of short-fiber-reinforced injection-molded composites: simulation and experimental results, *Polym. Compos.,* 1993 **14** 367–381.

Halpin, J.C. and Kardos, J.L. The Halpin–Tsai equations: a review, *Polym. Eng. Sci.,* 1976 **16** 344–352.

Hand, G.L. A theory of anisotropic fluids, *J. Fluid Mech.,* 1962 **13** 33–46.

Henry, D.P. Jr. and Bannerjee, P.K. A new boundary element formulation for two- and three-dimensional thermoelasticity using particular integrals, *Internat. J. Numerical Method Eng.,* 1988 **26** 2061–2078.

Henry, E., Kjeldsen, S. and Kennedy K. Fiber orientation and the mechanical properties of SFRP parts, *SPE ANTEC 94,* 1994 **1** 374–377.

Hinch, E.J. and Leal, L.G. The effect of Brownian motion on the rheological properties of a suspension of non-spherical particles, *J. Fluid Mech.,* 1972 **52** 683–712.

Hinch, E.J. and Leal, L.G. Constitutive equations in suspension mechanics. Part 2. Approximate forms for a suspension of rigid particles affected by Brownian rotations, *J. Fluid Mech.,* 1976 **76** 187–208.

Hookham, P.A. *Concentration and Velocity Measurements in Suspensions Flowing through a Rectangular Channel,* PhD Thesis, California Institute of Technology, 1986.

Ilic, V., Vincent, J., Zheng, R. and Phan-Thien, N. Sedimentation velocity of some complex-shaped particles in Newtonian and non-Newtonian liquids, *Proc. Eleventh Australian Fluid Mechanics Conference,* Hobart, Australia, Dec 1992, 447–450.

Ingber, M.S. and Mondy, L.A. A numerical study of three-dimensional Jeffery orbits in shear flow, *J. Rheol.,* 1994 **38** 1829–1843.

Isayev, A.I. (ed.) *Modeling of Polymer Processing,* Munich, Hanser, 1991.

Jeffery, G.B. The motion of ellipsoidal particles immersed in viscous fluid, *Proc. R. Soc. Lond. A,* 1922 **102** 161–179.

Joseph, D.D. and Liu, Y.J. Orientation of long bodies falling in a viscoelastic liquid, *J. Rheol.,* 1993 **37** 961–983.

Kamal, M.R. The coupling of crystallization kinetics and fountain flow in simulation of the injection molding of crystalline polymers, *SPE ANTEC 95,* 1995 613–618.

Karnis, A., Goldsmith, H.L. and Mason, S.G. The kinetics of flowing dispersions. I. Concentrated suspension of rigid particles, *J. Colloid Interface Sci.,* 1966 **22** 531–553.

Kennedy, P. *Flow Analysis Reference Manual*, Moldflow Shortrun Books. Melbourne, 1993.

Kim, S. and Karrila, S.J. *Microhydrodynamics: Principles and Selected Applications*, Butterworth-Heinemann, Boston MA, 1991.

Kitano, T., Kataoka, T. and Shirota, T. An empirical equation of the relative viscosity of polymer melts filled with various inorganic fillers, *Rheol. Acta*, 1981 **20** 207–209.

Koh, C.J., Hookham, P. and Leal, L.G. An experimental investigation of concentrated suspension flows in a rectangular channel, *J. Fluid Mech.*, 1994 **266** 1–32.

Krishnamachari, S.I. *Applied Stress Analysis of Plastics*, New York, Van Nostrand Reinhold, 1993.

Leal, L.G. and Hinch, E.J. The effect of weak Brownian rotations on particles in shear flow, *J. Fluid Mech.*, 1971 **46** 685–703.

Leal, L.G. and Hinch, E.J. The rheology of a suspension of nearly spherical particles subject to Brownian rotations, *J. Fluid Mech.*, 1972 **55** 745–765.

Leal, L.G. and Hinch, E.J. Theoretical studies of a suspension of rigid particles affected by Brownian couples, *Rheol. Acta*, 1973 **12** 127–132.

Leighton, D.T., *The Shear Induced Migration of Particles in Concentrated Suspensions*, PhD Thesis, Stanford University, 1985.

Leighton, D.T. and Acrivos, A. The shear-induced migration of particles in concentrated suspensions, *J. Fluid Mech.*, 1987 **181** 415–439.

Lipscomb II, G.G., Denn, M.M., Hur, D.U. and Boger, D.V. The flow of fiber suspensions in complex geometry, *J. Non-Newtonian Fluid Mech.*, 1988 **26** 297–325.

Milliken, W.J., Gottlieb, M., Graham, A.L., Mondy, L.A. and Powell, R.L. The viscosity-volume fraction relation for suspensions of rod-like particles by falling-ball rheometry, *J. Fluid Mech.*, 1989 **202** 217–232.

Mondy, L.A., Brenner, H., Altobelli, S.A., Abbott, J.R. and Graham, A.L. Shear-induced particle migration in suspensions of rods, *J. Rheol.*, 1994 **38** 444–452.

Papanastasiou, T.C. and Alexandrou, A.N. Isothermal extrusion of non-dilute fiber suspensions, *J. Non-Newtonian Fluid Mech.*, 1987 **25** 313–328.

Park, N. A. and Irvine, T.F., Jr. Measurements of rheological fluid properties with the falling needle viscometer, *Rev. Sci. Instrum.*, 1988 **59** 2051–2058.

Park, N.A., Cho, Y.I. and Irvine, T.F., Jr. Steady shear viscosity measurements of viscoelastic fluids with the falling needle viscometer, *J. Non-Newtonian Fluid Mech.*, 1990 **34** 351–357.

Petit, L. and Noetinger, B. Shear-induced structures in macroscopic dispersions, *Rheol. Acta*, 1988 **27** 437–441.

Phan-Thien, N. A nonlinear network viscoelastic model, *J. Rheol.*, 1978 **22** 259–283.

Phan-Thien, N. Constitutive equation for concentrated suspensions in Newtonian liquids, *J. Rheol.*, 1995 **39** 679–695.

Phan-Thien, N. and Graham, A.L. A new constitutive model for fibre suspensions: flow past a sphere, *J. Rheol.*, 1991 **30** 44–57.

Phan-Thien, N. and Kim, S. *Microstructure in Elastic Media: Principles and Computational Methods*, New York, Oxford University Press, 1994.

Phan-Thien, N. and Tanner, R.I. A new constitutive equation derived from network theory, *J. Non-Newtonian Fluid Mech.*, 1977 **2** 353–365.

Phan-Thien, N., Tran-Cong, T. and Graham, A.L. Shear flow of periodic arrays of particle clusters: a boundary-element method, *J. Fluid Mech.*, 1991a **228** 275–293.

Phan-Thien, N., Zheng, R. and Graham, A.L. The flow of a model suspension fluid past a sphere, *J. Stat. Phys.*, 1991b **62** 1173–1195.

Phan-Thien, N., Zheng, R. and Tanner, R.I. Flow along the centreline behind a sphere in a uniform stream, *J. Non-Newtonian Fluid Mech.*, 1991c **41** 151–170.

Phan-Thien, N., Jin, H. and Zheng, R. On the flow past a needle in a cylindrical tube, *J. Non-Newtonian Fluid Mech.*, 1993 **47** 137–155.

Phan-Thien, N. and Tullock, D.L. Completed double layer BEM in elasticity and Stokes flow: distributed computing through PVM, *Comput. Mech.*, 1994 **14** 370–383.

Phan-Thien, N., Altobelli, S.A., Graham, A.L., Abbott J.R. and Mondy, L.A. Hydrodynamic particle migration in a concentrated suspension undergoing flow between rotating eccentric cylinders, *Ind. Eng. Chem. Res.*, 1995 **34** 3187–3194.

Phillips, R.J., Armstrong, R.C., Brown, R.A., Graham, A.L. and Abbott, J.R. A constitutive equation for concentrated suspension that accounts for shear-induced particle migration, *Phys. Fluids A*, 1992 **4** 30–40.

Powell, R.L. Rheology of suspensions of rodlike particles, *J. Stat. Phys.*, 1991 **62** 1073–1094.

Rosenberg, J., Denn, M.M. and Keunings, R. Simulation of non-recirculating flows of dilute fibre suspensions, *J. Non-Newtonian Fluid Mech.*, 1990 **37** 317–345.

Ryskin, G. and Rallison, J.M., appendix to Ryskin, G. The extensional viscosity of a dilute suspension of spherical particles at intermediate microscale Reynolds numbers, *J. Fluid Mech.*, 1980 **99** 513–529.

Sherman, R. and Calder, A. Predicting fiber orientation distribution for the analysis of injection molded plastics, *Proceedings Structures Plastics '92 Conference and New Product Design Competition*, Dallas, Texas, April 5–8, 1992, 1–5.

Sun, J. *Some Developments and Applications of the Mixed Finite Element and Streamline Integration Method for Non-Newtonian Fluid Flow*, PhD thesis, University of Sydney, 1995.

Szeri, A.J. and Leal L.G. A new computational method for the solution of flow problems of microstructured fluids. Part 1. Theory, *J. Fluid Mech.*, 1992 **242** 549–576.

Szeri, A.J. and Leal L.G. A new computational method for the solution of flow problems of microstructured fluids. Part 2. Inhomogeneous shear flow of suspension, *J. Fluid Mech.*, 1994 **262** 171–204.

Tadmor, Z. and Gogos, C.E., *Principles of Polymer Processing*, New York, John Wiley & Sons, 1977.

Tanner, R.I., *Engineering Rheology*, 2nd edn., London, Oxford Press, 1988.

Tran-Cong, T. and Phan-Thien, N. Stokes problems of multiparticle systems: a numerical method for arbitrary flows, *Phys Fluids A*, 1989 **1**(3) 453–461.

Tucker, C.L. (ed.) *Fundamentals of Computer Modelling for Polymer Processing*, Munich, Hanser, 1989.

Tucker, C.L. Flow regimes for fiber suspensions in narrow gaps, *J. Non-Newtonian Fluid Mech.*, 1991 **39** 239–268.

Verlaye, V. and Dupret, F. Prediction of fiber orientation in complex injection molded parts, *Proc. ASME Winter Annual Meeting.*, Nov. 28–Dec. 3, 1993, New Orleans.

Walters, K. and Tanner, R. I. The motion of a sphere through an elastic fluid, in *Transport Processes in Bubbles, Drops and Particles,* eds Chhabra R.P. and De Kee D, Hemisphere Publication, New York, 1992, pp 73–86.

Yamane, Y., Kaneda, Y. and Doi, M. Numerical simulation of semi-dilute suspensions of rodlike particles in shear flow, *J. Non-Newtonian Fluid Mech.*, 1994 **54** 405–421.

Zheng, R., *Boundary Element Method for some Problems in Fluid Mechanics and Rheology*, PhD Thesis, University of Sydney, 1991.

Zheng, R. and Phan-Thien, N. Transforming the domain integrals to the boundary using approximate particular solutions: a boundary element approach for non-linear problems, *Appl. Numer. Math.*, 1992a **10** 435–445.

Zheng, R. and Phan-Thien, N. Boundary element simulation of the unsteady motion of a sphere falling along the axis of a cylindrical tube containing a viscoelastic fluid, *Rheol. Acta*, 1992b **31** 323–332.

Zheng, R., Phan-Thien, N. and Tanner, R.I. On the flow past a sphere in a cylindrical tube: limiting Weissenberg number, *J. Non-Newtonian Fluid Mech.*, 1990a **36** 27–49.

Zheng, R., Phan-Thien, N. and Tanner, R.I. Modelling the flow of a suspension fluid in injection moulding, in *Proc. Fifth National Society of Rheology*, ed. Y.L. Yeow and P.H.T. Uhlherr, Melbourne, 1990b, 141–144.

Zheng, R., Coleman, C.J. and Phan-Thien, N. A boundary element approach for non-homogeneous potential problems, *Comput. Mech.*, 1991a **7** 729–288.

Zheng, R., Phan-Thien, N. and Coleman, C.J. A boundary element approach for non-linear boundary-value problems, *Comput. Mech.*, 1991b **8** 71–86.

Zheng, R., Phan-Thien, N. and Tanner, R.I. The flow past a sphere in a cylindrical tube: effects of inertia, shear thinning and elasticity, *Rheol. Acta*, 1991c **30** 499–510.

Zheng, R., Phan-Thien, N. and Ilic, V. Falling needle rheometry for general viscoelastic fluids, *J. Fluid Eng.*, 1994 **116** 619–624.

Zheng, R., McCafferey, N., Winch, K., Yu H. and Kennedy, P. Predicting warpage of injection moulded fibre-reinforced plastics, *J. Compos. Thermoplastic Mater.*, 1996 **9** 90–106.

# 4

## Flow-induced alignment in injection molding of fiber-reinforced polymer composites

T D PAPATHANASIOU

## 4.1    Introduction

The orientation patterns observed in injection molded fiber-reinforced composites are the result of the response of the fibers to the flow in the cavity during filling and, to a lesser extent, the result of subsequent orientation during packing. Features of the filling flow affecting fiber orientation include: first, the shearing flow across the cavity thickness, second, the convective flow along the plane of the cavity, which is mainly determined by the geometry of the part and third, the fountain flow near the advancing free surface. Among geometrical factors, the gating arrangement plays a significant role by determining the amount of extension the fiber-loaded melt experiences upon entry into the cavity, as do inserts or contractions/expansions by altering the fiber orientation locally. There now exists a significant amount of published literature on the subject, from the pioneering experimental work of Darlington *et al.* (1976,1977) and Bright *et al.* (1978) to the more recent works of Bay and Tucker (1992a,b) and Gupta and Wang (1993), in which experimental measurements have been compared with model predictions, and the work of Chung and Kwon (1995) in which models for fiber orientation have been coupled to the kinematics of the flow during mold filling. Summaries of experimental and theoretical studies in this area are given in Tables 4.1 and 4.2. In spite of the expanding usage of injection molding in fabricating short and long fiber-reinforced parts, experimental and computational results on this subject have not been systematically presented and discussed in recent reviews or books dealing with injection molding. The aim of this chapter is to fill this gap and present a critical review of available experimental and computational studies in the area (the latter covered in some more detail in the preceding Chapters 2 and 3). A brief overview of the injection molding process and of the modeling of its filling stage are presented in Section 4.2. The main characteristics of the flow patterns during filling that are of relevance to fiber orientation are also presented in that section. The bulk of the chapter (Section 4.3) is concerned with a review of experimental observations of fiber orientation in injection molded parts. Where possible, these are related to computational results and trends are identified and stated. Finally

*Table 4.1.* Selected experimental studies on injection molding of fiber-reinforced composites

| Reference | Geometry | Fiber | Remarks |
|---|---|---|---|
| Akay and Barkley (1991) | Edge-gated plaque Multiple gates | SF/LF | SEM analysis across thickness. Mechanical properties. Knit lines. Edge effects. Fiber depletion. Effect of wall temperature and of injection speed |
| Akay and Barkley (1992) | Edge-gated strip | SF/LF | Jetting. Fiber-size degradation |
| Akay and Barkley (1993) | Edge-gated strip Single/double gate | SF/LF | Weld and knit lines, filling patterns, mechanical properties |
| Bailey and Rzepka (1991) | Edge-gated disk Fan-gated plaque | SF/LF | SEM analysis across thickness. Effect of packing and other processing conditions on core thickness. Fiber depletion |
| Bay and Tucker (1992a,b) | Film-gated strip Center-gated disk | SF | Use of orientation tensor $(\alpha)$. Gapwise variation of components of $(\alpha)$. Effect of gate. Comparison with model. Core thickness |
| Bouti *et al.* (1989) | Edge-gated strip | Glass Flake | Effect of cavity thickness. Effect of concentration. Mechanical properties |
| Bright *et al.* (1978) | Edge-gated strip | SF | Fiber orientation across thickness (contact microradiographs). Effect of injection speed. Orientation in the sprue and at the gate. |
| Darlington and Smith (1987) | Center-gated box, disk and bowl | SF | Fiber orientation across thickness (contact microradiographs). Core thickness. Effect of packing pressure. Fiber depletion |
| Fellahi *et al.* (1995) | Various | – | Review paper on weldlines |
| Fisa and Rahmani (1991) | Edge-gated plaque Two gates; inserts | SF | Formation and properties of weldlines. Effect of cavity thickness on core size. Free surface shape. |
| Hegler (1984) | Plaques, various gating schemes | SF | Skin–core structure, edge effects, weldlines |

(*continued*)

*Table 4.1. (continued)*

| Reference | Geometry | Fiber | Remarks |
|-----------|----------|-------|---------|
| Matsuoka *et al.* (1990) | Edge-gated plaque | SF | X-rays with metal tracer fibers. Thermal expansion coefficient. Effect of fiber volume fraction, edge effects. Comparison with model |
| McClelland and Gibson (1990) | Flow in nozzle, convergent dies and plaques | LF/SF | Visualization of fiber orientations (radiography). Rheology of LFR material. Core size. Fiber bending |
| Sanou *et al.* (1985) | Sprue-gated strip | SF | Short and full shots. Effect of fiber concentration, of cavity thickness and of injection speed |
| Sanschagrin *et al.* (1990) | Double-gated strip Inserts | SF/flake | Morphology (fractography) and properties of weld-lines. Shape of free surface. |
| Spahr *et al.* (1990) | Edge-gated plaque | SF/LF | Fractography, fracture toughness. Effect of fiber concentration. Effect of fiber aspect ratio on core thickness. Fiber depletion |
| Truckenmuller and Fritz (1991) | Edge-gated plaque | SF/LF | Filling patterns for LFR melts. Core size, fiber depletion, fiber bending and fiber clustering. Across-thickness fiber orientation via SEM. Mechanical properties |
| Vincent and Agassant (1986) | Center-gated disk | SF | Optical microscopy, core thickness, wall temperature |

SF, short fibers; LF, long fibers.

Section 4.4 is an outline of modeling techniques used in the prediction of fiber orientation in injection molding.

## 4.2    The injection molding process

### 4.2.1    Overview

Injection molding is an intermittent cyclic process used to produce uniform articles from a mold. It is widely used in plastics processing, particularly in the production of articles with a high degree of geometrical complexity. Injection

*Table 4.2.* Selected computational studies on fiber orientation in injection molding of fiber-reinforced composites

| Reference | Flow Studied | Method | Remarks |
|---|---|---|---|
| Bay and Tucker (1992a,b) | Filling of film-gated strip and of center-gated disk | uC | Orientation tensors (hybrid closure). Parametric studies (effect of inter-action coefficient, filling time, wall temperature, melt rheology, fountain flow). Comparison with experiment |
| Chung and Kwon (1995) | Filling of edge-gated tensile bar; multifaceted podium | C | Second order orientation tensor. Effect of fiber con-centration on clamping force and gapwise velocity profiles. Effect of interac-tion coefficient. Effect of orientation at the gate |
| Frahan et al. (1992) | Filling of multi-gated cavities incl. inserts. Podium | uC | Jeffery's theory and orien-tation tensor with quadra-tic closure. Orientations at the mid-plane. Isothermal flow. No fountain flow |
| Gillespie et al. (1985) | Filling of edge gated strip with insert | uC | Jeffery's orientations (un-coupled). Weldline. Edge-effects. 2D planar flow (no Hele–Shaw). Isothermal |
| Gupta and Wang (1993) | Filling of fan-gated plaque | uC | Second order orientation tensor including an inter-action coefficient. Effects of: initial fiber orientation, cavity thickness, injection speed, wall solidification, melt rheology. Non-iso-thermal |
| Matsuoka et al. (1990, 1991) | Filling of edge-gated plaque | uC | Folger–Tucker model. Hele–Shaw flow with gap-averaged velocities and orientations. Warpage pre-dictions |
| Ranganathan and Advani (1993) | Steady, isothermal flow in center-gated disk | C | Fourth order orientation tensor. Comparison of coupled vs. decoupled so-lution. Effect of the initial fiber orientation. |

C, coupled approach; uC, decoupled approach.

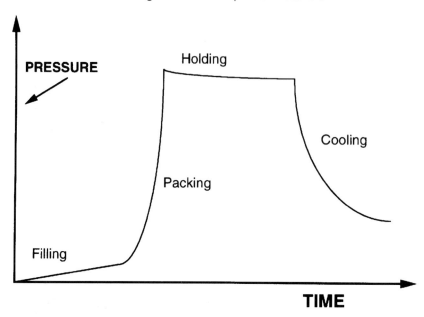

*4.1*  Pressure history in the mold cavity showing the stages of a typical injection molding cycle.

molded components range from large automobile, aerospace and computer parts to tiny gears or paper clips. Inherent advantages of the process are its high degree of reproducibility, its ability to produce parts of high geometrical complexity and its robustness in terms of feed material. Most polymers can be injection molded, including fiber-reinforced and engineering thermoplastics, thermosetting polymers and liquid crystal polymers. Variants of the process are also used in the shaping of 'green' ceramic preforms and in powder metallurgy. The typical injection molding process consists of three steps. During the filling step, the molten polymer is forced into the cavity whose walls are maintained at a temperature low enough to cause solidification (or, high enough to cause curing, in the case of thermosets). When the cavity is full, more material is packed into the mold to account for the shrinkage occurring during solidification; this elevated pressure is maintained for some time (packing/holding step) to enhance the dimensional stability of the product. After packing/holding is complete, the material is allowed to cool into the mold cavity until it is safe to be ejected without structural damage (cooling step). In terms of the variation of pressure in the cavity, the three stages outlined above can be summarized in Fig. 4.1. There is a substantial body of technical literature on injection molding. Besides sources discussed in this chapter, overviews of the process have been given by Isayev (1987) and Bernhardt (1983).

## 4.2.2  Modeling of mold filling

Filling is the most comprehensively studied stage of the injection molding process. This is not surprising since moldability and process optimization for both thermoplastics and thermosets hinge on the mold-filling step. Filling is of even greater importance in injection molding of fiber-reinforced thermoplastics, since the orientation assumed by the fibers in various sections of the part is almost entirely determined by the flow patterns during filling. Use of computational techniques in the analysis of flow during mold filling was pioneered in the 1970s by Kamal and co-workers (Kamal and Kenig, 1972a,b; Kuo and Kamal, 1976). Increased levels of sophistication have been introduced in subsequent analyses since the mid-1970s. In loose chronological order, these include the use of the finite element method (Hieber and Shen, 1980; Hieber *et al.*, 1983) as a means of discretizing three-dimensional thin complex-shaped parts and allowing for non-linear material properties (in particular, a viscosity that depends on shear rate, temperature and pressure), consideration of the three-dimensional transient nature of the energy equation to allow for a realistic description of heat transfer from the mold to the cavity (Shen, 1984), consideration of viscoelasticity (Kamal *et al.*, 1986,1988; Lafleur and Kamal, 1986; Mavridis *et al.*, 1988; Papathanasiou and Kamal, 1993), analysis of the fountain flow (Kamal *et al.*, 1986,1988; Lafleur and Kamal, 1986; Gogos *et al.*, 1986; Behrens *et al.*, 1987; Coyle *et al.*, 1987; Mavridis *et al.*, 1986,1988), the introduction of body-fitted curvilinear coordinates (Subbiah *et al.*, 1989; Papathanasiou and Kamal, 1993; Papathanasiou, 1995) and the coupling between filling and post-filling analyses along with prediction of effective product properties (Chiang *et al.*, 1991a,b,1994). Extensions of traditional mold-filling analyses to thermosetting polymers have also been reported (Macosco, 1986; Hayes *et al.*, 1991). A number of commercial packages for the simulation of the various stages of the injection molding process are available (Manzione, 1987). Some of these packages include modules for the prediction of fiber orientation. Such results are shown later in this chapter as well as in Chapter 3. In the following, the Hele–Shaw approach typically used in the modeling of mold filling is outlined and the basic features of typical mold-filling flows are presented.

### 4.2.2.1  *Planar flow and pressure drop*

The Hele–Shaw analysis is now widely accepted as appropriate for the modeling of the macroscopic characteristics of the filling stage of injection molding in cavities of general three-dimensional shape. The main requirement is that $L \gg 2b$, where $L$ is a characteristic length for the flow on the $x$–$y$ plane and $2b$ is the cavity thickness. Implicit in the Hele–Shaw approach are the following assumptions:

1   Viscous forces dominate during filling and thus, inertial forces are neglected in the momentum equation as are body forces.
2   No flow occurs in the transverse direction (i.e. across the cavity thickness).
3   Because of the geometrical proportions of thin cavities ($L \gg 2b$) and because of no-slip on the cavity walls, shear stresses across the thickness dominate and thus, the in-plane stresses are neglected in the momentum equation.
4   The flow is incompressible and inelastic.

Under these assumptions the flow is at quasi-steady state (i.e. time is only a stepping parameter) and two dimensional. With reference to Fig. 4.2, the Hele–Shaw model for mold filling is:

$$\frac{\partial P}{\partial x} = \frac{\partial}{\partial z}\{T^{xz}\} \qquad\qquad \frac{\partial P}{\partial y} = \frac{\partial}{\partial z}\{T^{yz}\} \qquad\qquad [1]$$

where $T$ is the deviatoric stress tensor, $x$ and $y$ define the plane of the flow and $z$ indicates the thickness direction. Neglecting viscoelasticity, the stresses in Eqn [1] are typically expressed using a generalized Newtonian constitutive model for the shear viscosity $\mu$:

$$\mu(\Theta, P, \gamma) = \frac{\mu_o(P, \Theta)}{1 + (\mu_o \gamma / \tau^*)^{1-n}} \qquad\qquad [2]$$

in which $\gamma$ is the shear rate and the parameters $n$ and $\tau^*$ determine the slope of the $\mu$ versus $\gamma$ curve and the width of the transition zone from the power-law to

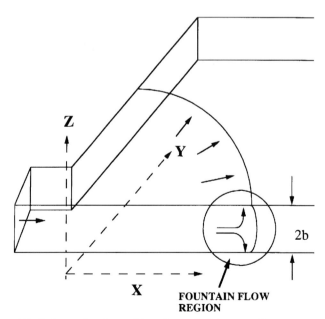

*4.2*  Schematic of a mold cavity and of the coordinate system adopted in this chapter.

the Newtonian (plateau) region, respectively. The zero-shear-rate viscosity $\mu_0$ is a function of temperature $\Theta$ (and pressure $P$) according to standard relationships, such as the Arhenius or the Williams–Landel–Ferry functional forms (Wang, 1992).

One first integration of Eqn [1] in the $z$-direction yields expressions for the gapwise variation of the in-plane velocities $u$ and $v$:

$$u(x, y, z) = -\left(\frac{\partial P}{\partial x}\right) \int_z^{b(x,y)} \frac{z\,dz}{\mu(x, y, z)} \qquad v(x, y, z) = -\left(\frac{\partial P}{\partial y}\right) \int_z^{b(x,y)} \frac{z\,dz}{\mu(x, y, z)}.$$

[3]

In Eqn [3], the cavity thickness is allowed to vary on the $x$–$y$ plane. Introducing the fluidity integral $S$,

$$S(x, y) = \int_0^{b(x,y)} \frac{z^2\,dz}{\mu(x, y, z)}$$

[4]

the gap-averaged velocities in the $x$ and $y$ directions ($\bar{u}$ and $\bar{v}$) can be related to the local pressure drop as follows:

$$\bar{u}(x, y) = -\frac{S(x, y)}{b(x, y)} \cdot \left(\frac{\partial P}{\partial x}\right) \qquad \bar{v}(x, y) = -\frac{S(x, y)}{b(x, y)} \cdot \left(\frac{\partial P}{\partial y}\right)$$

[5]

Differentiating the above equations with respect to $x$ and $y$ respectively and adding, making use of the incompressibility condition, yields an elliptic equation for the pressure:

$$\frac{\partial}{\partial x}\left(\frac{S}{b}\frac{\partial P}{\partial x}\right) + \frac{\partial}{\partial y}\left(\frac{S}{b}\frac{\partial P}{\partial y}\right) = 0$$

[6]

It can be seen in Eqn [5] that the local in-plane velocity vector is in the direction of $-\nabla P$ and can be calculated in a simple post-processing step once the pressure field is known. One implication of the Hele–Shaw formulation of the filling flow is the fact that the fluid is allowed to slip on the side walls of the cavity. Chung and Kwon (1995) have reformulated the Hele–Shaw model in the case of flow of fiber-reinforcing melts by considering the interaction between flow and fiber orientation. One of their basic findings is that the resultant velocities are no longer in the direction of $-\nabla P$; some more discussion on this will be given in Section 4.4.2. The gapwise variation of $u$ and $v$ can also be calculated through Eqn [3] once the temperature (and thus the viscosity) profile across the gap is known. Since Eqn [6] is the only one to be solved, only boundary conditions for the pressure need to be imposed on the boundaries of the domain. On the side walls (edges) of the cavity, the condition is that the normal pressure gradient vanishes

$$\frac{\partial P}{\partial n} = 0 \qquad [7]$$

implying zero melt flux across these boundaries (Eqn [5]). On the free surface the pressure is given an ambient (usually zero) value, while at the gate either the pressure or the pressure gradient are prescribed as functions of time (depending on whether the filling is under fixed pressure or fixed flowrate). For Newtonian flow the fluidity integral becomes:

$$S = \frac{b^3(x, y)}{3\mu} \qquad [8]$$

and thus, if the cavity is also of constant thickness, Eqn [6] reduces to the Laplace equation. Analytical solutions to the corresponding problem of isothermal filling in rectangular cavities of uniform cross-section have been presented by Kuo and Kamal (1976). Isothermal filling problems (also allowing for variable cavity thickness and power-law behavior (small $n$) of the melt) are also amenable to numerical solutions using an asymptotic distance method (Aronsson, 1994); this approach results in computationally efficient predictions of the filling patterns in cavities of practically arbitrary geometrical complexity.

The pressure equation (Eqn [6]) is coupled to the energy equation through the use of a temperature-dependent viscosity in the fluidity integral (Eqn [4]). The energy equation for injection mold filling can be written as:

$$\rho C_p \left[ \frac{\partial \Theta}{\partial t} + u \frac{\partial \Theta}{\partial x} + v \frac{\partial \Theta}{\partial y} \right] = \mu \cdot \Phi + \kappa \left[ \nabla^2 \Theta + \frac{\partial^2 \Theta}{\partial z^2} \right] \qquad [9]$$

in which $C_p$ is the heat capacity and $\kappa$ is the thermal conductivity of the polymer melt, the first term at the right hand side is the viscous dissipation term and the Laplacian is taken on the plane of the flow. Essential features of Eqn [9] are heat convection on the $x$–$y$ plane due to the advancement of the hot melt into the cavity and heat conduction in the thickness direction due to the temperature difference between the melt and the mold surfaces. The boundary conditions for Eqn [9] on the cavity surfaces are ideally obtained by coupling the heat transfer problem in the cavity to the heat transfer problem in the mold, including the cooling channels (Manzione, 1987). In the absence of such coupling, a heat flux condition based on an average heat transfer coefficient is usually imposed on the cavity surfaces.

Simultaneous solution of the Hele–Shaw flow model (Eqn [6]) and of the energy equation (Eqn [9]) allows for the determination of the effect of the thermal conditions in the mold on pressure drop, filling times and moldability. Prediction of filling patterns as well as prediction of the variation in temperature and pressure in typical mold cavities can be accomplished fairly accurately

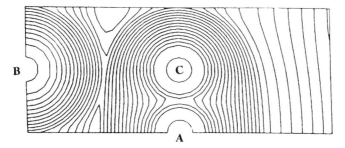

*4.3* Hele–Shaw flow predictions for isothermal filling patterns in a three-gated cavity (Frahan *et al.*, 1992, with permission). A, B and C indicated gate locations.

using the Hele–Shaw approach outlined previously. Existing commercial packages (Manzione, 1987; Chiang *et al.*, 1994; Chapter 3) have extended this capability to molds of great geometrical complexity. Typical filling patterns in a cavity with three gates, obtained from the Hele–Shaw model, are shown in Fig. 4.3.

## 4.2.3  Velocity profiles in mold filling

Understanding the general features of the mold-filling flow is essential for an at least qualitative understanding of the orientations assumed by fibers during injection molding of fiber-reinforced melts. It should be stated from the outset, however, that the velocity profiles during filling of a cavity with a fiber-filled melt are likely to deviate from those expected in molding of a neat resin under the same conditions. This is a result of the presence of the fibers (filler particles in general), which influences the rheology of the melt (Ranganathan and Advani, 1993; Tang and Altan, 1995; Chung and Kwon, 1995). Near walls, where the shear rate is highest, fibers tend to align faster than those nearer the centerplane where the shear rate is low. This leads to varying rheological properties across the cavity thickness and to a flatter velocity profile. More references on the subject of coupling between rheology and fiber orientation can be found in Chapter 3 of this book as well as in Section 4.4.2 of this chapter. In the following three sections we present the general features of the velocity profiles associated with mold filling using typical polymer melts.

### 4.2.3.1  *Gapwise velocity profile*

Since the viscosity of polymer melts is strongly temperature dependent and the surfaces of the mold are kept at a temperature well below the no-flow temperature of the melt (at least in the case of thermoplastic melts), the velocity across the gap deviates fundamentally from the fully developed profile corresponding to an isothermal melt flowing between parallel plates. The essence of this deviation is the existence of an inflection point whose location depends on the degree of non-isothermality. Because of this inflection point, the shear rate assumes its maximum value at some distance from the wall; this has significant implications in the morphology of injection molded parts. Tadmor (1974) proposed an empirical model which captures the qualitative features of the steady-state gapwise velocity profile under conditions of monotonically decreasing temperature from the center to the wall, that is, in the absence of viscous heating. The effect of the latter is present in most numerical simulations (e.g. Subbiah *et al.*, 1989; Chiang *et al.*, 1991a; Papathanasiou, 1995) and has been treated explicitly by Jansen and Vandam (1993). In a modified form, the expression proposed by Tadmor is:

$$u(z) = \bar{U}\frac{(J+3)}{2}\left[1 - \left(\frac{z}{b}\right)^s\right]^{(1+j)} \cdot \left[1 + (1+J)\left(\frac{z}{b}\right)^s\right], \quad s = \frac{1+n}{2n}$$

[10]

where $\bar{u}$ is the gap-averaged melt velocity. The value of the parameter $J$ expresses the degree of non-isothermality of the system. For isothermal conditions, $J = 0$ and Eqn [10] yields the expected power-law velocity profile with index $n$. Eqn [10] is able to reproduce quite well numerically predicted gapwise velocity profiles and thus can serve as a useful parametric expression for analysis purposes. Figure 4.4a shows the predictions of Eqn [10] for $J = 3$ as well as the corresponding isothermal ($J = 0$) power-law ($n = 0.264$) and Newtonian ($n = 1$) profiles obtained for the same melt flowrate. Evidently, non-isothermality reduces the 'flatness' of the velocity profile and this has significant implications in the morphology of injection molded short-fiber reinforced parts, as discussed in Section 4.3.1 below. The shear rate profile across the gap can be readily obtained by differentiation of Eqn [10]. As shown in Fig. 4.4b, when the non-isothermality of the flow increases the maximum shear rate is shifted away from the walls and towards the mid-plane of the cavity. In injection molding of fiber-reinforced melts, this is associated with an analogous shift of the oriented shell zone (Section 4.3.1).

### 4.2.3.2 *Planar velocities*

The planar flow in the mold, with gap-averaged velocities calculated from Eqn [5], is responsible for convecting the melt (and the fibers) from the gate into the cavity. It is also responsible for the orientation adopted by the fibers along the relatively shear-free region around the centerplane of the mold. The most interesting feature of the planar flow during filling is the expanding flow associated with a narrow edge-gate or a center-gate as well as the converging/ diverging flows induced by the geometry of the part (such as the flow around

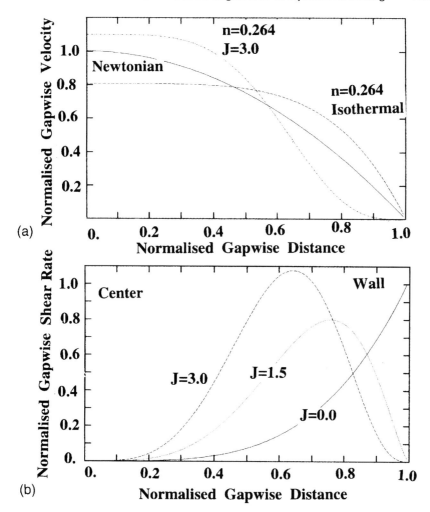

4.4 (a) Gapwise velocity profiles (normalized with respect to the Newtonian centerplane velocity) in the case of Newtonian flow, isothermal flow of a power-law fluid with $n=0.264$ and non-isothermal flow of a power-law fluid ($n=0.264$, $J=3$). (b) Effect of non-isothermality on the gapwise distribution of shear rate (normalized with respect to the wall shear rate corresponding to isothermal flow). Power-law fluid with $n=0.264$.

obstacles, around corners, etc.). As an example, the prediction of fiber orientation for flow around a cylindrical obstacle is covered in Chapter 3 of this book. Figure 4.5 shows predicted velocity profiles and streamlines in the region around an edge-gate; the presence of deceleration in the main direction of flow and of stretching transverse to it are evident.

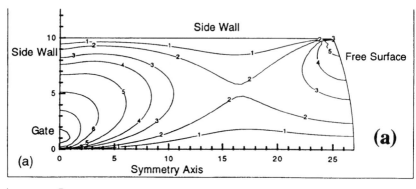

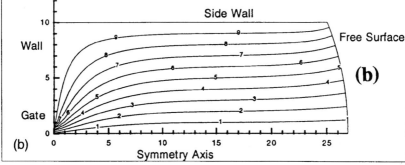

4.5 Contours of the transverse planar velocity component $\bar{v}$ (a), and of the stream function (b) in a rectangular cavity at some instant during filling. Velocities are normalized with respect to the maximum value. Contour levels are: (a) 0.00, 0.01, 0.02, 0.03, 0.05, 0.1, 0.2, 0.5; (b) 0.1–0.9 in increments of 0.1.

### 4.2.3.3    Fountain flow

The Hele–Shaw approximation has been found to be accurate in predicting filling patterns and is considered adequate in determining the pressure required for filling. Improvements in the quantitative nature of these predictions are being sought by using more realistic material data and refining the numerical models rather than by reformulating the basic Hele–Shaw equations. However, since the gapwise transverse flow is ignored in the Hele–Shaw analysis (in order to derive a two-dimensional representation with the ensuing economy in computational requirements) this approximation cannot give any information on the single most important feature of the filling process, namely the fountain flow (Rose, 1961). This is the term that has been used to describe the gapwise transverse flow immediately behind an advancing melt/air interface. As the term implies, fountain flow is characterized by deceleration of the fluid elements as they approach the (slower moving) interface. As a result of incompressibility, these fluid elements acquire a transverse velocity, spilling outward towards the wall.

Fountain flow has been the subject of extensive experimental and theoretical work in the context of injection mold filling (Kamal *et al.*, 1986,1988; Behrens *et al.*, 1987; Coyle *et al.*, 1987; Mavridis *et al.*, 1986,1988). Through these, it is now known that fountain flow is primarily responsible for the rearrangement of the material entering the cavity. The tracer experiments of Schmidt (1974) have shown that material entering the mold first ends up deposited on the surface of the cavity closer to the injection gate, while subsequently injected melt ends up on the surface at longer distance from the injection point. The fountain flow has also been found to be responsible for the formation of oriented polymer layers at the surface of the molding (Kamal and Tan, 1979; Moy and Kamal, 1980; Isayev, 1983; Mavridis *et al.*, 1988). The calculations of Mavridis *et al.* (1986) have shown that the kinematics of fountain flow are primarily governed by conservation of mass and are only affected by viscoelasticity in a secondary manner (Bay and Tucker, 1992a). Coyle *et al.* (1987) have presented a comprehensive experimental and computational study of the fountain flow in Newtonian and shear-thinning fluids, in the presence and absence of gravity. Figure 4.6 shows simulation results concerning the deformation of three tracer lines (indicated as 1,2,3), initially perpendicular to the direction of flow and at a location one gap-width behind the contact point between the free surface and the walls of the mold. Without loss of generality, a frame moving with the free surface has been considered in this simulation; the walls therefore appear to move in the $-x$ direction. Evidently fountain flow operates by stretching and orienting fluid elements during molding. The V-shaped deformation patterns appearing near the walls in Fig. 4.6 have been confirmed with tracer experiments (Coyle *et al.*, 1987) and are in qualitative agreement with the injection molding tracer experiments of Schmidt (1977) and Faulker and Schmidt (1977). These flow patterns have been confirmed in other numerical studies and comprise an

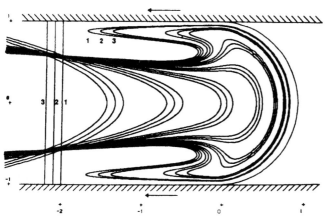

*4.6* Schematic diagram of the deformation of fluid elements by the fountain flow (Coyle *et al.*, 1987, with permission).

accepted model for the local flow of polymer melts behind the advancing free surface during injection mold filling.

As anticipated, fountain flow does play a role in the development of the fiber orientation distributions observed in injection molded fiber-reinforced parts (Bay and Tucker, 1992a,b). Its influence, however, is frequently overwhelmed by that of the shear and extensional flows outlined previously.

## 4.3    Experimental observations of fiber orientation in injection molding

### 4.3.1   Filling patterns with fiber-reinforced melts

Because of the nature of the Hele–Shaw flow during mold filling (highly viscous melts flowing in thin cavities), the melt under usual molding conditions will fill the cavity by spreading in a smooth manner (Fig. 4.3). Typically, the melt front assumes a continuous smooth shape which is maintained as the cavity is filled. These patterns have also been observed during injection molding of short-fiber-reinforced thermoplastics. However, in the filling of cavities with short-fiber-reinforced melts at high fiber concentrations or with long-fiber-reinforced melts, peculiar filling patterns are known to occur. The main observations are that the free surface ceases to be smooth and continuous but appears irregular and broken up, with the melt front progressing faster in the regions adjacent to the side walls of the cavity and that, sometimes, jetting occurs after an initial circular melt front has been established. These phenomena are schematically shown in Fig. 4.7.

Truckenmuller and Fritz (1991) found that short shots of (30% w/w) long-glass-fiber-reinforced (GFR) polyamide-6,6 exhibited a free surface which appeared discontinuous and fragmented. This was the case regardless of the manner of incorporation of the long fibers in the resin (through pultrusion compounding or through direct addition into the molding machine). In contrast, short GFR polyamide-6,6 (also 30% w/w) exhibited a smooth filling pattern, much like the neat resin. Similar irregular filling patterns (with the free surface advancing faster along the side walls of the cavity) were observed by Akay and Barkley (1993) in 50% w/w short- and long-GFR polyamides. According to Akay and Barkley (1993) and Truckenmuller and Fritz (1991), this filling pattern is the result of the reduced resistance to melt flow in the regions adjacent to the side walls of the cavity, where the fibers are predominantly aligned in the flow direction (Section 4.3.3.1) and thus offer less resistance to melt flow compared with the (interlocking) fibers in the central part of the mold. Possible breakage of fibers near the side walls leads to further reduction in local viscosity. This explanation is not based on direct measurements or computations; however, it is in agreement with the observation that lower fiber content SFR melts (30% and

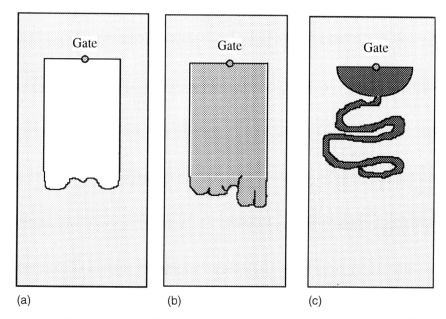

(a)                                (b)                                (c)

**4.7**  Schematic of filling patterns typical of concentrated- or long-fiber-reinforced melts. (a) Preferential progression along the edges; (b) fragmented free surface; (c) jetting (after Akay and Barkley 1985; Truckenmuller and Fritz, 1991).

40% w/w) exhibit smooth filling patterns and that the irregular patterns described above are more prominent for long-fiber-reinforced melts. Akay and Barkley (1985,1992) also observed that jetting occurred in 50% w/w short-fiber GFR polyamide as well as in the long-fiber-reinforced melt. No jetting was observed for the neat resin or for short-GFR material at fiber concentrations up to 30%
w/w and the propagation of the free surface was smooth and similar to that observed with a neat resin. The anisotropic viscosity of fiber dispersions is the main reason behind the filling patterns observed in concentrated and/or long-fiber-reinforced melts. These patterns have not yet been predicted theoretically or computationally and are beyond the predictive capabilities of existing commercial computer aided design (CAD) packages for the injection molding process.

## 4.3.2  Skin-core structure

Injection molded fiber-reinforced thermoplastics exhibit a distinct laminate structure across their thickness. Kenig (1986) talks about a nine-layer structure, including a very thin fiber-depleted layer at the surface, a sequence of three shell layers with alternating aligned–transverse–aligned (to the main direction of flow,

MDF) orientations and a core with orientation transverse to the MDF. Bay and Tucker (1992b) observe (and predict computationally) a seven-layer structure, including two shell layers with differing levels of orientation, the least oriented (and smaller) of which is attributed to the action of the fountain flow. If thin transitional layers are ignored, the various shell layers can be described as one continuous zone with orientation changing gradually but being primarily along the MDF. This five-layer structure (two skin layers, two shell zones and one core region, for a total of five layers) is supported by the overwhelming majority of microstructural analyses in fiber-reinforced injection molded parts and is adopted in the following discussion.

In conditions typical of commercial injection moldings in edge-gated cavities, the fibers in the core region are aligned transversely to the MDF, while those in the shell zone are aligned parallel to the MDF. The fibers immediately adjacent to the walls of the molding (skin layer) show a random-in-plane orientation. Figure 4.8 shows contact microradiographs of slices obtained from injection molded rectangular strips fed by a point gate located at the narrow end of the strip. The material used was a GFR polypropylene, containing 20% by weight of glass fibers of aspect ratio 60 (before injection) and the injection time was 0.2 s (fast injection). The formation of a skin–core structure with the fibers in the core aligned transversely to the MDF and the fibers in the shell zone aligned along the MDF is evident. Because of the short injection time, the thickness of the skin layer is insignificant and not visually identifiable in this figure. Similar observations have been reported in numerous other studies (Lhymn and Schultz, 1985; Spahr et al., 1990; Akay and Barkley, 1991; Darlington and Smith, 1987; Gupta and Wang, 1993; Bay and Tucker, 1992b).

The formation of a skin layer, in which the fibers assume a random-in-plane orientation, is the result of the fountain flow which moves material (including fibers) from the core to the cavity walls where it freezes before gapwise shearing is able to align the fibers in the flow direction (Bay and Tucker, 1992a,b; Gupta and Wang, 1993). The thickness of this layer depends on cooling rate and filling

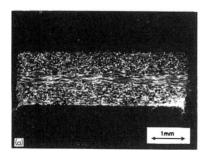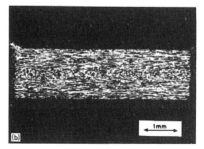

*4.8* Contact microradiographs of sections across the part thickness cut perpendicularly (a) and parallel (b) to the main flow direction (Bright *et al.*, 1978, with permission). Fast injection ($t_{fill} = 0.2$ s).

time (Kamal and Papathanasiou, 1993). Bay and Tucker (1992b) have suggested that the Graetz number ($Gz$) can be used as a measure of the anticipated thickness of the skin layer. $Gz$ is defined as the ratio of the characteristic time for heat conduction to the characteristic time for mold filling:

$$Gz = \frac{4b^2}{\alpha \cdot t_{fill}}$$ [11]

where $2b$ is the cavity thickness, $\alpha$ the thermal diffusivity of the melt and $t_{fill}$ the mold filling time. Higher values of $Gz$ indicate the formation of thinner skin layers. Bay and Tucker (1992b) suggest that no skin will form if $Gz > 100$.

The formation of a shell zone, in which the fibers are strongly oriented along the main direction of flow, is the result of shearing. The shell zone develops around the location of the maximum shear rate across the cavity thickness, as suggested by the shear rate profiles of Fig. 4.4b. Bay and Tucker (1992b) have reported experimental and computational results in a center-gated disk using various GFR melts. Simulations, including an approximate but realistic treatment of the fountain flow, predict that immediately below the skin layer there exists a thin sub-skin layer in which the orientation is between that of the shell and of the core zones. Bay and Tucker (1992b) attribute the presence of this region to the action of the fountain flow, in a manner similar to that used to explain the orientation of macromolecules (Isayev, 1983; Mavridis et al., 1988) near the skin of injection molded neat resins (polystyrene, Kamal and Tan, 1979 or polyethylene, Moy and Kamal, 1980). This sub-skin layer is shown to be erased by the action of the gapwise shear flow when the injection speed is high (Bay and Tucker 1992b).

The formation of a transversely oriented core region in the final molding is the result of two factors:

1   The presence of an extensional flow in the vicinity of the gate. This can be the result of a small (point) edge gate and the consequent existence of a strong expanding flow in its vicinity as the melt spreads to fill the cavity or the presence of a center gate, as is the case in top-gated cavities. As indicated by Jeffery's model (Chapter 2), fibers in a diverging flow will align in the direction of extension, which is in this case perpendicular to the main direction of flow. This transverse orientation is achieved irrespective of the orientation possessed by the fibers upon entry into the cavity.

2   The convection of the thus oriented fibers downstream into the cavity by a relatively shear-free flow. Apart from the centerplane, shear flow across the cavity thickness tends, in general, to reorient the fibers along the MDF (Chapter 2). However, the existence of a flat (blunted) velocity profile across the thickness of the cavity helps maintain the transverse fiber orientation acquired as outlined in 1) above and convey it downstream into the cavity without significant reorientation.

A transversely oriented core is thus associated with extensional flow (edge-gated cavities with small gates, center-gated cavities irrespective of gate size) and 'flat' gapwise velocity profiles (fast injection speeds, thin cavities and/or highly shear-thinning melts). If one or both of 1) or 2) above are not realized in a particular application, the fibers in the core will be to a larger or lesser extent aligned with the MDF. In the following, some of the factors affecting fiber orientation in the core region are briefly expanded upon.

### 4.3.2.1  Influence of the injection gate

Gating arrangements which create a diverging flow in their vicinity cause the formation of a core in which the fibers are aligned in the direction of elongation (which is perpendicular to the MDF) irrespective of their orientation upon entry into the cavity (Bay and Tucker 1992b; Vincent and Agassant 1986; Gupta and Wang 1993). Gating arrangements which do not create such a flow (as in film-gated or sprue-gated moldings) facilitate the formation of a core in which the fibers are primarily aligned in the flow direction. Bay and Tucker (1992b) confirmed this by comparing experimentally determined fiber orientations at the core of parts made of GFR nylon and polypropylene (using the second order orientation tensor to quantify measurements obtained through optical micro-scopy) in a center-gated disk and in a film-gated strip. They found that while the fiber orientation at the core was strongly transverse to the main direction of flow in the disk, the fibers at the core of the strip were aligned with the MDF. Sanou et al. (1985) report similar observations (using optical microscopy) in molding GFR polypropylene through a sprue-type edge-gated cavity. As a result of the converging flow from the sprue-gate, the fibers align in the flow direction and maintain this orientation as they are transported downstream into the cavity. The presence of some fibers at the core whose orientation is transverse to the MDF can be traced to fibers maintaining the orientations they had in the sprue prior to entering the cavity.

The influence of the gating arrangement on the relative size of the (transverse) core and (aligned) shell zones has been demonstrated in the work of Darlington and Smith (1987), who studied the orientation of fibers in sections taken from a washing bowl (in the shape of a thin-walled cylinder open at the top and gated at the center of its base) injection molded from GFR nylon 66. The flow in the circular base is radial while the flow in the side walls of the cylindrical bowl is one-dimensional, as if caused by a film gate. Their results (obtained using contact microradiography) show that the core thickness in sections taken from the side walls (where the flow is primarily one-dimensional) is smaller than along the base (where an extensional component tends to orient the fibers transversely to the main (radial) flow direction). Evidently, the shear flow during filling of the side walls has resulted in a reorientation of some part of the core along the MDF. In conclusion, available experimental evidence on the effect of the gating arrangement on the fiber orientation at the core can be summarized as follows:

> Point- or center-gate ⟷ Transverse core
>
> Film- or sprue-gate ⟷ Aligned core

### 4.3.2.2  Effect of injection speed

Transversely oriented fibers in the core have been commonly associated with point-gated cavities. However, it has been shown that, even in point-gated cavities, low injection speeds can result in the formation of a core aligned, to a larger or lesser extent, with the main direction of flow. Using contact microradiography, Bright *et al.* (1978) have presented microstructural analysis of GFR polypropylene strips molded under slow and fast filling conditions. The results for the latter case have been presented in Fig. 4.8. The effect of lower injection speed on fiber orientation is clearly seen in Fig. 4.9 which shows the fiber orientations, observed in the same cavity and for the same material as in Fig. 4.8, when the injection time was increased from 0.2 s to 11 s.

Evidently, the shell and core regions observed in Fig. 4.8 have now blended in one zone in which the fibers are aligned along the MDF (which is perpendicular to the page in Fig. 4.9a and from right to left in Fig. 4.9b). Because of the long injection times, significant solidification on the walls of the molding occurs during slow filling (Section 4.3.2, Eqn [11]). This explains the formation of quite a substantial skin layer (Fig. 4.9) in which the fibers have maintained the random-in-plane orientation acquired through the action of fountain flow. The experiments of Gupta and Wang (1993) in a fan-gated plaque have also indicated that reducing the injection speed results in a core region which is more aligned along the MDF. Akay and Barkley (1991) reported that the size of the (transversely oriented) core in injection molded plaques of GFR polypropylene and polyamide increased roughly linearly from 20% to 60% of the plaque thickness when the ram speed increased from 10 mm s$^{-1}$ to 25 mm s$^{-1}$.

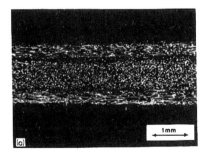 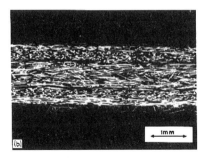

**4.9** Contact microradiographs of sections across the part thickness cut perpendicularly (a) and parallel (b) to the main flow direction (Bright *et al.*, 1978, with permission). Slow injection ($t_{fill} = 11$ s).

Bright *et al.* (1978) attributed the change in the orientation of the core region at lower filling rates (from transverse to the MDF towards aligned along the MDF) to the modification of the velocity profile across the gap caused by the corresponding lower shear rates. As outlined in Section 4.3.2.6, if the shear rate is low enough so that the viscosity of the melt is in the plateau region, the corresponding velocity profile will be steeper (closer to parabolic) compared to what would be observed in the high shear rate (power-law) region. This will result in high shear rates across a larger section across the cavity thickness and thus in a reorientation of some (or even all) of the initially transverse core along the MDF. However, this mechanism will only operate for certain combinations of melt rheology and injection speeds. If the shear rates corresponding to what is considered 'fast' filling fall in the power-law region of the viscosity curve and those corresponding to 'slow' filling fall in the transition or plateau region, then this mechanism will contribute to realigning an initially transversely-oriented core along the MDF. This appears to be the case in the work of Bright *et al.* (1978) as can be confirmed by careful examination of their data on injection speed and melt rheology. However, this mechanism will not operate if both 'fast' and 'slow' injections correspond to the power-law region, as has been clearly shown in the computational work of Gupta and Wang (1993). By choosing a very small value of the parameter $\tau^*$ ($\tau^* = 8.665$ Pa) in the shear viscosity model:

$$\mu = \frac{\mu_0}{1 + (\mu_0 \gamma / \tau^*)^{1-n}} \qquad [12]$$

they made the melt behave effectively as a power-law fluid in the entire range of shear rates examined ($\gamma > 0.1$ s$^{-1}$). Consequently, the gapwise velocity profiles were independent of injection speed (and typical of a pseudoplastic melt). Under these conditions, decoupled (Section 4.4.2) and isothermal filling/orientation simulations predicted that the fiber orientation at the core is identical for all injection speeds examined. In conclusion, available evidence on the effect of injection speed on the relative size of the core can be summarized as follows:

> Faster injection $\longleftrightarrow$ Thicker core (transverse to MDF)
> Slower injection $\longleftrightarrow$ Thinner core (transverse to MDF) or
> core aligned in the MDF

### 4.3.2.3  *Effect of wall solidification*

In parallel with the influence of the injection speed on the nature of the gapwise velocity profile, Bright *et al.* (1978) have argued that the observed orientations for slow injection speed into edge-gated cavities could be explained as the consequence of the reduction in cross-sectional area available for flow caused by extensive wall solidification. In this case, the expanding flow at the gate is not so strong and thus the transverse orientation assumed by the fibers upon entry into the cavity is not prominent. Numerical experiments by Gupta and Wang (1993) have suggested that wall solidification may be the key in inducing a fiber orientation along the MDF even in point-gated cavities. When the filling simulations for a power-law fluid outlined in the previous section were run under non-isothermal conditions (considering a cold wall and including viscous dissipation in the model), the fibers in the core were predicted to be more aligned in the MDF in the slow-injection case than in the case of fast injection. This has been interpreted by Gupta and Wang (1993) as the result of the formation of solidified polymer layers on the surface of the cavity whose thickness grows with distance from the gate. This creates a slightly converging flow which helps orient the fibers along the MDF. The thickness and therefore the relative importance of these layers become more prominent at low injection speeds. The existence of such a converging flow under conditions of non-isothermal filling and wall solidification had been predicted analytically by Richardson (1983,1986) and in subsequent computational work by Kamal et al. (1988). Such a solidification pattern on the $x$–$y$ plane under conditions of uniform cooling on the surfaces of the cavity is shown in Fig. 4.10. Motivated from their result, Gupta and Wang (1993) investigated the orientation of fibers in cavities with a converging cross-section. Their simulations predicted that in such cavities the core will be to some degree aligned along the MDF, even under isothermal conditions and using a point-gated cavity.

Considering that the local thickness of the solidified layer depends on the local cooling rate (Papathanasiou, 1995), the above result implies that fiber

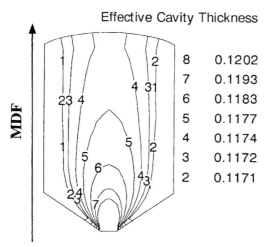

4.10 Predicted spatial variation of effective cavity thickness in a fan-gated rectangular cavity filled at constant rate and under uniform cooling conditions. The formation of a converging channel near the gate can be seen, judging by the spacing of the contour lines in its vicinity. The initial cavity thickness is 0.13 and the contour values represent the thickness of the cavity available for flow (Papathanasiou, 1995).

orientation could be influenced, to some degree, by judicious choice of the local cooling rate.

### 4.3.2.4 Effect of cavity thickness

For comparable injection speeds, thinner cavities are associated with higher shear rates and thus, all other parameters (cooling rate, material properties) being equal, with flatter velocity profiles. These in turn will help maintain the orientation imparted to the fibers upon entry (e.g. transverse to the MDF orientation near point gates) as they are transported downstream into the cavity. Bouti et al. (1989) presented a study of the microstructure and the mechanical properties of glass-flake reinforced polypropylene, injection molded in point-gated rectangular plaques of thickness between 1.6 mm and 6.4 mm. As in fiber-reinforced polymers, a skin–core structure was observed as a result of the orienting action of the flow field on the flakes, which were found to be aligned with the flow in the shell layer, while they assumed a transverse orientation in the core. The relative size of the core (transverse to the MDF) was found to increase with increasing cavity thickness ($2b$), representing about 15% of $2b$ for a molding with $2b = 1.6$ mm, between 15–30% when $2b = 3.2$ mm and between 40–50% when $2b = 6.4$ mm. This resulted in higher elastic tensile moduli in the thin samples and was attributed to the higher shear rate associated with thin moldings (since all samples were molded under the same injection speed). Not surprisingly, a similar increase in the relative thickness of the core as the cavity thickness increased was observed by Fisa and Rahmani (1991) in the case of GFR polypropylene (40% w/w). Bouti et al. (1989) also found that the size of the transversely oriented core increases with flake concentration, the latter ranging between 2% and 26% by weight, and that this increase was less pronounced for thinner cavities. In transfer molding experiments of a 33% w/w short-glass fiber reinforced thermoset resin, Gillespie et al. (1985) found (using optical microscopy) that in thin cavities ($2b = 1.6$ mm) the orientation at the core was slightly in the main direction of flow, while in thicker cavities ($2b = 3.2$ mm) the fibers in the core were oriented transverse to the MDF as would be expected in the edge-gated cavity employed. Simulation results presented by Gupta and Wang (1993) also indicate that the relative thickness of the transversely oriented core increases when the thickness of the cavity increases (from $2b = 0.79$ mm to $2b = 1.59$ mm). They argue that due to the low thermal conductivity of typical polymer melts, the thickness of the solidified layer is independent of ($2b$) and thus, factors favoring fiber orientation along the MDF will not operate effectively in thick cavities since the reduction in cross-sectional area due to solidification is proportionally smaller. In summary:

> Thicker cavity $\longleftrightarrow$ Thicker core (transverse to the MDF)

### 4.3.2.5 Effect of cavity wall temperature

The influence of the temperature at the cavity wall on the relative size and orientation of the core and shell layers can be traced to the shape of the gapwise velocity profile (Fig. 4.4) as well as to the amount of wall solidification and thus, is qualitatively similar to the effect of injection speed and to the effect of cavity thickness (Sections 4.3.2.3 and 4.3.2.4). Experiments by Vincent and Agassant (1986) on a center-gated disk indicate that raising the wall temperature from 80°C to 150°C results in a better oriented shell layer and a substantially thinner skin layer, while the thickness and orientation of the core remain unaffected. As shown in Fig. 4.4b, higher mold temperatures (lower $J$, Eqn. [10]) tend to shift the region of high shear rate towards the walls of the cavity, thus reducing the thickness of the skin layer, in agreement with the results of Vincent and Agassant (1986). However, a decrease in the core thickness at lower wall temperatures would be anticipated, since in that case the maximum shear rate is shifted towards the center of the cavity. The results of Vincent and Agassant do not indicate such a trend. Simulations performed by Gupta and Wang (1993) do indeed show that increasing the thermal conductivity and/or decreasing the mold temperature results in a reduction in the thickness of the transversely oriented core. This in turn means that the thermal conductivity of the fibers themselves (usually significantly higher than that of the melt) has a role to play in the microstructure of molded parts. Addition of 20% glass fibers (with thermal conductivity $k=0.0024$ cal cm$^{-1}$s$^{-1}$K$^{-1}$) into a polystyrene resin ($k=0.00029$ cal cm$^{-1}$s$^{-1}$K$^{-1}$, Sanou et al. (1985)) will result in a composite with $k=0.000324$ cal cm$^{-1}$s$^{-1}$K$^{-1}$ (if the model of Nielsen (1974) is used to predict the effective $k$ of the mixture):

$$k = k_p \frac{1 + ABf}{1 - \xi Bf} \qquad B = \frac{(k_f/k_p) - 1}{(k_f/k_p) + A} \qquad \xi = 1 + \left(\frac{1 - \phi_m}{\phi_m^2}\right)f \qquad [13]$$

where $k_p$ and $k_f$ are the thermal conductivities of polymer and fibers, respectively, $f$ is the fiber volume fraction, $\phi_m$ is the fiber volume fraction at maximum packing and $A$ is a constant depending on the orientation of the fibers and the direction of heat flow ($A=0.5$ for heat flux normal to uniaxially oriented fibers; Papathanasiou et al., 1995). Since the orientation of the fibers changes across the thickness, detailed investigation of this effect can be significantly more complicated.

Results of simulations performed by Bay and Tucker (1992b) show no effect of wall temperature on the thickness and orientation of the core and skin regions; however, this most probably reflects the choice of the conditions examined (wall temperature variation of only 30K). In agreement with the predictions of Gupta and Wang (1993), Akay and Barkley (1991) reported experimental data in injection-molded GFR polypropylene and polyamide plaques which show that the size of the core increased roughly linearly from 20% to 60% of the plaque thickness when the wall temperature increased from 60°C to 100°C. It should be pointed out that the fiber orientations measured in actual molded samples always depend on the interplay between a number of factors, such as the amount of viscous dissipation, the thermal and rheological properties of the fiber-reinforced melt and the filling rate. Discrepancies between various experimental investigations can very well be due to such factors which are not quantifiable. However, it appears that a general rule of thumb for the effect of wall temperature on microstructure in fiber-reinforced injection molded parts can be:

> Colder walls ⟷ Thinner core (of transverse to MDF orientation)
> ⟷ Thicker skin layers

### 4.3.2.6    Effect of melt rheology

The existence of a 'flat' velocity profile across the gap is the result of the shear thinning character of polymer melts. The smaller the power-law index of the melt, the flatter the gapwise velocity profile and the greater its ability to convey the orientation obtained by the fibers at the gate further downstream into the cavity. Convection of a transversely oriented core into the cavity without reorientation is further facilitated by the presence of the fibers themselves; it has been shown that the presence of filler particles enhances the plug-flow nature of the velocity profile (Ranganathan and Advani 1993; Tang and Altan 1995; Chung and Kwon 1995) which can be 'flat' even under isothermal conditions.

The effect of a flat velocity profile in conveying and maintaining the transverse orientation imparted to the fibers at the gate has been explicitly demonstrated in the experimental work of Darlington and Smith (1987). GFR polypropylene (GFRp) and GFR nylon (GFRn) were injected at constant rate in a center-gated disk cavity. Rheological measurements for these polymers indicate that the GFRn grade is less shear thinning than the GFRp grade. The velocity profile across the gap is therefore expected to be flatter in the case of GFRp and, thus, the transverse orientation acquired by the fibers in the core upon entry into the cavity is expected to be less disturbed by shear in the case of GFRp than in the case of GFRn. This has indeed been found to be the case (Darlington and Smith 1987). This effect of matrix rheology on the form of the skin–core structure has also been predicted by the computations of Bay and Tucker (1992b). Using material data for polypropylene (PP), polycarbonate (PC) and nylon and non-isothermal filling analysis followed by determination of the components of the second order orientation tensor (decoupling flow kinematics from fiber orientation but considering the effect of fiber–fiber interactions through a non-zero value of the fiber interaction coefficient ($C_I = 0.01$)) they predicted that the material with the 'flatter' velocity profile (PP) resulted in a molding with a thicker core, while the material with the more pointed velocity profile (PC) gave the thinnest core. These observations can be summarized as follows:

> Lower power-law index ⟷ Thicker core (transverse to the MDF) (more shear-thinning)

## 4.3.3  Other influences

### 4.3.3.1  Edge-effects

Kenig (1986) proposed that the filling flow in point-gated rectangular cavities assumes a converging character when an initially semi-circular flow front meets the side walls of the cavity and is progressively transformed into a flat free surface. He suggested that this will result in alignment of the fibers along the edges of the mold. Akay and Barkley (1991) have presented extensive experimental results concerning the distribution of fiber orientations in edge-gated strips with a point gate, using short- and long-fiber-reinforced polypropylene and polyamide. Besides determining the extent of the skin–core structure across the thickness of molded specimens using scanning electron microscopy, they also determined the orientation of fibers along the edges of the molding. In agreement with the arguments put forth by Kenig (1986) they observed that the core does not occupy the entire width of the molding but evolves from virtual non-existence at its edges to its full thickness in the middle of the cavity width. For a 73 mm wide molding of long-fiber-reinforced polyamide they found that the core only started featuring at 10 mm from the edges; between the edges and the core the fibers were clearly oriented along the main direction of flow (Fig. 4.11), contrary to the orientation of the fibers in the core which was transverse to the MDF (as expected in the point-gated cavities employed).

The orientation of the fibers along the edges of the molding was found to be mainly along the MDF for both short- and long-fiber composites. Figure 4.12 shows the evolution of the skin–core structure from the middle to the edges of the cavity (Fig. 4.12b and 12c) as well as the orientation of the fibers on the edge of the molding (Fig. 4.12a). Similar observations, that is, the formation of aligned-with-the-flow edge layers along the side walls of an edge-gated strip which surround a transversely oriented core, have been reported by Gillespie et al. (1985) and by Matsuoka et al. (1990) who also predicted the formation of these edge layers using the Folgar–Tucker model to describe the evolution of fiber orientation.

In injection molding experiments with a GFR polyester resin (30%w/w) in a fan-gated rectangular cavity, Gupta and Wang (1993) also observed that near the edges of the mold the fibers were aligned in the flow direction across the entire thickness of the part. This was found to be the case for a range of cavity thicknesses (0.79 mm and 1.59 mm) and filling times (0.33 s, 0.43 s and 1.76 s). Gupta and Wang (1993) also attribute this orientation to the shearing

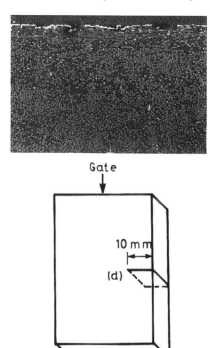

*4.11* Scanning electron micrograph of long-fiber-reinforced polyamide, showing fiber alignment along the MDF near the edges of the molding (Akay and Barkley 1991, with permission).

nature of the flow in the *x–y* plane. These observations are backed by simulations based on the (decoupled) solution of the Hele–Shaw mold-filling equations and of an evolution equation for the second order orientation tensor, including fiber–fiber interactions ($C_I = 0.001$). The rheological and thermal properties used in the simulation were those of the actual materials and this adds to the realism and relevance of the model predictions. Besides predicting the formation of edge layers aligned with the MDF, the simulations of Gupta and Wang also predicted that the extent of these edge layers increases with decreasing injection speed (or, equivalently, that the width of the transversely oriented core decreases with decreasing injection speed).

### 4.3.3.2  *Fiber depletion and fiber segregation*

Formation of fiber-depleted layers close to the surface of injection molded top-gated disks and bowls made of GFR polypropylene and nylon has been observed by Darlington and Smith (1987). The depletion patterns they observed appear similar to those reported by Bright *et al.* (1978) in fiber-reinforced

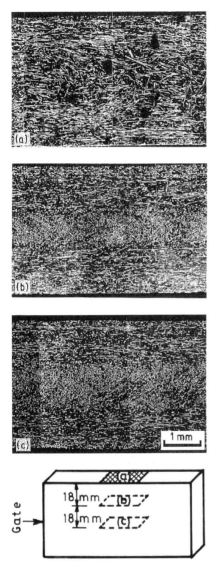

**4.12** Scanning electron micrograph of long-fiber-reinforced polyamide, showing the transition in orientation from the edges to the middle of the molding (Akay and Barkley 1991, with permission).

polypropylene (see Fig. 4.9) and are also similar to the patterns observed by Schmidt (1977) (in glass-bead-filled polypropylene injection molded into a top-gated plaque), in which a polymer-rich line appears to trail from the top side of the gate into the cavity until it blends with the rest of the (macroscopically homogeneous) mixture. Burn-off data from injection molded specimens reported by Akay and Barkley (1991) indicate that there is a gradual depletion of fibers from core to skin, which is more pronounced in long- (from 58% w/w in the core to 40% w/w near the skin) than in short-fiber-reinforced polyamide composites (from 53% w/w in the core to 47% w/w near the skin). Determining variations in fiber volume fraction from local density measurements, Spahr *et al.* (1990) have also found that the concentration of fibers at the skin was lower than in the core, for both long- and short- GFR polypropylene (the difference being significantly greater for the long-fiber composite). Similar results have been reported by Bailey and Rzepka (1991) for GFR Verton resins, Hegler (1984) and Singh and Kamal (1989). Besides fiber depletion from the skin to the core, Hegler and Mennig (1985) observed a change in local glass volume fraction with distance from the gate in injection molded glass-bead and glass-fiber reinforced polymers. Similar segregation has been observed in glass-bead-reinforced polystyrene (Ogadhoh and Papathanasiou, 1996) and has been attributed to lateral migration across the cavity thickness in conjunction with the fountain flow at the melt front. It was found that the extent of particle segregation was more pronounced for larger particles. However well documented experimentally, fiber segregation during injection molding has not been analysed theoretically and its dependence on processing parameters is not well established.

### 4.3.3.3  Fiber orientation in the sprue

The fiber orientation in the sprue is important because it affects the state at which the fibers enter the cavity. Since a sprue is a diverging channel, flow in it has both shear and elongational components. Analysis of the motion of a single fiber in a sprue with angle $\alpha$ (Vincent and Agassant, 1986) shows that, depending on the location of the fiber, its motion will be dominated by either shear or elongation. If $\beta$ is the angle formed between the axis of the sprue and the line connecting the center of the fiber to the origin of the sprue, the motion of fibers located in a ring $\beta > \beta_0$ is characterized by tumbling, similar to that usually associated with shear flow. The size of this ring depends on sprue angle $\alpha$ and fiber aspect ratio $r_e$:

$$\cos 2\beta_0 = r_e^2 \cos 2\alpha + \sqrt{(1 - r_e^2)(1 - r_e^2 \cos^2 2\alpha)} \qquad [14]$$

For $\beta > \beta_0$ the fibers remain aligned with the main direction of flow for the largest part of time, periodically rotating through 180°. The motion of centrally located ($\beta < \beta_0$) fibers is governed by elongation; these fibers will tend to align transversely to the main direction of flow. These results are independent of the flow rate through the sprue and depend only of fiber aspect ratio and sprue angle. Even though derived for the simple case of isolated fibers in a Newtonian fluid, the above conclusions have been found to be in qualitative agreement with experimental observations. Figure 4.13 shows a contact microradiograph of a slice cut perpendicular to the main flow direction near the narrow end of an

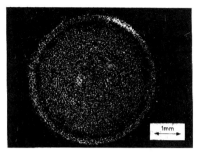

**4.13** Contact microradiograph of a slice taken at the narrow end of a sprue and perpendicular to the MDF (Bright *et al.*, 1978, with permission).

injection-molding sprue (short GFR polypropylene). It is evident that the fibers near the walls are aligned with the flow, while those at the center have assumed a transverse orientation, in qualitative agreement with the previous analysis. Similar results have been reported by McClelland and Gibson (1990) for long-fiber-reinforced nylon. In this case, significant fiber bending has been shown to occur in the interior of the sprue.

### 4.3.3.4  Effect of packing

Because of the temperature dependent density of polymer melts, injection molded parts tend to shrink upon cooling. In unreinforced materials this can lead to sink marks – surface imperfections manifested by depressions at the skin of the molding – especially in regions of larger part thickness. In fiber-reinforced polymers (in particular those incorporating long fibers) the formation of stiff skin layers (due to the presence of fibers) will not usually allow sink marks to form; instead, microvoids can form in the core of the part (Akay and Barkley, 1991,1993). To compensate for this shrinkage, more material is packed into the cavity during the packing and holding stages. The effect of the extra flow associated with packing/holding on the orientation of fibers has not been investigated in detail. Malzahn and Schultz (1986) have presented fiber orientation results in short shots and full moldings. They observe that while the fibers in the core are aligned along the MDF in the short shots, the fiber orientation in the core of full moldings is transverse to the MDF; this change was attributed to the effect melt flow during packing may have on fiber orientation. Sanou *et al.* (1985) have also reported fiber orientations in short and full shots; no significant difference was observed between the orientation in the full part and the short shots. However, this could be due to the use of a sprue gate, which will induce a more or less unidirectional melt flow during filling. The work of Bailey and Rzepka (1991) and Darlington and Smith (1987) has shown that the thickness of the transversely oriented core is reduced upon packing and holding. Figure 4.14 is a schematic of a mechanism that has been suggested by Bailey

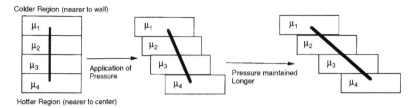

**4.14** Schematic of the mechanism suggested by Bailey and Rzepka (1991) by which application of packing pressure can lead to fiber alignment. Due to non-isothermality, $\mu_1 > \mu_2 > \mu_3 > \mu_4$ where $(\mu_i)$ indicate local viscosities.

and Rzepka (1991) to explain this reduction (observed in injection moldings made from a number of long-fiber-reinforced Verton resins). It can be seen that due to the temperature (and thus, viscosity) profile established across the cavity thickness at the end of filling, the effect of packing/holding pressure will be to align some of the fibers not already aligned with the MDF. Evidently, the mechanism shown in Fig. 4.14 will be more pronounced at the boundary between the skin and core regions where the viscosity gradient will be highest and its results will be more visible in systems incorporating long fibers. This could explain the observations of Bailey and Rzepka (1991) (in GFR polyamides) that moldings obtained without applying a packing pressure exhibited a thicker core than moldings obtained under standard conditions; evidently, misaligned fibers located at the boundary between the shell and core regions are being aligned as a result of packing-induced flows.

The effect of packing on fiber orientation has not been analysed theoretically. Even though certain commercial packages incorporate a unified treatment of filling and post-filling flows (Chiang *et al.* 1994), reported fiber orientations have usually been obtained during or at the end of the filling stage.

### 4.3.3.5  Weldlines

Weldlines have been termed the 'inescapable by-products' of the injection molding process. More specifically, weldlines are the result of the geometrical complexity of injection molded parts (and/or of the use of multiple gates) and of the complex filling patterns, frequently involving multiple and/or splitting free surfaces, associated with them. Weldlines are generated by the merging or impingement of separate melt flow fronts and are encountered in all moldings of practical significance. Distinct variations in part properties have been associated with weldlines. However, due to the flow-induced fiber orientation in the weldline and the drastic effect this orientation has on properties, these variations are much more pronounced in fiber- and flake-reinforced composites than in neat resins (Sanschagrin *et al.* 1990). An excellent review on the subject of weldline formation in injection molded parts, encompassing neat polymers, fiber-reinforced composites and polymer blends has been recently compiled by Fellahi *et al.* (1995). We will review here the orientation of fibers observed in and around weldlines in fiber-reinforced molded parts. Ways to improve the strength of weldlines have been reviewed by Fellahi *et al.* (1995) and Tomari *et al.* (1995). As with every aspect of fiber orientation in injection molding, the orientations observed in weldlines have their origins in the melt flow associated with their formation. Since weldlines are associated with colliding or merging flow fronts, the orientations assumed by the fibers around them will represent the orientational state of the fibers in the vicinity of the free surface at the moment the two free surfaces meet.

Two types of weldlines are usually encountered in injection molded parts: weldlines formed when two flow fronts impinge head-on (Type I) and those formed when two flow fronts merge as a result of flow around an insert or of flow from adjacent gates (Type II). In either case, what is usually termed a weld-*line* is actually a weld-*zone* of finite dimensions in which the orientation of fibers is fundamentally different from the orientation in the rest of the part. In Type-I weldlines, the orientation of fibers has been found to be along the main direction of the weld across the entire width and thickness of the weldline. This represents a discontinuity in the typical skin–core structure of injection molded parts and has been blamed for differential shrinkage of molded parts in this region (Akay and Barkley 1993). Figure 4.15 is a typical SEM (scanning electron micrograph) of fiber orientation along the plane of a Type-I weldline. A schematic of fiber orientations around a Type-I weldline and on a plane perpendicular to the weld is shown in Fig. 4.16; it is evident that the skin–core structure has been disrupted and that fibers in the weld-*zone* are aligned with the weld and are perpendicular to the main direction of flow.

Type-II weldlines are characterized by fiber orientation parallel to the weld, which now coincides with the main direction of flow. A significant difference between the formation of weldlines of Type-II in neat (or bead reinforced) and in fiber-reinforced polymers is the fact that weldlines formed in the latter persist visually (in the form of a V-shaped free surface with the point of the V located at the weldline) for a longer length behind the obstacle than do weldlines in the former case. The reason for this has to be found in the anisotropy of fiber-loaded melts, as a result of which the material in the weld zone flows more slowly than the material in the rest of the cavity, thus resisting the homogenization of the free surface (Fisa and Rahmani, 1991). As a result of the local shape of the free surface at the weldline, the latter behaves effectively as a wall: fibers are deposited on both sides by the expanding free surfaces, thus generating a unidirectional microstructure in its vicinity.

A schematic of fiber orientations typical of Type-II weldlines is shown in Fig. 4.17. As in Type-I weldlines, the parallel-to-the-weld orientation extends across

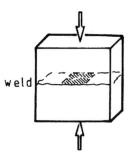

*4.15* Scanning electron micrograph of a section taken along a Type-I weldline (Akay and Barkley, 1993, with permission).

the entire thickness of the part and represents a disruption in the transversely oriented core usually found in edge-gated moldings. The difference between the orientation at the weld-zone and the orientation in the rest of the sample will persist downstream from the insert. Uniformity in fiber orientation along the cavity width will only be established when (and if) the rest of the transversely oriented core is aligned along the MDF by the action of the shear flow across the cavity thickness. Keeping in mind the discussion in the preceding section on the factors causing the formation of a skin–core structure, this homogenization will be facilitated by lower injection speeds or thinner cavities. The latter influence has been demonstrated in the work of Gillespie *et al.* (1985), in which a weldline formed downstream of a molded-in hole was clearly discernible in a thick (3.2 mm) molding while it was not so evident in a thin (1.6 mm) molding (also Section 4.3.2.4).

Weldlines formed in long-fiber-reinforced composites are weaker than those corresponding to short-fiber composites and their strength decreases with fiber

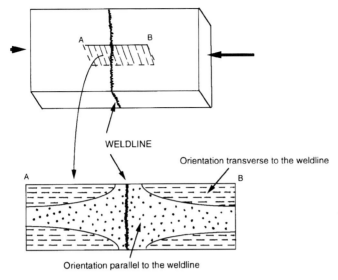

**4.16** Schematic diagram of fiber orientations typically observed on a plane perpendicular to a Type-I weldline (after Akay and Barkley, 1993).

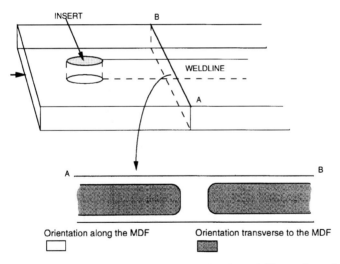

**4.17** Schematic diagram of the distribution of fiber orientations in a Type-II weldline (after Fisa and Rahmani, 1991)

concentration. This has been attributed to incomplete homogenization and to the presence of fiber bundles (Spahr et al., 1990; Truckenmuller and Fritz, 1991), the formation of microvoids as discussed in the following Section (4.3.3.6) and to fiber bending (Fellahi et al. 1995). Fiber orientation around weldlines of Types I and II can and has been predicted using (decoupled) models for fiber

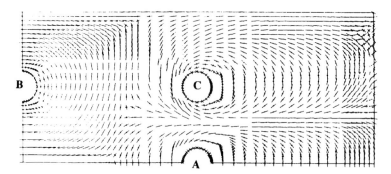

4.18 Predicted fiber orientation distributions in a triple-gated mold (shown are the eigenvectors of the orientation tensor), showing orientations typical of weldline regions (Frahan *et al.*, 1992, with permission). A, B and C indicate gate locations.

orientation during filling. Figure 4.18 is taken from the work of Frahan *et al.* (1992) who used dilute suspension orientation models and an isothermal Hele–Shaw filling analysis and shows the predicted fiber orientations in a triple-gated cavity in which both Type-I and Type-II weldlines form.

### 4.3.3.6 Long fibers

Traditional fiber-reinforced injection molding compounds have been available since at least the 1960s. These are made by extrusion compounding, in which the polymer and chopped strands of fibers are mixed in an extruder, passed through a suitable die and the extruded product pelletized. As a result of fiber attrition during extrusion compounding the length of fibers is reduced, typically down to between 200 $\mu$m and 400 $\mu$m, with a consequent reduction in the mechanical properties of the parts generated by injection molding the thus produced pellets. Pultrusion compounding has recently enabled the production of injection molding compound which contains fibers of large size, typically of the order of the size of the pellet, exceeding significantly the aspect ratios found in conventional short-fiber-reinforced thermoplastics. Injection molded parts produced using such pultrusion-compounded pellets are characterized by improved stiffness, strength and, in some cases, toughness and fatigue crack propagation resistance (Bailey and Rzepka 1991). Still, fiber degradation during processing is a problem (Khanh *et al.*, 1991) and alternative processes, such as the direct addition of roving strands have been investigated (Truckenmuller and Fritz 1991).

Considering the extreme length of fibers used, long-fiber-reinforced (LFR) materials are not very difficult to process by injection molding, their shear viscosity being not much higher than that of short-fiber-reinforced melts. McClelland and Gibson (1990) found that fibers in LFR composites are protected from breakage during processing by remaining locally parallel to each other. For this reason, their flow appears to involve convection of locally oriented domains which sometimes appear in the finished product as clusters of fibers (McClelland and Gibson, 1990). Overall, the skin–core structure typical of traditional short-fiber-reinforced moldings is also observed in LFR injection molded parts. This is anticipated since, qualitatively, the mechanisms responsible for fiber orientation in short-fiber moldings operate in the presence of long fibers as well. However, the increased fiber length results in decreased fiber mobility, especially at the melt front, increased fiber–wall interactions, fiber bending and/or break-up and poor fiber dispersion manifested in the presence of fiber bundles in the molding (Truckenmuller and Fritz, 1991). It has been found that the thickness of the core increases when fibers of higher aspect ratio are used (Spahr et al., 1990; Bailey and Rzepka, 1991). Also the extent of fiber depletion from the core to the skin increases with fiber aspect ratio (Spahr et al., 1990; Blanc et al., 1987), even though this most probably reflects the segregated state at which long fibers enter the cavity. The orientation in both the shell and the core layers was found to be higher for the short-fiber composites (Spahr et al., 1990; McLelland and Gibson, 1990); this could be explained as a result of steric interactions which inhibit the motion and reorientation of longer fibers. Microphotographs or SEM pictures of sections cut through molded specimens containing long fibers indicate that significant bending of the fibers can occur (Spahr et al., 1990; Akay and Barkley, 1991; Truckenmuller and Fritz, 1991). Akay and Barkley (1991) also found that microvoiding, particularly in the (denser) core region, was a problem in long-fiber-reinforced polyamide but not in the corresponding short-fiber material. Akay and Barkley (1992) and Truckenmuller and Fritz (1991) reported that significant jetting and irregular filling patterns occurred during molding of long-fiber-reinforced polyamides (Fig. 4.7 and Section 4.3.1).

### 4.3.3.7   Effect of fiber concentration

Spahr et al. (1990) have presented a comprehensive investigation of the properties of short ($r_e = 70$) and long ($r_e = 320$) GFR polypropylene moldings. Even though the emphasis was on fracture behavior, a detailed comparison between samples containing various amounts of (short and long) fibers, namely at concentrations of 10%, 20%, 30% and 40% w/w, was made. It was found that the thickness of the core increased with fiber concentration (from 13% of the total thickness of the molding in the sample containing 10% w/w of long fibers to 21% of the total thickness of the molding in the sample containing 40% w/w of long fibers). The corresponding numbers for the short-fiber composites were 3% and 11% of the total thickness, respectively. Fiber concentration was also found to affect the extent of fiber depletion from the core to the skin; increasing the fiber concentration resulted in increased variation between the local fiber concentrations in the core and the skin of the molding. These effects are not restricted to fiber-reinforced melts. In injection molding experiments with glass-flake-filled polypropylene, Bouti *et al.*(1989) found that the size of the transversely oriented core increases with flake concentration, the latter ranging between 2% and 26% by weight, and that this increase was more pronounced for thicker cavities. The effect of fiber concentration on the observed orientational state of the fibers in injection molding can only be predicted in the context of coupled filling models (Section 4.4.2 below). Using such an analysis of mold filling, Chung and Kwon (1995) predicted that the clamping force will increase quadratically with fiber volume fraction and also, that the gapwise velocity profile will be more blunted as the fiber concentration increases. Flatter velocity profiles have been previously associated with a thicker core (Sections 4.3.2.2 and 4.3.2.6). The effect of fiber concentration on the morphology of injection molded parts cannot be predicted by existing commercial packages for mold filling.

## 4.4    Prediction of fiber orientation in injection molding

The prediction of fiber orientation is a critical step in extending existing capabilities of CAD packages for injection molding (usually restricted to prediction of filling patterns, temperature and pressure profiles) to the dimensional stability and the anticipated mechanical properties of fiber-reinforced parts. This extension is seen as an essential requirement in applying concurrent engineering practices in the injection molding of fiber-reinforced composites. Provided that the distribution of fiber orientation can be predicted from a filling analysis, the mechanical and thermophysical properties of a fiber-reinforced composite can be estimated, with acceptable accuracy, from the fiber orientation tensor (Gupta and Wang, 1993; Chiang *et al.*, 1994; see also Chapters 3 and 8). The rheology of fiber dispersions as well as the interaction between flow and fiber orientation has been the subject of extensive study (e.g. Powell, 1991) and is covered in Chapters 2 and 3 of this book. The present section will briefly review the state of the art in predicting the evolution of fiber orientation in injection mold filling.

### 4.4.1    Modeling strategies: prediction of fiber orientation

*4.4.1.1    Jeffery's model for non-interacting fibers*

The first attempts to predict the orientation of fibers in flows related to injection molding were based on solution of Jeffery's equation (Chapter 2) for a number of test fibers along streamlines determined from a separate (decoupled) determination of the flow kinematics. Statistical averages for the orientation in each point of the domain were obtained by suitably averaging the orientations of individual test fibers. This approach is documented in the works of Givler (1983); Givler *et al.* (1983); Gillespie *et al.* (1985); Vincent and Agassant (1986); Subbiah *et al.* (1989) and Frahan *et al.* (1992). Even though conceptually simple, the above approach has the obvious disadvantage of requiring significant computational resources (Bay and Tucker, 1992b). Most importantly, the effect of interfiber interactions on orientation (let alone the effect of the fibers on the rheology of the melt) cannot be included in this context. However, the first predictions for the orientation of fibers in flows related to injection molding, which elucidated the effect of the expanding flow at the gate and the effect of the fountain flow, were based on this approach and have been found to be in qualitative agreement with experimental evidence (Givler, 1983; Subbiah *et al.*, 1989). Figure 4.19 shows predicted fiber orientations in an expanding planar flow for two initial orientations (random and normal to the MDF). Figure 4.20 shows similar orientation predictions in the fountain flow region. Since fiber interactions and their influence on rheology are ignored, this approach cannot be relied upon to fine-tune the distribution of fiber orientation in a molded part by manipulation of the processing conditions.

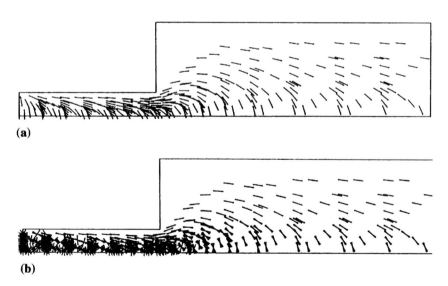

*4.19*  Predicted fiber orientation in a planar expansion for initial fiber orientation perpendicular to the MDF (a) and random (b). In each case, the fibers in the expanding section of the flow assume orientations predominantly transverse to the MDF. High fiber aspect ratios and flow from left to right (from Givler, 1983).

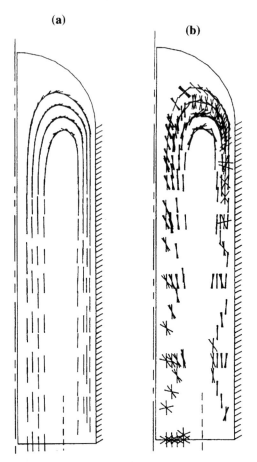

*4.20* Predicted fiber orientation in the fountain flow region for initial fiber orientation along the MDF (a) and random (b). In each case, the fibers near the wall assume orientations predominantly parallel to the wall. High fiber aspect ratios and flow upwards (from Givler, 1983).

#### 4.4.1.2 Rotary diffusion models

The effect of interfiber interactions on fiber orientation in flow has been included in models which incorporate the randomizing influence of these interactions through the addition of a diffusive term in the evolution equation for the orientation distribution function (ODF, $\psi$) (Folgar and Tucker, 1984; Kamal and Mutel, 1989). According to the Folgar–Tucker model, the evolution equation for the orientation distribution function $\psi$ is:

$$\frac{\mathrm{D}\psi}{\mathrm{D}t} = -\nabla(\psi \cdot \Omega) + C_I \gamma \nabla^2 \psi \qquad [15]$$

where the angular velocity $\Omega$ is still described by Jeffery's equation and where $C_1$ is the interaction coefficient, expressing the strength of interfiber interactions. Such models have extended the predictions for fiber orientation into the semi-concentrated regime (Tucker and Advani, 1994) the main uncertainty being the value chosen for the interaction coefficient (Chapter 2). Matsuoka et al. (1990,1991) have presented a model for fiber orientation in injection mold filling using the Folgar–Tucker equation. In their filling analysis an 'equivalent Newtonian behavior' of the fiber-reinforced melt was assumed; however, this is not necessary and does not preclude the use of the more general non-linear Eqn. [6] for the flow in the cavity. The evolution of the ODF was determined in a gap-averaged sense, since the objective was comparison with experimental measurements of the thermal expansion coefficient and thus no results concerning the morphology across the thickness of the plaque were given in that study. Predictions for the gap-averaged fiber orientation were in qualitative agreement with X-ray images of molded plaques seeded by tracer metal fibers. Predictions for the warpage of ribbed moldings were also found to be in reasonable agreement with measurements. This application notwithstanding, use of Eqn [15] has been limited, mainly due to the introduction of orientation tensors as more efficient descriptors of fiber orientation in multifiber systems.

### 4.4.1.3  Use of orientation tensors

The use of evolution equations for orientation tensors appears to be the method of choice in more recent works. The main reason for the use of orientation tensors (instead of the complete ODF) in describing fiber orientation in injection molding is computational economy: Bay and Tucker (1992a) have argued that for an injection molding simulation in a model with 5000 nodes, description of the fiber orientation through the ODF (considering 20 increments in the $\theta$ and 40 increments in the $\phi$ directions, $\phi$ and $\theta$ being the fiber orientation angles) would result in a problem with $4.0 \times 10^6$ degrees of freedom. Adoption of the second order orientation tensor $\boldsymbol{\alpha}^{(2)}$ requires only five extra unknowns (the five independent components of $\boldsymbol{\alpha}^{(2)}$) to be determined at each node. This reduces the extra number of unknowns to 25 000, less than 1% the size of the problem corresponding to full use of the ODF.

The evolution equation for the components of the second order orientation tensor $\boldsymbol{\alpha}^{(2)}$ can be obtained from the ODF through the definition:

$$\alpha_{ij}^{(2)} = \langle r_i r_j \rangle = \int \psi(\mathbf{r}) r_i r_j \mathrm{d}\mathbf{r} = \int\limits_{\phi=0}^{2\pi} \int\limits_{\theta=0}^{\pi} r_i r_j \psi(\phi, \theta) \sin(\theta) \mathrm{d}\theta \mathrm{d}\phi \qquad [16]$$

and the interaction between fibers can be included in the same sense as in the Folgar–Tucker model (Eqn [15]) through the interaction coefficient $C_I$. The resultant evolution equations for the components of $\boldsymbol{\alpha}^{(2)}$ are (Tucker and Advani, 1994; Hinch and Leal, 1976):

$$\frac{D\alpha_{ij}^{(2)}}{Dt} + \frac{1}{2}(W_{ij}\alpha_{kj} - W_{kj}\alpha_{ik}) = \frac{\beta}{2}(D_{ik}\alpha_{kj} + D_{kj}\alpha_{ik} - 2D_{kl}\alpha_{ijkl})$$

$$+ 2C_{I}\gamma(\delta_{ij} - 3\alpha_{ij}) \qquad [17]$$

where $W$ is the vorticity tensor, $D$ is the shear rate tensor and $\beta = \dfrac{r_e^2 - 1}{r_e^2 + 1}$. In Eqn. [17] the left hand side represents the corotating or Jauman derivative of $\boldsymbol{\alpha}^{(2)}$, which accounts for the fact that the fiber orientation is convected with the bulk fluid motion and $\delta_{ij}$ is the Kronecker delta function. It is evident from Eqn [17] that the fourth order orientation tensor $(\alpha_{ijkl})$ is required in order to carry out the integration and determine the evolution of the five independent components of $\boldsymbol{\alpha}^{(2)}$. This is common in all tensorial representations of the orientation state and necessitates the use of a closure, that is, of an approximation of a higher order tensor (fourth order in this case) in terms of tensors of lower order (second, in this case). Various such closure approximations have been developed and tested (Tucker and Advani, 1994; Advani and Tucker, 1987,1990). Bay and Tucker (1992a) suggest that a hybrid closure yields best steady-state orientation results. This closure is:

$$\alpha_{ijkl} = (1 - f)\left[ -\frac{1}{35}(\delta_{ij}\delta_{kl} + \delta_{ik}\delta_{jl} + \delta_{il}\delta_{jk}) + \cdots + \frac{1}{7}(\alpha_{ij}\delta_{kl} + \alpha_{ik}\delta_{jl} \right.$$

$$\left. + \alpha_{il}\delta_{jk} + \delta_{ij}\alpha_{kl} + \delta_{ik}\alpha_{jl} + \delta_{il}\alpha_{jk}) \right] + f\alpha_{ij}\alpha_{kl} \qquad [18]$$

where the term multiplying $(1 - f)$ is the linear closure approximation, while the term multiplying $(f)$ is the quadratic approximation. The scalar $(f)$ is a measure of orientation which equals zero for random orientation and unity for full alignment:

$$f = 1 - 27 \det\{\alpha_{ij}\} \qquad [19]$$

Through the use of a closure it is possible to integrate Eqn [17] and determine the evolution of orientation in a general three-dimensional flow with $(C_I > 0)$ or without $(C_I = 0)$ interfiber interactions. It should be remembered that Eqn [17] is an exact transformation of Jeffery's model, with the added advantage of allowing for interfiber interactions when $C_I > 0$. However, the use of a closure is an approximation which may affect the quantitative accuracy of the predictions of Eqn [17]. The quantitative accuracy of Eqn [17] in the presence of interfiber interactions will be further affected by the value chosen for the interaction coefficient $C_I$. More details on this can be found in Chapters 2 and 3 of this book and references therein and in Section 4.4.3.2 of this chapter with reference to injection molding.

Gupta and Wang (1993) used Eqn [17], coupled with a standard non-isothermal Hele–Shaw filling analysis, for the prediction of fiber orientation in filling of a fan-gated plaque. The three-dimensional orientation of fibers was determined through the evolution equations for the in-plane components of $\boldsymbol{\alpha}^{(2)}$ using the hybrid closure of Advani and Tucker (1987,1990) and $C_I = 0.001$. The out-of-plane components of $\boldsymbol{\alpha}^{(2)}$ were given empirically chosen constant values. Gupta and Wang (1993) presented fiber orientation predictions at various planes across the thickness of the plaque and carried out parametric studies which elucidated the effect of processing conditions (such as wall temperature and injection speed), cavity thickness, material properties and the effect of wall solidification on the strength and direction of fiber orientation in the core and skin layers. These predictions were found to be in qualitative agreement with experimental results, as outlined in the preceding Section 4.3. Frahan *et al.* (1992) modeled isothermal Hele–Shaw filling flows in various geometries, including double-gated cavities with inserts and thin-walled three-dimensional parts, and determined the orientation of fibers along the mid-plane using both Jeffery's theory and the second order orientation tensor (with a quadratic closure) in the absence of fiber–fiber interactions ($C_I = 0$). Their results for fiber orientation around weldlines are in qualitative agreement with experimental results in this field (Section 4.3.3.5). Fiber orientation predictions, based on the use of the second order orientation tensor with a hybrid closure approximation have also been reported by Bay and Tucker (1992a) (for essentially one-dimensional, non-isothermal filling of a center-gated disk and of a film-gated rectangular strip) and by Chung and Kwon (1995) who also coupled the kinematics of the filling flow to the orientation of the fibers (Section 4.4.2).

The second order orientation tensor approach appears to be the method of choice in commercial CAD packages for the injection molding process. Figures 4.21.a and b show a wire-mesh geometry (a) of a car radio faceplate along with predictions of fiber orientation (b) at the end of the filling stage obtained through the commercial code C-MOLD. It is evident that the predicted fiber orientations at various sections of the part comply with the filling patterns associated with a point-gated cavity, namely an expanding flow near the gate, converging flows behind the two circular and the rectangular inserts and roughly one-dimensional flow along the narrow ribs. These results, along with examples presented in Chapter 3 define the state of the art in the capabilities of contemporary commercial codes in predicting the distribution of fiber orientation in injection molded fiber-reinforced composites. In all such codes as well as in the studies outlined in previous sections, the effect of fibers on the kinematics of the flow has been neglected.

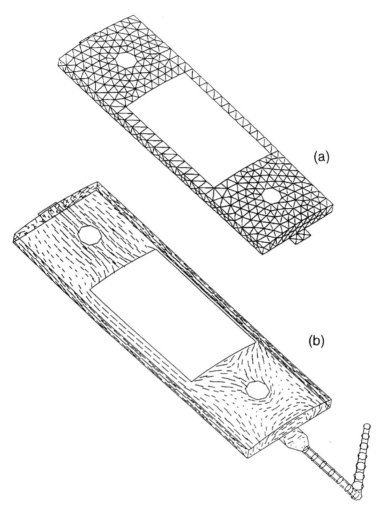

*4.21* Wire-mesh drawing along with finite element mesh of a car radio face plate (a) and predicted fiber orientation distributions (b). In part (b), the orientation of the lines indicates the first principal direction of the orientation tensor and the length of each line represents the degree of orientation (with permission, AC Technology Europe b.v)

## 4.4.2    Modeling strategies: effect of fibers on flow kinematics

Unlike dispersions of particulates, fiber dispersions are anisotropic. This means that the flow kinematics of a fiber dispersion are governed not only by the fiber volume fraction but also by the orientation and aspect ratio of the fibers. Irrespective of the method used for predicting fiber orientation, two main approaches, usually termed the uncoupled and the coupled approach, have evolved in modeling flows of fiber dispersions. In the former, it is assumed that the orientation of the fibers does not alter the rheological behavior of the melt. Thus, the fiber orientation problem is decoupled from the flow kinematics and the calculations can be carried out in a sequential manner, by first computing the flow kinematics through, for example, Eqns [3]–[6] and then finding the corresponding fiber orientation using any of the models outlined in Section 4.4.1. Interfiber interactions can be included in this approach, as outlined in the previous section. The validity of decoupling the kinematics of the flow from the orientation of the fibers in flows related to injection molding has been investigated by Tucker (1991) who concluded that the flow and fiber orientation problems can be considered as decoupled if the cavity thickness is small enough (compared to the fiber size) so that the fiber orientation is effectively in-plane, having no effect on the gapwise shear stress and thus the gapwise velocity profile. This implies that the decoupled analysis will be more appropriate for long- than for short-fiber-reinforced composites and for thinner rather than thicker cavities.

In the coupled approach, fiber orientation and flow kinematics are coupled through a suitable constitutive relationship which expresses the influence of the fibers (their concentration, aspect ratio and orientation) on the stress tensor $T$. This influence is usually expressed as:

$$T = T^{(v)} + T^{(p)} \qquad [20]$$

where the superscripts (v) and (p) indicate the contributions of the viscous suspending fluid and that of the fibers (respectively) on the stress tensor. In the coupled approach a constitutive relationship must be assumed in order to calculate the fiber-related part ($T^{(p)}$) of the stress tensor as a function of the fiber aspect ratio, concentration and orientation. This is an area of considerable current interest; more information on this subject can be found in Chapter 3 of this book as well as in Tang and Altan (1995) and Ranganathan and Advani (1993) and references therein.

The overwhelming majority of published simulation results dealing with the prediction of fiber orientation in injection molding, including the state-of-the-art in commercial CAD packages, have adopted the decoupled approach. Coupled flow-orientation calculations in the center-gated disk geometry (ignoring factors such as the presence of an advancing free surface, the non-isothermality and the pseudoplasticity of the fluid) have been presented by Ranganathan and Advani (1993), while the problem of isothermal developing flow of a fiber dispersion between two parallel plates has been treated by Tang and Altan (1995). As of the time of this writing, the only work in which the coupled approach has been applied in actual mold filling simulations is that of Chung and Kwon (1995). Assuming that the gapwise shearing stress dominates (this may not be a good assumption in LFR melts which exhibit very high extensional viscosities), the filling flow problem can still be treated as a Hele–Shaw type of flow. When Eqn [20] is considered, the gapwise shear stress can be written as:

$$T_{zj} = \mu \cdot u_{zj} + (\mu N) \cdot [u_{1,z}\alpha_{3j13} + u_{2,z}\alpha_{3j23}] \qquad j = 1, 2 \qquad [21]$$

where the standard notation is used ($u_{ij} = \partial u_i/\partial x_j$) and the direction $z$ (corresponding to index 3 in the components $\alpha_{ijkl}$ of the fourth order orientation tensor (Section 4.4.1.3) above) is taken in the gapwise direction. The coefficient ($N$) typically depends on the concentration and aspect ratio of the fibers. According to the Dinh–Armstrong model (1984), it is:

$$N = \frac{\pi m L^3}{6 \ln(h/R)} \qquad [22]$$

where $m$ is the number density of fibers, $h$ the average interfiber spacing (which in turn depends on fiber orientation) and $R$ and $L$ indicate the radius and length of the fibers, respectively.

Following Eqn [21], the momentum equation in the $x$ direction becomes:

$$\frac{\partial P}{\partial x} = \frac{\partial}{\partial z}\left[\mu\frac{\partial u}{\partial z} + \mu N\left\{\alpha_{3113}\frac{\partial u}{\partial z} + \alpha_{3123}\frac{\partial v}{\partial z}\right\}\right] \qquad [23]$$

with a similar equation holding for the pressure gradient in the $y$ direction. Based on Eqn [23] and through suitable redefinition of the fluidity integral (Eqn [4]), a non-linear Poisson equation for the pressure can be formulated (Chung and Kwon, 1995). By thus considering the effect of fiber concentration and orientation on rheology, Chung and Kwon (1995) were able to obtain results unavailable through the traditional decoupled approach, such as predictions of the effect of fiber concentration on the required clamping force and on the shape of the gapwise velocity profiles. The versatility of their approach was also demonstrated by presenting fiber orientation predictions for various test cases, such as a molded tensile specimen and a multifaceted podium. Coupling of the flow kinematics to fiber orientation is the only approach that can possibly allow for an (at least semi-quantitative) determination of the effect of processing conditions and fiber characteristics on the morphology (and thus, the properties) of fiber-reinforced injection molded composites (as outlined in Section 4.3 of this chapter). Notably, this includes the effect of processing conditions and dispersion characteristics on the relative thickness of the core and shell zones (Section 4.3.2) as well as of the zones associated with weldlines (Section 4.3.3.5) or with cavity edges (Section 4.3.3.1). These can only be qualitatively predicted through the decoupled approach. Furthermore, models for the prediction of fiber segregation during processing can only be incorporated in the context of a coupled flow-orientation analysis.

## 4.4.3  Parametric sensitivity studies

Even though computer-aided analysis lends itself ideally to sensitivity studies as a means of identifying the mechanisms and/or the physical phenomena affecting fiber orientation, very little has been done in this direction to date. Much of the results of computational sensitivity studies are presented and discussed in Section 4.3 since they help create a framework for the explanation of available experimental evidence. In the following we focus on sensitivity studies that are particularly important in a modeling context, namely the effect of considering the fountain flow in a filling model, the effect of the numerical value of the interaction parameter $C_I$ on fiber orientation predictions and the effect of the (assumed) orientation of fibers at the gate.

### 4.4.3.1  *Effect of fountain flow on fiber orientation predictions*

Fountain flow is not included in the Hele–Shaw model for the mold-filling process. In order to account for its effect on fiber orientation, approximate local treatments have been proposed (Crochet *et al.*, 1994). These can be seen as variants of earlier thermal models for the fountain flow region (Subbiah *et al.*, 1989, Dupret and Vanderschuren, 1988), the essence of which is that material approaching the free surface around the centerplane is convected outwards while at the same time it is stretched and/or reoriented by the fountain flow. Bay and Tucker (1992a) included, in an approximate sense, the effect of fountain flow on temperature and fiber orientation during mold filling. The detailed effect of fountain flow on fiber motion and orientation was also investigated by the same authors using the commercial software package FIDAP in conjunction with a particle tracing scheme. Numerical predictions were found in good agreement with fiber visualization experiments and the observed discrepancies were attributed to the interaction between fibers and wall and between fibers and the free surface. Since the magnitude of these interactions is expected to drop with decreasing fiber size, such schemes are more likely to be accurate for small fibers. Bay and Tucker (1992b) also found that the fountain flow is responsible for the formation of a dip in the orientation of fibers at a narrow region below the skin. The location and width of that region were predicted well, but the exact orientation in it was not, possibly as a result of the finite size of fibers and/or non-isothermal effects in the fountain flow region. This local dip in the orientation below the surface is not predicted by models which do not consider the fountain flow. Previous treatments of the fountain flow (Bay and Tucker, 1992b; Crochet *et al.*, 1994) have ignored the interaction between fibers and flow kinematics (decoupled approach). The effect of fountain flow on fiber orientation has not yet been investigated in the context of coupled models.

*4.4.3.2  Effect of the interaction parameter on fiber orientation predictions*

At this stage, this parameter cannot be derived from first principles and can, at the best, be determined by fitting experimental data (Chapter 2). In introducing the rotary diffusion model, Folgar and Tucker (1984) suggested that the value of $C_I$ should increase with the aspect ratio and/or the concentration of fibers, since in this case the strength and frequency of interfiber interactions increases. Ranganathan and Advani (1991) have demonstrated that this premise holds only in the semi-dilute concentration regime, where $\phi r_e < 1$, $\phi$ being the fiber concentration and $r_e$ the fiber aspect ratio. Beyond that point, they found that the interaction coefficient decreases exponentially with the product $\phi r_e$. This has been attributed to steric inhibition effects, which, in concentrated systems, force the fibers preferentially to adopt orientations aligned with their neighbours, thus yielding smaller apparent values of the interaction coefficient. In summary, while in semi-dilute systems the presence of interfiber interactions prevents the fibers from becoming completely aligned with the flow and introduces some randomness in the orientation distribution, in concentrated systems interfiber interactions lead to enhanced alignment.

Matsuoka *et al.* (1990) attempted to determine $C_I$ from experimentally measured values of the thermal expansion coefficient in injection molded plaques of GFR polypropylene. Since the coefficient of thermal expansion depended linearly on calculated values of a scalar measure of fiber orientation, it was argued that the numerical value of the interaction coefficient could be determined as the one giving the best linear fit between the predicted orientation parameter and the measured thermal expansion coefficient. However, their results indicate that the actual fiber orientation was not very sensitive to the chosen value of $C_I$ and was probably governed by flow kinematics. For example, the correlation of the linear fit between the thermal expansion coefficient and orientation parameter was 0.884 for $C_I = 0.01$ and 0.873 for $C_I = 0.1$ (this for a 10% w/w GFR polypropylene). In a detailed examination of the effect of the interaction parameter on predictions of fiber orientation during injection molding, Bay and Tucker (1992b) found that as $C_I$ decreased, both shell and core regions became more aligned (along and transverse to the MDF, respectively). It was also found that while the $\alpha_{11}$ component of the orientation tensor was predicted quite well, the components $\alpha_{33}$ and $\alpha_{13}$ were not, the quality of the fit depending on the value chosen for $C_I$. Since experimental fiber orientation results through the shell zone for flow in a strip could be fitted well using the alternative orientation distribution function approach, Bay and Tucker (1992b) concluded that the observed discrepancies were probably due to inaccuracies introduced by the use of a closure approximation. In the context of the coupled approach, the simulations of Chung and Kwon (1995) predicted that increasing the value of $C_I$ will result in an increased value of the required clamping force.

### 4.4.3.3  Choice of inlet orientation

Figure 4.22 is a contact microradiograph of a section of a molded edge-gated rectangular strip taken 0.3 mm from the gate. The center of the picture reveals that fibers emerging from the gate are very strongly aligned in the flow direction. However, experimental data like this are rare and the orientation of the fibers at the gate is usually unknown, even though it is an essential initial condition for the integration of the fiber orientation model (Eqn [17]) in the context of a filling simulation. The effect of the fiber orientation at the gate on the final fiber orientations in the part has been the subject of some study through parametric analyses. Some results obtained from Jeffery's model, indicating that the transverse orientation in a point-gated cavity and the alignment of fibers with the walls through the action of the fountain flow are independent of initial fiber orientation, have been given in Figures 4.19 and 4.20. Gupta and Wang (1993) found that the fiber orientation at the gate has only a small effect on the value of the $\alpha_{11}^{(2)}$ component of $\boldsymbol{\alpha}^{(2)}$ in a small region (roughly 20% of the total plaque length) from the gate; this effect was found to be somewhat larger for slow injection speeds. Similar conclusions were drawn by Chung and Kwon (1995) using a non-isothermal coupled model for the filling process (Section 4.4.2). The calculations of Frahan et al. (1992) (using decoupled, isothermal filling with kinematics computed on the cavity mid-plane) also indicate that the predicted orientation distributions in the cavity are insensitive to the fiber orientation at the gate. In the absence of experimental data, these are somewhat reassuring results. The effect of fiber orientation at the gate was also investigated by Ranganathan and Advani (1993) in the context of the coupled approach for steady isothermal spreading flow of a Newtonian fiber-filled fluid in a disk (excluding free surface effects). That study concluded that the difference between the predictions corresponding to isotropic and aligned initial orientations was maximum at some point halfway between the centerplane and the cavity wall; this difference was also found to diminish with distance from the injection point.

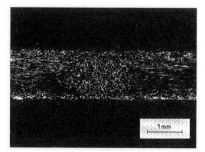

**4.22**  Contact microradiograph of a section cut 0.3 mm from the gate and perpendicular to the main direction of flow, indicating that the fibers enter the cavity aligned with the MDF (Bright et al., 1978, with permission).

## 4.5    Conclusion

Since about 1980 significant progress has been made in the science and technology of injection molding of fiber-reinforced composites. Besides continuing applications in forming short-fiber-reinforced materials (with the proven advantages of automation, reproducibility and ability to generate complex shapes), the introduction of pultrusion technology for the generation of pellets containing long fibers has led to a renewed interest in the process as a means of fabricating long-fiber-reinforced components with superior mechanical properties. Use of feed materials containing higher volume fractions of fibers is also desirable for the same reason. The interplay between processing conditions and product morphology has been the subject of extensive experimenal work and is now fairly well understood qualitatively. A review of relevant studies has been presented in this chapter and is summarized in Table 4.1. The key challenge in the continued and expanded usage of the process to more demanding applications is the ability of the designer to predict quantitatively the distribution and orientation of the fibers in the finished article and thus tune the process to achieve a desirable set of properties. In this direction, significant progress has also been made. Current state-of-the-art simulation codes can predict, semi-quantitatively, the distribution of fiber orientation in injection molded parts of arbitrarily complex shape, including the orientations near gates and around weldlines. Estimates for the anticipated effective properties of the part can also be derived from these predictions. However, there are a number of issues that need to be considered in more detail before such predictions can be relied upon quantitatively and before fine-tuning of the process can be guided by model predictions. One development which will be critical for the quality of fiber orientation predictions in short-fiber-reinforced parts at high fiber concentrations is the consideration of the interaction between fiber orientation and flow kinematics during filling. Besides implementation issues, progress in this area is directly linked to the development and further testing of constitutive models for concentrated fiber dispersions. Other areas where improvements can be made include the consideration of the effect of packing on fiber orientation, the effect of melt elasticity on fiber orientation and the consideration of longer fibers, whose size can be comparable to the cavity thickness.

## References

Akay, M. and Barkley D. (1985) Processing-structure-property interaction in injection moulded glass-fibre reinforced polypropylene, *Compos. Sruct*, **3**, 269–293.

Akay, M. and Barkley, D. (1991) Fibre orientation and mechanical behaviour in reinforced thermoplastic injection mouldings, *J. Mater. Sci.*, **26**, 2731–2742.

Akay, M. and Barkley, D. (1992) Jetting and fibre degradation in injection moulding of glass-fibre reinforced polyamides, *J. Mater. Sci.*, **27**, 5831–5836.

Akay, M. and Barkley, D. (1993) Flow aberrations and weld-lines in glass-fibre reinforced thermoplastic injection moulding, *Plastics Rubber Compos. Processing Applic.*, **20**, 137–149.

Advani, S.G. and Tucker III, C.L. (1990) Closure approximations for three-dimensional structure tensors, *J. Rheol.*, **34**, 367–386.

Advani, S.G. and Tucker III, C.L. (1987) The use of tensors to describe and predict fiber orientation in short fiber composites, *J. Rheol.*, **31**, 751–784.

Aronsson, G. (1994) On p-harmonic functions, complex duality and an assymptotic formula for injection mold filling, Linkoping University Report LiTH-MAT-R-94-27.

Bailey, R. and Rzepka, B. (1991) Fibre orientation mechanisms for injection molding of long fibre composites, *Internat. Polym. Process.*, **6**(1), 35–41.

Bay, R.S. and Tucker III, C.L. (1992a) Fiber orientation in simple injection moldings. Part 1: Theory and numerical methods, *Polym. Compos.*, **13**, 317–321.

Bay, R.S. and Tucker III, C.L. (1992b) Fiber orientation in simple injection moldings. Part 2: Experimental results, *Polym. Compos.*, **13**, 332–341.

Behrens, R.A., Crochet, M.J., Denson, C.D. and Metzner, A.B. (1987) Transient free surface flows: motion of a fluid advancing in a tube, *AIChE J.*, **33**(7), 1178–1186.

Bernhardt, E.C. (ed.) (1983) *Computer-Aided Engineering for Injection Molding*, Hanser Verlag, Munich.

Blanc, R., Philipon, S., Vincent, M., Agassant, J.F., Alglave, H. Muller, R. and Froelich, D. (1987) Injection molding of reinforced thermosets, *Internat. Polym. Process.*, **2**(1), 21–27.

Bouti, A., Vu-Khan, T. and Fisa, B. (1989) Injection molding of glass flake reinforced polypropylene: flake orientation and stiffness, *Polym. Compos.*, **10**(5), 352–359.

Bright, P.F., Crowson, R.J. and Folkes, M.J. (1978) A study of the effect of injection speed on fiber orientation in simple mouldings of short glass fibre-filled polypropylene, *J. Mater. Sci.*, **13**, 2497–2506.

Chiang, H.H., Hieber, C.A. and Wang, K.K. (1991a) A unified simulation of the filling and post-filling stages in injection molding. Part I: Formulation, *Polym. Eng. Sci.*, **31**, 116–124.

Chiang, H.H., Hieber, C.A. and Wang, K.K. (1991b) A unified simulation of the filling and post-filling stages in injection molding. Part II: Experimental verification, *Polym. Eng. Sci.*, **31**, 125–139.

Chiang, H.H., Santhanam, N., Himasekhar, K. and Wang, K.K. (1994) Integrated CAE analysis for fiber-filled plastics in injection molding, in *Advances in CAE of Polymer Processing*, eds Himasekhar, K., Prasad, V., Osswald, T.A. and Batch, G. ASME, New York.

Chung, S.T. and Kwon, T.H. (1995) Numerical simulation of fiber orientation in injection molding of short-fiber-reinforced thermoplastics, *Polym. Eng. Sci.*, **35**(7), 604–618.

Coyle, D.J., Blake, J.W. and Macosko, C.W. (1987) The kinematics of fountain flow in mold-filling, *AIChE J.*, **33**, 1168–1177.

Crochet, M.J., Dupret, F. and Verleye, V. (1994) Injection molding, in *Flow and Rheology in Polymer Composites Manufacturing*, ed. Advani, S.G., Elsevier, Amsterdam, pp. 415–463.

Darlington, M.W. and Smith, A.C. (1987) Some features of the injection molding of short fiber reinforced thermoplastics in center sprue-gated cavities, *Polym. Compos.*, **8**(1), 16–21.

Darlington, M.W., McGinley, P.L. and Smith, G.R. (1976) Structure and anisotropy of stiffness in glass-reinforced thermoplastics, *J. Mater. Sci.*, **11**, 877–886.

Darlington, M.W., Gladwell, B.K. and Smith, G.R. (1977) Structure and mechanical properties in injection molded disks of glass-fiber reinforced polypropylene, *Polymer*, **18**, 1269–1275.

Dinh, S.M and Armstrong, R.C. (1984) A rheological equation of state for semi-concentrated fiber suspensions, *J. Rheol.*, **28**, 207–227.

Dupret, F. and Vanderschuren, L. (1988) Calculation of the temperature field in injection molding, *AIChE J.*, **34**, 1959–1972.

Faulkner, D.L. and Schmidt, L.R. (1977) Glass bead filled polypropylene Part I: Rheological and mechanical properties, *Polym. Eng. Sci.* **17**, 657–662.

Fellahi, S., Meddad, A., Fisa, B. and Favis, B.D. (1995) Weldlines in injection molded parts: A review, *Adv. Polym. Technol.*, **14**(3), 169–195.

Fisa, B. and Rahmani, R. (1991) Weldline strength in injection molded glass fiber-reinforced polypropylene, *Polym. Eng. Sci.*, **31**(18), 1330–1336.

Folgar, F. and Tucker III, C.L. (1984) Orientation behavior of fibers in concentrated suspensions, *J. Reinf. Plastics Compos.*, **3**, 98–119.

Frahan, H.H., Verleye V., Dupret F. and Crochet, M.J. (1992) Numerical prediction of fiber orientation in injection molding, *Polym. Eng. Sci.*, **32**(4), 254–266.

Gillespie, J.W., Vanderschuren, J.A. and Pipes, R.B. (1985) Process induced fiber orientation: numerical simulation with experimental verification, *Polym. Compos.*, **6**(2), 82–86.

Gogos, C.G., Huang, C.F. and Schmidt, L.R. (1986) The process of cavity filling including fountain flow in injection molding, *Polym. Eng. Sci.*, **26**, 1457–1466.

Givler, R.C. (1983) Numerical techniques for the prediction of flow-induced orientation, *Technical Report CCM-83-11*, Center for Composite Materials, University of Delaware.

Givler, R.C., Crochet M.J. and Pipes, R.B. (1983) Numerical prediction of fiber orientation in dilute suspensions, *J. Compos. Mater.*, **17**, 330–343.

Gupta, M. and Wang, K.K. (1993) Fiber orientation and mechanical properties of short-fiber-reinforced injection-molded composites: simulated and experimental results, *Polym. Compos.*, **14**(5), 367–382.

Hayes, R.E., Dannelongue, H.H. and Tanguy, P.A. (1991) Numerical simulation of mold filling in reaction injection molding, *Polym. Eng. Sci.*, **31**(11), 842–848.

Hegler, R.P. (1984) Faserorientierung beim Verarbeiten kurzfaserverstarkter Thermoplaste, *Kunststoffe*, **74**(5), 271–277.

Hegler, R.P. and Mennig, G. (1985) Phase separation effects in processing of glass-bead and glass-fiber-filled thermoplastics by injection molding, *Polym. Eng. Sci.*, **25**(7), 395–405.

Hieber, C.A. and Shen, S.F. (1980) A finite element/finite difference simulation of the injection mold filling process, *J. Non-Newtonian Fluid Mech.*, **7**, 1–31.

Hieber, C.A., Socha, L.S., Shen, S.F., Wang, K.K. and Isayev, A.I. (1983) Filling thin cavities of variable gap thicknesses: A numerical and experimental investigation, *Polym. Eng. Sci.*, **23**, 20–26.

Hinch, E.J. and Leal, L.G. (1976) Constitutive equations in suspension mechanics. Part 2: Approximate forms for a suspension of rigid particles affected by Brownian rotations, *J. Fluid Mech.*, **76**, 187–208.

Isayev, A.I. (1983) Orientation development in the injection molding of amorphous polymers, *Polym. Eng. Sci.*, **23**(5), 271–284.

Isayev, A.I. (ed.) (1987) *Injection and Compression Molding Fundamentals*, Marcel Dekker, New York.

Jansen, K.M.B. and Vandam, J. (1993) An analytical solution for the temperature profiles during injection molding, including dissipation effects, *Rheol. Acta.*, **31**(6), 592–602.

Kamal, M.R. and Kenig, S. (1972a) The injection molding of thermoplastics Part I: Theoretical model, *Polym. Eng. Sci.*, **12**(4), 294–301.

Kamal, M.R. and Kenig, S. (1972b) The injection molding of thermoplastics Part II: Experimental test of the model. *Polym. Eng. Sci.*, **12**(4), 302–308.

Kamal, M.R. and Mutel, A.T. (1989) The prediction of flow and orientation behaviour of short fiber reinforced melts in simple flow systems, *Polym. Compos.*, **10**(5), 337–343.

Kamal, M.R. and Papathanasiou, T.D. (1993) Filling of a complex-shaped mold with a viscoelastic polymer. Part II: Comparison with experimental data, *Polym. Eng. Sci.*, **33**, 410–417.

Kamal, M.R. and Tan, V. (1979) Orientation in injection molded polystyrene, *Polym. Eng. Sci.*, **19**(8), 558–563.

Kamal, M.R., Chu, E., Lafleur, P.G. and Ryan, M.E. (1986) Computer simulation of injection mold filling for viscoelastic melts with fountain flow, *Polym. Eng. Sci.*, **26**(3), 190–196.

Kamal, M.R., Goyal, S.K. and Chu, E. (1988) Simulation of injection mold filling of viscoelastic polymers with fountain flow, *AIChE J.*, **34**(1), 94–106.

Kenig, S. (1986) Fiber orientation development in molding of polymer composites, *Polym. Compos.*, **7**(1), 50–55.

Khanh, T.V., Denault, Habib, P. and Low, A. (1991) The effects of injection molding on the mechanical properties of long-fiber-reinforced PBT/PET blends, *Compos. Sci. Tech.*, **40**, 423–435.

Kuo, Y. and Kamal, M.R. (1976) The fluid mechanics and heat transfer of injection mold filling of thermoplastic materials, *AIChE J.*, **22**(4), 661–669.

Lafleur, P.G. and Kamal, M.R. (1986) A structure-oriented computer simulation of the injection molding of viscoelastic crystalline polymers Part I: Model with fountain flow, packing, solidification, *Polym. Eng. Sci.*, **26**(1), 92–102.

Lhymn, C. and Schultz, J.M. (1985) Fracture of glass-fibre reinforced poly(phenylene sulphide), *J. Mater. Sci. Letters*, **4**, 1244–1248.

Macosko, C.W. (1986) *Fundamentals of Reaction Injection Molding*, Carl Hanser Verlag, Munich.

Malzahn, J.C. and Schultz, J.M. (1986) Transverse core fiber alignment in short-fiber injection molding, *Compos. Sci. Tech.*, **25**, 187–192.

Manzione, L.T. (ed.) (1987) *Applications of Computer-Aided Engineering in Injection Molding*, Hanser Verlag, Munich.

Matsuoka, T., Takabatake, J., Inoue, Y., and Takahashi, H. (1990) Prediction of fiber orientation in injection molded parts of short-fiber-reinforced thermoplastics, *Polym. Eng. Sci.*, **30**, 957–966.

Matsuoka, T., Takabatake, J., Koiwai, A., Inoue, Y., Yamamoto, S. and Takahashi, H. (1991) Integrated simulation to predict warpage of injection molded parts, *Polym. Eng. Sci.*, **31**(14), 1043–1050.

Mavridis, H., Hrymak, A.M. and Vlachopoulos, J. (1986) Finite element simulation of fountain flow in injection molding, *Polym. Eng. Sci.*, **26**(7), 449–454.

Mavridis, H., Hrymak, A.M. and Vlachopoulos, J. (1988) The effect of fountain flow on molecular orientation in injection molding, *J. Rheol.*, **32**, 639–663.

McClelland, A.N. and Gibson, A.G. (1990) Rheology and fibre orientation in the injection moulding of long fibre reinforced nylon 66 composites, *Compos. Manufact.*, **1**, 15–25.

Moy, F.H. and Kamal, M.R. (1980) Crystalline and amorphous orientation in injection molded polystyrene, *Polym. Eng. Sci.*, **20**(14), 957–964.

Nielsen, L.E. (1974) The thermal and electrical conductivity of two-phase systems, *Ind. Eng. Chem. Fundam.*, **13**(1), 17–20.

Ogadhoh,S.O. and Papathanasiou, T.D. (1996) Particle rearrangement during processing of glass-reinforced polystyrene by injection moulding, *Composites Part A*, **27A**(1), 57–63.

Papathanasiou, T.D. (1995) Modelling of injection mold filling: Effect of undercooling on polymer crystallisation, *Chem. Eng. Sci.*, **50**(21), 3433–3442.

Papathanasiou, T.D. and Kamal, M.R. (1993) Filling of a complex-shaped mold with a viscoelastic polymer. Part I: The mathematical model, *Polym. Eng. Sci.*, **33**(7), 400–409.

Papathanasiou,T.D., Soininen, R. and Caridis, K.A. (1995) Internal microstructure and the thermal response of functionally gradient metal matrix composites, *Scand. J. Metallurgy*, **24**, 159–167.

Powell, R.L. (1991) Rheology of suspensions of rodlike particles, *J. Stat. Phys.*, **62** (5/6), 1073–1093.

Ranganathan, S. and Advani, S.G. (1991) Fiber–fiber interactions in homogeneous flows of non-dilute suspensions, *J. Rheol.*, **35**, 1499–1522.

Ranganathan, S. and Advani, S.G. (1993) A simultaneous solution for flow and fiber orientation in axisymmetric, diverging radial flow, *J. Non-Newtonian Fluid Mech.*, **47**, 107–136.

Richardson, S.M. (1983) Injection moulding of thermoplastics: freezing during mould filling, *Rheol. Acta*, **22**, 223–236.

Richardson, S.M. (1986) Injection moulding of thermoplastics: freezing of variable viscosity fluids. II. Developing flows with very low heat generation, *Rheol. Acta*, **25**, 308–318.

Rose, W. (1961) Fluid–fluid interfaces in steady motion, *Nature*, **191**, 242.

Sanou, M., Chung, B. and Cohen, C. (1985) Glass fiber-filled thermoplastics. II. Cavity filling and fiber orientation in injection molding, *Polym. Eng. Sci.*, **25**(16), 1008–1016.

Sanschagrin, B., Gauvin, R., Fisa, B. and Vu-Khan, T. (1990) Weldlines in injection molded polypropylene: effect of filler shape, *J. Reinf. Plastics Compos.*, **8**, 194–208.

Schmidt, L.R. (1974) A special mold and tracer technique for studying shear and extensional flows in a mold cavity during injection molding, *Polym. Eng. Sci.*, **14**(11), 797–801.

Schmidt, L.R. (1977) Glass-bead filled polypropylene Part II: Mold filling studies during injection molding, *Polym. Eng. Sci.*, **17**(9), 666–670.

Shen, S.F. (1984) Simulation of polymeric flows in the injection molding process, *Internat. J. Numer. Method Fluids*, **4**, 171–183.

Singh, P. and Kamal, M.R. (1989) The effect of processing variables on microstructure of injection molded short fiber reinforced polypropylene composites, *Polym. Compos.*, **10**, 344–351.

Spahr, D.E., Friedrich, K., Schultz, J.M. and Bailey, R.S. (1990) Microstructure and fracture behavior of short and long fiber-reinforced polypropylene composites, *J. Mater. Sci.*, **25**, 4427–4439.

Subbiah, S., Trafford, D.L. and Guceri, S.I. (1989) Non-isothermal flow of polymers into two-dimensional, thin cavity molds: a numerical grid generation approach, *Int. J. Heat Mass Transfer*, **32**(3), 415–434.

Tadmor, Z. (1974) Molecular orientation in injection molding, *J. Appl. Polym. Sci.*, **18**, 1753–1772.

Tang, L. and Altan, M.C. (1995) Entry flow of fiber suspensions in a straight channel, *J. Non-Newtonian Fluid Mechanics*, **56**, 183–216.

Tomari, K., Takashima, H. and Hamada, H. (1995) Improvement of weldline strength of fiber reinforced polycarbonate injection molded articles using simultaneous composite injection molding, *Adv. Polym. Technol.*, **14**(1), 25–34.

Truckenmuller, F. and Fritz, H.G. (1991) Injection molding of long-fiber-reinforced thermoplastics: A comparison of extruded and pultruded materials with direct addition of roving strands, *Polym. Eng. Sci.*, **31**(18), 1316–1329.

Tucker III, C.L. (1991) Flow regimes for fiber suspensions in narrow gaps, *J. Non-Newtonian Fluid Mech.*, **39**, 239–268.

Tucker III, C.L. and Advani, S.G. (1994) 'Processing of short-fiber systems', in *Flow and Rheology in Polymer Composites Manufacturing*, ed. Advani, S.G., Elsevier, Amsterdam, pp. 147–197.

Vincent, M. and Agassant, J.F. (1986) Experimental study and calculations of short glass fiber orientation in center gated molded disks, *Polym. Compos.*, **7**(2), 76–83.

Wang, K.K. (1992) Injection molding of polymers and composites: process modelling and simulation, *MRS Bulletin*, April 1992, 45–49.

# Control and manipulation of fibre orientation in large-scale processing

PS ALLAN AND MJ BEVIS

## 5.1    Introduction

The flow fields that apply during the filling of mould cavities in the injection moulding of discontinuous fibre-reinforced thermoplastic matrix composites lead to non-uniform distribution of microstructure and physical properties, both along the length and through the thickness of moulded parts. This is exemplified in the paper by Chivers *et al.*[1] that reports on the influence of flow length on the stiffness and toughness of some engineering thermoplastics. Stiffness properties were measured at several locations around rectangular cross-section picture frame mouldings. The modulus anisotropy varied with flow path length and could be correlated with the observed flow pattern in the mould tool. The orientation distribution of short fibres in injection moulding or extrusions of complex geometry vary significantly with position as revealed in comprehensive reviews.[2,3]

The principal topic of this chapter relates to the application of macroscopic shears to solidifying polymer matrix composites, with the objective of inducing favourable properties in the finished part. The structures which develop in injection moulded composites are well understood, and are a result of the various interactions of the moving composite melt with the mould. The orientation of the fibres in the central core region within mouldings is usually controlled by the orientation imposed by the gate. The orientation in the outer skin layers is caused by fountain flow at the flow front and in general aligns fibres along the flow direction. Under some circumstances a third layer develops, termed the shear or shell layer, which is located between the core and the skin layers and is a result of shearing flow between the faster moving core region and the often frozen skin. The shearing flow results in preferential alignment of fibres approximately along the flow direction. The balance between the different layers and their resulting fibre orientations depends on a range of factors, including the injection speed, gate configuration, mould geometry and mould thickness.

The role of shear controlled orientation technology (SCORTEC) is to influence the balance between the different layers, through the purposeful imposition of macroscopic shearing actions to the solidifying melt within a mould cavity or extrusion die. Macroscopic shears of specified magnitude and direction, applied at the melt–solid interface provide several advantages.

1   Enhanced fibre alignment by design in fibre-reinforced polymer matrices and in ceramic and metal matrices processed with the aid of sacrificial polymer binders.

2   Elimination of mechanical discontinuities that result from the initial mould filling process, including internal weld lines.

3   Reduction in the detrimental effects of a change in moulded section thickness on the mechanical performance of fibre-reinforced polymers.

4   Elimination or reduction in defects resulting from the moulding of thick sectioned components.

The principle of SCORTEC as applied to injection moulding (SCORIM)[4,5] is introduced in Section 5.2, with reference to the enhancement of the strength of internal weld lines in injection moulded thermoplastic and thermosetting matrix composites. The use of SCORIM for the enhancement and control of tensile strength and Young's modulus, and the control of thermal expansion in injection moulded thermoplastics is presented in Section 5.3. The control of the orientation of reinforcing short fibres and porosity in thick section mouldings is presented in Section 5.4, and includes the use of oscillating packing pressures to control porosity, using as an example thick-section blocks representing an actuation support structure. The application of SCORIM for the control of fibre orientation and/or porosity in a selection of mould geometries is presented in Section 5.5, that includes moulded rings and a tapered fin moulding. The application of the SCORTEC principle for the production of laminated microstructures and the *in-situ* formation and management of fibres in moulded thermoplastics, to large mouldings and continuous profile extrusion (SCOREX)[6] are presented in Section 5.6.

## 5.2   Application of SCORIM for weldline strength enhancement

To apply macroscopic shears to a solidifying melt within a cavity requires the provision of a multiplicity of live feeds to the cavity, where the pressure applied to each of the feeds can be independently controlled. Another requirement is to be able to displace sufficient molten material within the mould to create a macroscopic shearing of the melt.

A simple embodiment of the concept in the form of a two live-feed device, is illustrated in Fig. 5.1, located between the injection moulding machine screw/barrel and the mould cavity. Initially, the pistons 1 and 2 are positioned to allow the melt to enter the mould from either one or both of the feeds. When the mould is full, pistons 1 and 2 may be moved back and forth to operate on the solidifying melt as follows:[4,5]

1   Pistons 1 and 2 can be pumped back and forth at the same frequency but with a phase difference of 180°.

2   Pistons 1 and 2 can be pumped back and forth at the same frequency and in phase.
3   Pistons 1 and 2 can be held down under a static pressure.

The operation of the first of the three modes induces a repeated shearing of the molten material within the mould cavity. As the melt–solid interface propagates from the outer to the inner regions of the cavity section, the alignment of the fibres within the matrix which results from the shearing of the melt is frozen into the solid. The frequency and working pressure of the SCORIM pistons can be programmed through a series of stages of operation to give the optimum conditions for the fibre orientation and moulding cycle time. A specific example of the detailed selection of the SCORIM process parameters is presented in Section 5.3. An alternative to the use of auxiliary pistons (Fig. 5.1) for applying macroscopic shears to solidifying melts is to use the action of reciprocating injection screws on multi-barrelled moulding machines, as proposed by Becker et al.[7] for the 'push-pull' process and as applied to industrial products[8].

The most pronounced increase in the strength of injection mouldings that results from the application of SCORIM is achieved when the process is applied to mouldings that contain internal weldlines. A weldline occurs in any moulded plastic part in which two flow fronts come together. This impingement may result from the meeting of multiple flows injected through separate gates or from recombination of a single flow divided by passage around an internal pin or obstruction. It is well known that the weldline represents a potential source of weakness, which is attributed to the lack of mechanical mixing or diffusion at

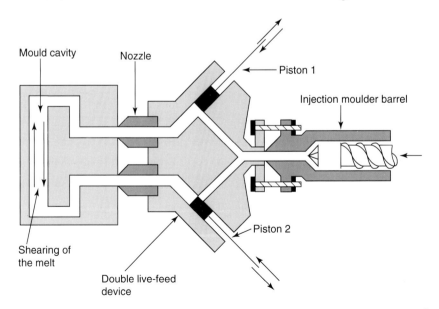

5.1   An arrangement for a SCORIM double live-feed moulding process.[4]

the weld–melt interface, and the fountain flow pattern, which tends to produce at the weldline, a flow orientation at an angle of 90° from the direction of main flow. In reinforced plastics, the local variation in fibre orientation in the weld-line region is the major cause of weakness. Several studies in the 1970s and 1980s focused on relationships between processing, structure and properties of weld lines in thermoplastics, and showed that internal weldline integrity can be improved marginally by selection of injection moulding process parameters.

SCORIM provides a route for the substantial enhancement of internal weld-line strength. The macroscopic shears generated at the melt–solid interface by the technique during the solidification of a moulded part, result in the re-orientation of polymer and any other alignable constituents, in the vicinity of an internal weldline, thereby erasing the weldline. The application of SCORIM[5.9] for the modification of the microstructure of internal weldlines is analogous to the approaches described by Hamada et al.[10] and by Piccarolo et al.,[11] that, respectively, cause a more favourable microstructure at the internal weld, in part by appropriate designs of gate geometry. The application of SCORIM for the enhancement of the internal weldline strength of injection moulded composites is demonstrated by two case studies: a short glass fibre-filled copolyester thermotropic liquid crystal polymer (TLCP)[12] and a high temperature cure polyester dough moulding compound (DMC).[13]

## 5.2.1 Copolyester Liquid crystal polymer containing glass fibres

The 3 mm and 6 mm thick moulded bars of glass fibre filled copolyester (Hoechst–Celanese Vectra A130 grade) were produced[12] using the following procedures:

1  A–conventional moulding with double gate fill.
•  B–asymmetric operation of SCORIM pistons (double live-feed moulding) with two gate fill.
2  C-conventional moulding with single gate fill

These conditions were used to provide for the comparison of the tensile strength of mouldings without a weldline, with a weldline, and with a weldline erased using SCORIM. Figures 5.2(a), (b) and (c) show contact X-ray microradiographs taken from sections of the 3 mm thick bars, produced according to procedures A, B and C, respectively. Microradiography revealed that in the conventionally moulded double gated bars there were clearly defined weld regions where, in the weld region, the glass fibres were oriented predominantly normal to the injection direction. Figure 5.2b shows that in the double gated SCORIM bar there was no evidence of a weldline in the longitudinal cross-section, apart from the outer skin region where there was

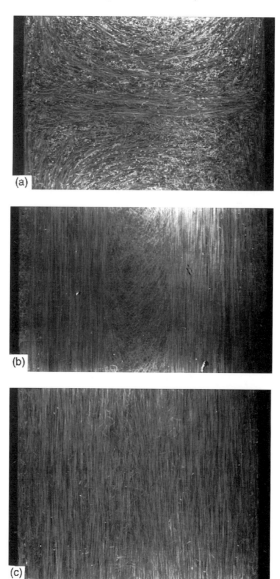

5.2 Contact microradiographs showing fibre orientation in 6 mm thick moulded bars produced by: a) conventional moulding, double gate fill; b) double gate fill followed by SCORIM; c) conventional moulding single gate fill.[12]

*Table 5.1.* Tensile properties of glass fibre filled TLCP (thermotropic liquid crystal polymers) moulded bars

| Test specimen production method | Thickness (mm) | $E$ (GPa) | $\sigma_{max}$ (MPa) |
|---|---|---|---|
| A | 3 | 10.4 | 23.9 |
| B | 3 | 18.9 | 205.2 |
| C | 3 | 16.9 | 198.3 |
| A | 6 | 9.0 | 25.2 |
| B | 6 | 19.5 | 190.2 |
| C | 6 | 14.0 | 160.5 |

evidence of randomly oriented glass fibres. Overall there was pronounced preferred orientation of glass fibres along the injection direction. The cooling rate within the 6 mm thick moulded bar was less than in the 3 mm bar, and consequently the prolonged higher temperature within the core provided for more prolonged application of the macroscopic shearing action, and hence a more complete uniaxial alignment of fibres through the thickness of mouldings. However, the microradiographs gained from the conventional single gated bars, (see Figure 5.2(c)) showed, as expected, that the 3 m thick bar exhibited a greater proportion of preferred orientation from the skin to the core than the 6 mm thick bars. Typical stress–strain curves of the double gated glass fibre filled bars produced by conventional injection moulding and by SCORIM are shown in Fig 5.3(a) and (b), and the corresponding tensile modulus and strength data are summarized in Table 5.1.

## 5.2.2  High temperature cure DMC

The study materials selected[13] for an investigation of the application of SCORIM to the injection moulding of thermosetting polymer matrix composites included a high temperature cure polyester that exhibited low shrink characteristics. The SCORIM process was shown to result in substantial enhancement of tensile and flexural strength in $20 \times 20$ mm square cross-section moulded bars, as produced using single or double gates to give mouldings without and with internal weldlines respectively. The substantial enhancement of tensile and flexural strengths that results from the use of SCORIM are shown in Fig 5.4(a) and (b) which represent the influence of the number of oscillations applied during curing, with a position across the width of single and double gated moulding, respectively. One conclusion from this investigation was that SCORIM has a potential application in thermoset matrix composites, particularly with respect to the enhancement of internal weldline strength. There was a need, however, to optimize the design of the SCORIM pistons in order to avoid seizure in operation due to the presence of inorganic filler. This involved the use of thermoplastic PEEK pistons[13].

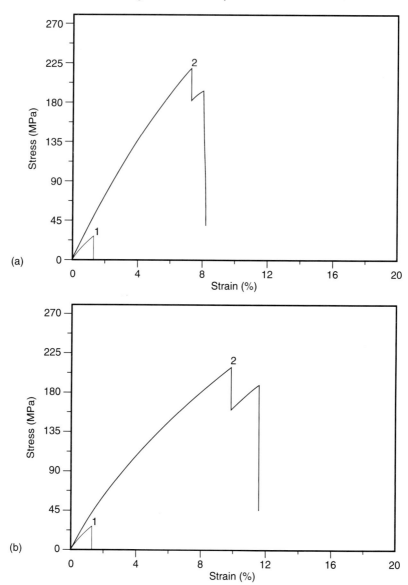

(a)

(b)

5.3  Stress–strain curves of glass fibre filled TLCP: a) 3 mm and b) 6 mm
      thick double gated bars produced conventionally (1) and SCORIM
      (2) respectively.[12]

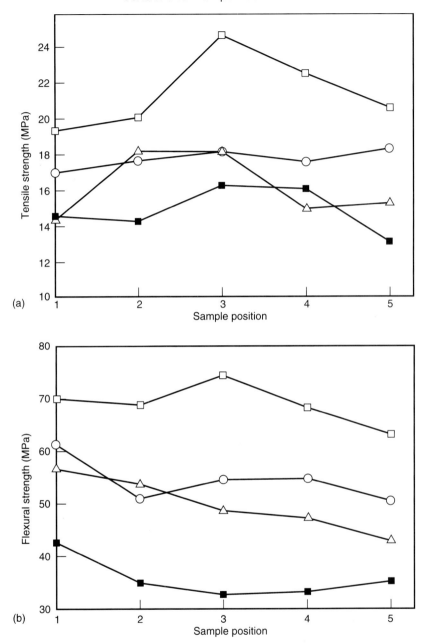

5.4 (a) Tensile strength dependence, and (b) flexural strength dependence, on process conditions and position through the thickness of
double gated DMC mouldings. Process conditions: ■, conventional;
△ 10 oscillations; ○, 20 oscillations; □; 30 oscillations.[13]

## 5.3    Application of SCORIM for physical property enhancement

Two materials have been selected to demonstrate the potential of SCORIM to manage the morphology of short fibre reinforced thermoplastics, throughout the volume of a rectangular plaque $84 \times 84 \times 6$ mm. They are, first, a polypropylene matrix, reinforced by very closely controlled fibres of length 5 mm at 20 vol.%, and second, a glass fibre-reinforced thermoplastic supplied by ICI in 1988 as an injection moulding grade of copolyester, Victrex 1500GL30. In the latter example the matrix polymer was synthesized from residues: $p$-hydroxy benzoic acid, isophthalic acid and hydroquinone, and contained 27 wt % glass fibre.

### 5.3.1  Glass fibre-reinforced polypropylene

A detailed analysis of the fibre orientation distribution in $84 \times 84 \times 6$ mm plaques produced by conventional injection moulding and SCORIM has been reported.[14] The study was based on a polypropylene matrix, reinforced by fibres with a uniform length of 5 mm. The experiments were designed to determine accurately the three-dimensional fibre orientation distributions resulting from the conventional and SCORIM processes, and correlate these with the processing conditions and the resulting mechanical properties. Theoretical models were also used to support the results.

The mechanical properties and fibre orientation distributions were measured at a number of different positions on the samples, although only measurements at the centre position, designated $\delta_2$ in Fig. 5.5, are presented here. The reader is referred to the original publication[14] for a description of the procedures adopted for the measurement of mechanical properties and fibre orientations, and the use of theoretical simulations to link the measured fibre orientation distributions and the mechanical properties. The mechanical properties were measured using an ultrasonic velocity technique. The measurement of fibre orientation distributions utilized a transputer controlled image analyser which enables accurate orientation data to be collected at high speed. The instrument works directly on images from a polished and etched section taken from the composite, where each fibre at the surface image appears as an ellipse. The composition of the conventional and SCORIM injection mouldings investigated[14] resulted in the following conclusions:

1   The SCORIM process dramatically improves the fibre alignment along the injection direction changing the orientation distribution from random-in-plane to preferentially aligned. This is as a result of the development of a thick highly aligned shell or shear region that is associated with the region in the mould cavity that responds to the macroscopic shearing action of the SCORIM process.

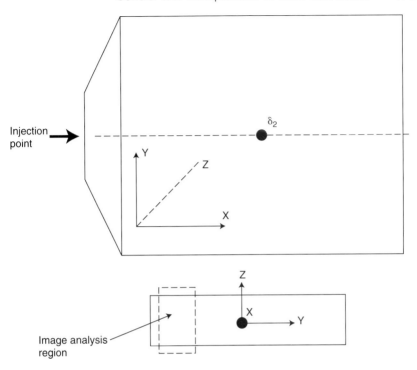

*5.5*  Sample details and measurement positions.[14]

2   The large changes in fibre orientation structure seen in the two moulding
    types only affect the stiffnesses along the two in-plane axes ($X$ and $Y$). The
    other stiffness constants are largely unaffected by fibre orientation.

Conclusion (1) here is illustrated in part in Fig 5.6(a) and (b), which show the
orientation averages $\langle \cos^2 \Theta \rangle$ along the three axes $X$, $Y$ and $Z$ for conventional
and SCORIM mouldings respectively and measured for the strips with the long
directions parallel to the $Y$ axis as shown in the insets. The higher the value of
the orientation average the better the alignment. For perfect alignment along one
axis, the orientation average would be equal to 1 with the other two averages
equal to zero. For random alignment in one plane two averages would be equal
to 0.5 with the other zero. At the edges of the conventionally moulded sample
(Fig.5.6(a)), normally termed the skin region, the fibres were preferentially
aligned along the injection direction ($X$), while in the centre of the specimen,
termed the core, the fibres were aligned more perpendicular to the injection
direction ($Y$). At the very centre of the sample there was a clear region of
appreciable out-of-plane orientation ($Z$ axis). In general, though, the fibres lie in
the plane of the sample. In the SCORIM moulding (Fig. 5.6(b)) it is first
noticeable that all the orientation structure is symmetrical about the centreline
and that the out-of-plane orientation is again very low. There is a central core

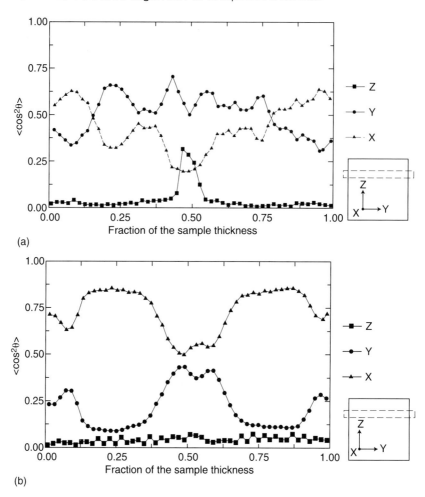

(a)

(b)

5.6  Fibre alignment through sample thickness for a) conventional and
b) SCORIM mouldings.[14]

where the fibres are about random in-plane, a shell region where the fibres are
aligned along the injection direction and an outer skin where the fibres are still
predominantly aligned along the injection direction but with a lower orientation
factor.

The substantial differences in fibre orientation distribution described above
are responsible for the differences in the mechanical properties of conventional
and SCORIM mouldings, and with respect to conclusion (2). Table 5.2
summarizes the measured stiffness constants, $C_{ij}$, as determined by ultrasonic
measurements, and the engineering constants. It is interesting to note[14] that the
stiffness constants $C_{22}$, $C_{13}$, $C_{23}$, $C_{12}$, $C_{44}$, $C_{55}$ and $C_{66}$ are little affected by the
large changes in orientation seen in the two different mouldings. This agrees
with recent work on other oriented systems,[15] where the latter six constants were

Table 5.2. Mechanical properties (stiffness constants, $C_{ij}$, Young's constants modulus, $E_{ij}$, Poisson's ratio, $v$, shear modulus, $G_{ij}$) of glass fibre-reinforced polypropylene conventional and SCORIM moulding with respect to position $\delta_2$ and definition $X=3$, $Y=1$, $Z=2$ (Fig. 5.5)

|          | Conventional moulding | SCORIM moulding |          | Conventional moulding | SCORIM moulding |
|----------|------------|------------|----------|------------|------------|
| $C_{33}$ | 10.0  | 17.5 | $E_{33}$ | 7.43  | 15.0 |
| $C_{22}$ | 7.37  | 8.12 | $E_{22}$ | 5.61  | 5.88 |
| $C_{11}$ | 12.2  | 9.28 | $E_{11}$ | 9.50  | 6.69 |
| $C_{13}$ | 4.59  | 4.20 | $v_{13}$ | 0.263 | 0.306 |
| $C_{23}$ | 3.73  | 3.88 | $v_{23}$ | 0.375 | 0.316 |
| $C_{12}$ | 3.65  | 4.32 | $v_{12}$ | 0.192 | 0.409 |
| $C_{44}$ | 2.07  | 2.46 | $G_{13}$ | 3.59  | 2.68 |
| $C_{55}$ | 3.59  | 2.68 | $G_{23}$ | 2.07  | 2.46 |
| $C_{66}$ | 1.94  | 2.03 | $G_{12}$ | 1.94  | 2.03 |

shown to be largely independent of orientation. The two effects of the SCORIM process, are first to exchange the order of $C_{33}$ and $C_{11}$ by aligning more fibres in the injection direction and second, to produce a higher maximum stiffness.

An early application of SCORIM[4] demonstrated that SCORIM provides for the effective alignment of short glass fibres in thick sectioned polypropylene matrix mouldings. The uniaxial alignment of glass fibres that may be achieved in 20 mm thick moulding is illustrated in Fig. 5.7(b), and may be compared with the equivalent X-ray radiograph obtained from a conventional moulding (Fig. 5.7(a)). A substantial enhancement of mechanical properties was present for the mouldings produced by the SCORIM route.

## 5.3.2  Glass fibre filled copolyester

In the moulding of the glass fibre filled copolyester,[16] a melt temperature of 340°C was used with a mould temperature of 85°C. An injection rate of 36 cm$^3$ s$^{-1}$ was employed for the production of conventional and SCORIM plaques. Two injection pressures were used for the production of the SCORIM plaques. For each condition, the SCORIM pistons were programmed to produce a reciprocating shear flow of the melt during the time the material was solidifying. The length of time for which an effective shearing of the melt was operating was termed the 'shear time' for the mouldings. The following parameters were set for each of the SCORIM stages, and typify the profiles that generally apply for achieving optimum microstructure and physical properties, though will differ according to study material and mould cavity geometry.

1  Stage duration (seconds).
2  Move time (seconds): the time given for one stroke of a SCORIM piston, i.e. with a move time of 2 s a complete cycle of oscillation will take 4 s.

**I.D.**

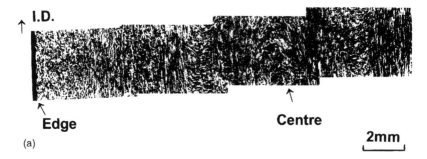

**Edge**                    **Centre**

(a)                                      **2mm**

**I.D.**

**Edge**                    **Centre**

(b)                                      **2mm**

*5.7* Contact *X*-ray radiographs showing the difference between the levels of glass fibre orientation achieved in 20 mm thick conventional (a) and SCORIM (b) mouldings.[4]

3    Compression pressure (C-Pres): the pressure limit on the hydraulic oil delivered to drive the SCORIM piston down the bore.
4    Relaxation pressure (R-Pres): the pressure limit on the hydraulic oil delivered to the SCORIM piston to return it to the back position in the bore.

The initial stage of each of the profiles was set to promote the maximum displacement of material through the mould cavity which was possible for the equipment used (i.e. the full stroke of the pistons was utilized at the start of the hold pressure stage). The second stage of the profiles was set to allow the moulding to solidify in a controlled way while maintaining a constant shearing of the remaining molten material. The total shearing times were 64 and 56 s for the glass fibre filled plaques as produced at low and medium pressures. A typical cavity pressure trace recorded during the conventional moulding cycle is shown in Fig.5.8(a). The use of the SCORIM process produces an oscillating response in the cavity pressure profile, as represented in Fig. 5.8(b) and (c) for the low and medium pressures respectively. These profiles indicate the effective application of the reciprocating shear flow in the cavity. The cavity pressure profiles also reveal that, as expected, an increase in the applied pressure causes the corresponding pressure pulse amplitude and the mean cavity pressure to increase and the total shearing time to decrease.

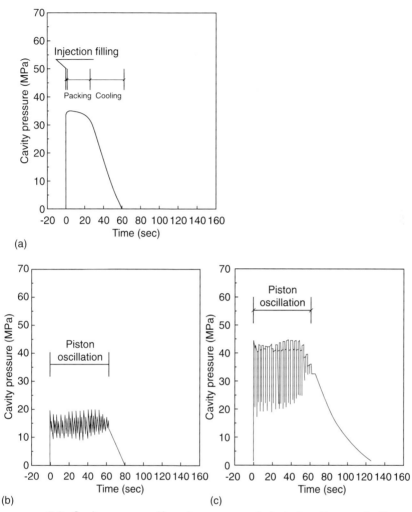

*5.8* Cavity pressure/time traces recorded during the production of conventionally moulded (a), and SCORIM low (b) and medium (c) pressure mouldings.[16]

### 5.3.2.1   Fibre orientation in SCORIM plaques

Contact microradiography shows that four different layers of preferred glass fibre orientation occur in the conventionally moulded plaques namely a skin with a more or less random glass fibre orientation, a shear layer with pronounced orientation in the injection direction that extends about 400 μm from the skin, a transition layer with unoriented glass fibres and a core with preferred glass fibre orientation parallel to the transverse direction. In contrast to the conventional moulding, the two SCORIM mouldings exhibit only three distinct regions: a skin, macroscopic shear (shell) and core regions. Figure 5.9(a) and (b) show the

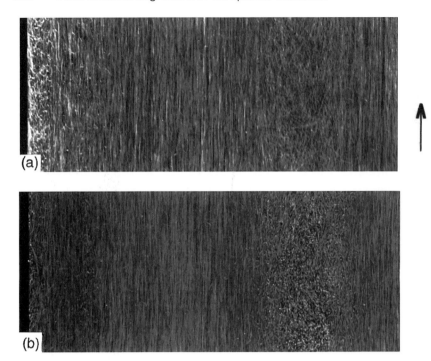

**5.9** CMR showing glass fibre orientations in longitudinal cross sections of, (a) SCORIM low pressure moulding, (b) medium pressure moulding. Arrows indicate injection or shear direction.[16]

extent of preferred glass fibre orientation development in low and medium pressure mouldings, respectively.

### 5.3.2.2 Young's modulus measurements

The dependence of Young's modulus on the position of test bars cut longitudinally (L) and transversely (T) to the processing direction, as shown in Fig 5.10, is presented in Fig 5.11(a) and (b) respectively. Figures 5.11(a) and (b) show that the SCORIM mouldings possess greater anisotropic properties than the conventional mouldings, which is attributed to the extended shear regions in the SCORIM mouldings. The SCORIM plaques also have a more uniform Young's modulus than conventional mouldings in both the longitudinal and transverse directions. The measured values of Young's modulus were much greater in the longitudinal direction than in the transverse direction for the SCORIM mouldings than for the conventional mouldings. This effect is reversed for measurements in the transverse direction. The more uniform Young's modulus distribution in the SCORIM mouldings is attributed to the repeated macroscopic shear flows that occurred along the whole length of the moulding

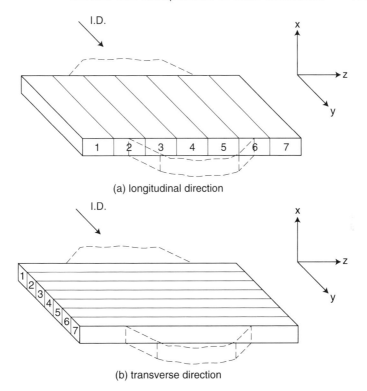

(a) longitudinal direction

(b) transverse direction

*5.10*  Sampling from moulded plaques for tensile modulus measurements.[16]

during solidification. The repeated shearing produced an aligned fibre microstructure through a larger cross-section, when compared with conventional moulding where the fountain flow effect during mould filling causes preferential alignment near to the gate.

### 5.3.2.3  Linear thermal expansion of mouldings

The linear thermal expansion behaviour of the conventionally moulded and SCORIM glass fibre filled plaques is shown in Figs. 5.12(a) and (b). The SCORIM and conventional plaques have low and high coefficients of thermal expansion in the longitudinal and the transverse directions, respectively. The thermal expansion in the longitudinal direction for the SCORIM mouldings is less than that for the conventional mouldings, whereas the reverse behaviour applies for the transverse expansion. The SCORIM mouldings which have the largest sheared layer exhibit the highest and lowest recorded thermal expansions in the transverse and longitudinal directions, respectively. A decrease in the extent of the sheared layer causes the thermal expansion in the transverse direction to decrease, and to increase in the longitudinal direction. The two

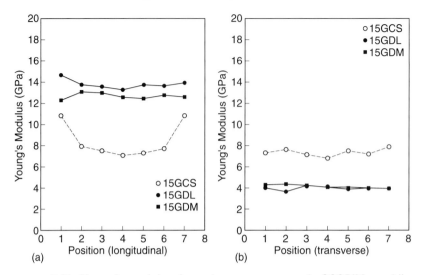

5.11 Young's modulus dependence on pressure in SCORIM moulding, where ○, ● and ■ represent conventional, low and medium pressure SCORIM moulding respectively: (a) measured parallel to the macroscopic shear direction across the width of the plaques; (b) measured transverse to the macroscopic shear direction along the length of the plaques.[16]

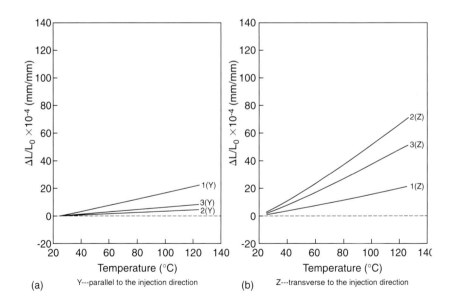

5.12 Thermal expansion of glass fibre filled SCORIM mouldings which show with increasing temperature: (a) a decrease in the longitudinal direction; (b) an increase in the transverse.[16]

examples referred to above illustrate the potential of SCORIM for the purposeful control of the physical properties of moulded composites. The extension of the SCORIM concept to some generic cavity geometries is presented in Section 5.5.

## 5.4    Control of porosity in thick-section mouldings

The principal reason for using an oscillating packing pressure in injection moulding is to ensure that thick-sectioned mouldings have sufficient time to freeze off in the mould cavity before the gate, sprue, runner or any other narrower section separating the thick section of the moulding cavity from the nozzle has frozen off. This is achieved by the periodic compression and expansion of the melt in the thick sections which causes enhanced shearing of the melt in the narrow mould sections. The internal heat generated by this shearing delays the freezing of the narrow sections, until the thick section of the mould has solidified. Using this technique, therefore, additional melt can be added to the thick cavity section to compensate for the shrinkage as the material solidifies. Thick moulding of fibre-reinforced thermoplastics can be produced without the presence of internal defects such as voids and cracks which are associated with an unsatisfactory packing of the mould.

The application of macroscopic shears caused by the out-of-phase operation of the two pistons, as in Fig. 5.1, provides an efficient alternative to the in phase operation of two pistons or the operation of a single piston,[17] for the suppression of porosity. The examples cited below illustrate both routes, the selection in practice being determined by the complexity of mould geometry and the requirement to manage fibre orientation distribution.

### 5.4.1  Control of porosity and fibre orientation in a variable cross-section bar

Changes in moulded section thickness can have a marked effect on the mechanical properties of fibre-reinforced plastics components. The detrimental effects of changes in moulded section thickness may be substantially avoided by the use of SCORIM, and illustrated here by reference to a test bar[18] that contained a change of section from 10 mm to 20 mm at the centre. Figure 5.13 is a diagram of the test bar, including the gating arrangement used for the production of 30 wt% short glass fibre-reinforced polypropylene bars. Three processing methods were used to mould bars, which were subsequently used for characterization of porosity and fibre orientation by contact X-ray microradiography according to the section illustrated in Fig. 5.13(b). Radiographs were obtained[18] from representative mouldings produced by conventional injection moulding with single gate fill, symmetric operation of SCORIM pistons with

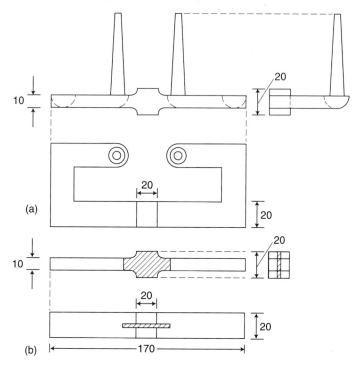

*5.13* The mould geometry used to investigate the influence of SCORIM on the microstructure of a variable thickness moulding.

single gate fill (oscillating packing pressure), and asymmetric operation of SCORIM pistons with single gate fill. The macroporosity that was present in the thick central section of the conventional moulding was removed by the use of an oscillating packing pressure. The use of the asymmetric operation of the SCORIM pistons during solidification results in a reduction of the transversely oriented fibres present within the thicker section and, as with the use of an oscillating packing pressure, leads to the elimination of porosity.

Tensile tests were made on the bars after the 20 mm section was machined off to match the remaining 10 mm thick sectioned bar. The respective tensile strength of void-free bars produced by oscillating packing pressure was 43 MPa and of the SCORIM bars with the preferred fibre orientation parallel to the injection direction was 61 MPa. This example illustrates the useful role of SCORIM in the elimination of porosity in thick-section moulding and in influencing the preferred orientation of reinforcing fibres with substantial enhancement of mechanical properties. It also indicates the relative importance of porosity and fibre orientation in determining the tensile strength of moulded composites.

## 5.4.2  Moulding and characterization of a thick-section moulding representing an actuation support structure

An evaluation of the oscillating packing method for the production of void-free thick-sectioned injection moulded components has been reported.[19] The report refers to a short glass fibre-reinforced thermoplastics moulding of a block representing the main features of an actuation support structure. The study materials were 30 wt.% glass fibre-reinforced polyethersulphone and poly-ethermide, respectively. Figure 5.14 and 5.15 are schematic diagrams of an actuation support structure and a simplified structure used for moulding and characterization work, respectively. Figure 5.16 provides detail of the gating finally adopted for the reported moulding trials, which comprised:

- mouldings produced using a static packing pressure as in conventional moulding
- mouldings produced using the oscillating packing pressure technique.

A single injection sprue located in the centre of the base together with a diaphragm gate was used to produce the mouldings. The presence of internal weldlines, at positions indicated in Fig. 5.15, could not be avoided by altering sprue and gate geometries. To enable an evaluation of the oscillating packing pressure method to be made, a series of comparative analyses were carried out with the two types of mouldings as follows:

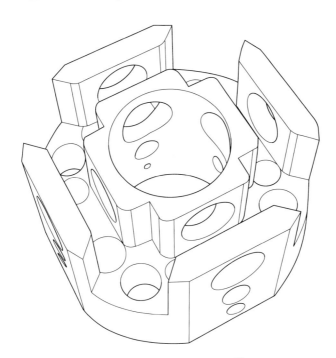

*5.14*  An actuation system support structure.[19]

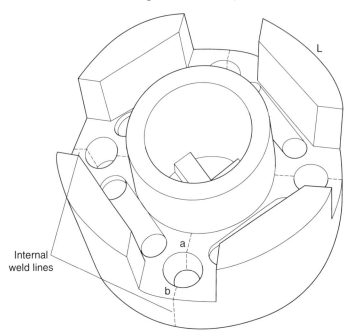

*5.15* The simplified structure used for moulding and characterization work. The maximum diameter of the block is approximately 151 mm.[19]

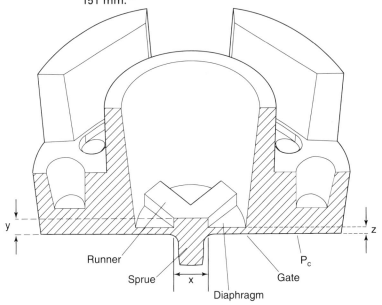

*5.16* Detail of the gating finally adopted for moulding trials, where $X = 25$ mm, $Y = 8.75$ mm, $Z = 3$ mm, $P_c$ identifies the location of the cavity pressure transducer for monitoring the response in the mould cavity to the application of an oscillating packing pressure.[19]

- Mass analysis – to quantify the amount of material present in the mouldings.
- Dimensional analysis – to investigate shrinkage of the mouldings.
- Sectional analysis – to examine the mouldings for internal voids.
- Strength analysis – to indicate the variation in material strengths obtained throughout the volume of a thick-sectioned structure and of the effects of using the oscillating pressure technique.

The mass of the moulding in glass fibre reinforced polyethersulphone (PES) was about 1.3 kg, and several times the conventional shot capacity of the moulding machine used for the trials. This meant that the mould had to be filled by flow moulding. The injection moulding machine was effectively used as an extruder for the initial stage of mould filling. A relatively poor surface finish was exhibited by the mouldings resulting from this practice, because the time for filling the mould cavity was between 2 min and 3 min. The poor surface finish would be alleviated by the use of an injection moulding machine with shot capacity to match the mould.

### 5.4.2.1  Moulding procedures

Two methods were used to pack the mould cavity after initial filling. For the first method, a static packing pressure was applied to the plastic in the mould cavity and the part was removed when the recorded cavity pressure dropped to zero. For the second method, an oscillating packing pressure was applied via an auxiliary mould packing device. The two methods are subsequently referred to as static and dynamic packing, respectively. Figure 5.17 shows the four stages in the flow moulding cycle based on the use of a mould packing device.

1  Screw rotation without movement along the barrel causing melt to be extruded into the mould cavity,
2  Mould cavity becomes full and the screw begins to move down the barrel against the screw back-pressure,
3  When the flow time expires, the screw stops rotating and an oscillating packing pressure is applied to the melt in the mould cavity by oscillation of the piston in the mould packing device in response to computer control. This piston otherwise stays in the up position, unless it is used as a shut-off valve,
4  When the packing time expires, the screw rotates and charges the barrel up to the position shown in Fig. 5.17(i). After a short cooling time, the moulding is ejected.

The packing time was determined by monitoring the cavity pressure $P_c$. When the oscillations in the cavity were seen to decay, it meant that either the cavity section or the sprue section had frozen and the cooling period has started. Two typical moulding cycles corresponding to different packing pressures, as recorded by the cavity pressure transducer $P_c$, are shown in Fig. 5.18. The packing time can be extended by increasing the oscillating piston pressure which

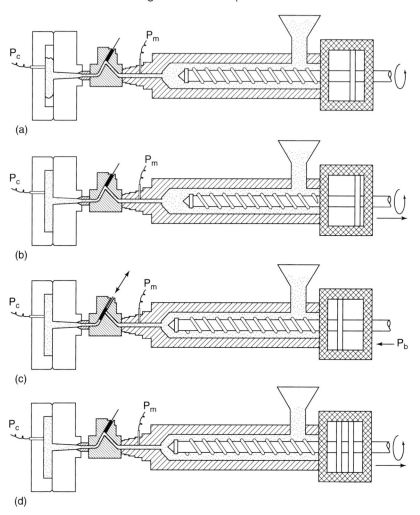

*5.17* Schematic representation of 4 stages in the flow moulding cycle based on the use of the mould packing device.[19]

caused an increase in the cavity pressure at the end of the packing stage. This can cause the part to be overpacked so care must be taken to avoid this.

### 5.4.2.2 *Characterization of mouldings*

Mouldings were produced using a range of packing pressures. The resultant mouldings were weighed and then examined by X-ray radiography as well as visual examination of sections for evidence of porosity. The optimum packing pressure, which gave mouldings free of porosity with minimum cycle time, was identified. Measurement of the dimensions were carried out on mouldings

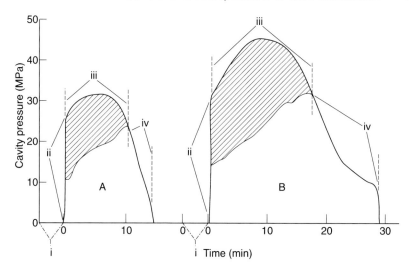

5.18 Typical mould cavity pressure recordings during the cycle illustrated in Fig. 5.17. The pressure responses during the 4 phases shown in Fig. 5.17 have been indicated.[19]

produced by dynamic and static packing, respectively. The mouldings selected for examination were categorized according to recorded packing profile, as identified from cavity pressure traces. To give some indication of the material strengths obtained, a simple strength analysis was carried out on selected mouldings. This consisted of a series of tensile tests. Three regions of the moulding were selected for tensile testing as follows:

1   The lugs that would be used to support loaded shafts,
2   The thickest region of the block, between each pair of 14 mm diameter holes,
3   The region around the 18 mm diameter holes.

A set of 32 tensile specimens according to ASTM-D638 and designated:

- Top lug longitudinal,
- Bottom lug transverse,
- Outer thick-section transverse,
- Top lug transverse, and
- Inner thick-section transverse

and covering the three regions referred to above, were taken from each moulding selected for testing. The influence of dynamic packing pressure on packing time and, principally, the extension of packing time resulting from an increase in oscillating packing pressure is illustrated in Fig. 5.18 for two (polyetherimide) (PEI) mouldings. Figure 5.19 is a record of the packing times for individual

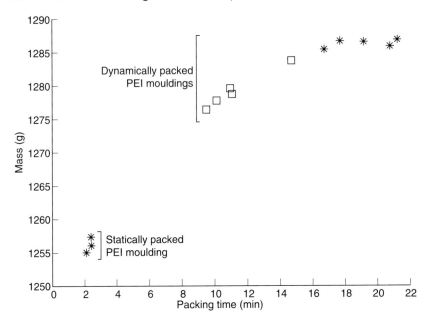

5.19 Record of packing times and corresponding masses of PEI moulding – mouldings selected for mechanical analysis *, mouldings dynamically packed for intermediate time ☐.[19]

mouldings and their masses. The mouldings packed with static pressure have the lowest recorded masses (about 1257 g) and possess substantial porosity; illustrated, for example, in the radiographs shown in Fig. 5.20 (a) and (b). Dynamically packed mouldings with packing times in the range 15–22 min exhibited no detectable porosity, as illustrated in Fig. 5.20(b), and masses of about 1285 g. Mouldings with an intermediate packing time of 9–11 min contained microporosity and has a mass of 1277 g. The mass and dimensions of the glass-reinforced thermoplastics mouldings were shown to be dependent on the duration of dynamic packing. The packing pressure profile is controlled by computer which ensures reproducible dimensions and mass provided that reproducibility of melt temperature and the functioning of the hydraulic valves are maintained.

Overall, the reported tensile strengths and moduli showed that manufacturers' data based on the testing of thin-section mouldings is not realised in practice in thick-sectioned mouldings. This result was attributed to the relatively poor preferred fibre orientation occurring in thick-sectioned mouldings. Porosity, when coupled with the transverse orientation of fibres, was the major cause of weakness. Porosity can be eliminated by the application of dynamic packing. However, the presence of transverse orientation of fibres in the absence of porosity results in tensile strengths of 55% of manufacturers' data for PEI. The

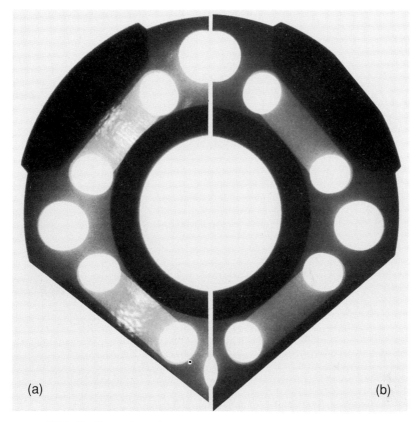

(a)                                                              (b)

*5.20* Radiographs of one half the cross-section taken through the thickness of PEI mouldings: a) statically packed; b) dynamically packed are representative of a 15 min packing time.[19]

most favourable conditions within the support structure, in terms of mould filling geometry and wall thickness, result in tensile strengths equalling 82% of manufacturers' data for PEI. The tensile strengths attainable, in the longitudinal top lug measurements on glass-fibre reinforced PEI, closely approached those required for the intended application. The adoption of SCORIM technology for the production of the actuation support structure, utilizing strategically placed live feeds, would result in significantly higher tensile strength and modulus in measurements on longitudinal top lug test pieces and would also provide for the more effective elimination of porosity.

## 5.5    Control of fibre orientation in a selection of mould geometries

One embodiment of a four live-feed arrangement[20] utilizes an injection moulding machine fitted with twin conventional injection units, each providing

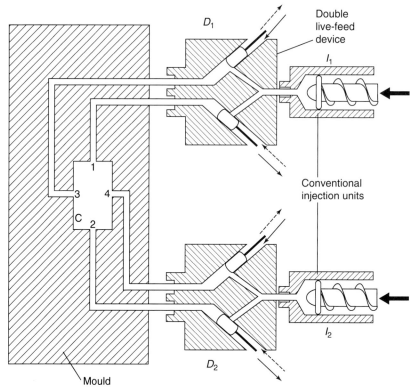

**5.21** A SCORIM four live-feed arrangement.[20]

melt for a double live-feed device. The schematic diagram shown in Fig. 5.21, illustrates a four live-feed arrangement based on twin injection units $I_1$ and $I_2$ and two double live-feed devices $D_1$ and $D_2$. On completion of mould filling of the mould cavity (C) through one or more of the gates (1)–(4) as required, and in place of the conventional mould packing stage, the pistons 1–4 are caused to oscillate according to a specific sequence and thereby influence fibre orientation during solidification of the melt.

One four live-feed arrangement in the Wolfson Centre for Materials Processing at Brunel University is based on a Negri Boss 130-90-90 two colour injection moulding machine. The two double live-feed devices were designed, constructed and integrated with the NB 130-90-90 injection moulding machine. Subsequently a four live-feed system was installed on a DEMAG 60 single barrelled injection moulding machine, with complete integration of the SCORIM process control and hydraulic actuation into the machine's microprocessor control and hydraulic supply. Two mould geometries are used to illustrate the benefits that result from the application of more than two live-feeds.

## 5.5.1  Injection moulded plaque

Figure 5.22 is a photograph of a 6 mm thick, 85 × 85 mm plaque moulding, featuring four independently controlled fan gates 1–4. The out-of-phase operation (mode A) of the live feeds at gates 1 and 2 after mould filling tends to cause alignment of the fibres to form with a preferred orientation in the direction shown in Figure 5.22. A multilayer laminated structure may be formed through the thickness of the plaque according to the sequencing and duration of the operation of pairs of live feeds, as the melt solidifies from the outside to the centre of the plaque cavity. Figure 5.22d indicates the positions used for contact microradiography on: 1) single-gated mouldings packed with a static pressure and four-gated mouldings packed using the four live-feed arrangement according to Fig. 5.22(b) and (c). Figure 5.22(e) is a microradiograph corresponding to the moulding produced by four live feeds, respectively. The latter clearly shows the layered structures exhibiting different preferred orientations (X and Y) of fibres, produced by alternate operation of live feeds 1–2 and 3–4, and concluding with prolonged operation of live feeds 3–4, to give the uniaxially aligned fibres in the centre of the moulding.

## 5.5.2  Injection moulded fin

Figure 5.23 is a photograph of an injection moulded fin tapering in thickness from 5–2 mm. In the example selected for presentation the initial injection of 30% glass fibre-reinforced polyarylamide was through gate 4 as illustrated in Figure 5.23. This resulted in a preferred orientation of fibres parallel to 1–4 in the surfaces of the fin. On completion of cavity filling, the live feeds 3 and 2 were caused to operate in mode A to produce transverse alignment parallel to 3–2 through the thickness of the moulding, as illustrated in Fig 5.23. The extent of the transversely orientated fibres through the thickness of the moulding was determined by the mould temperature and the start time for the operation of live feeds 3 and 2. Typically, the application of SCORIM resulted in a 22% increase in flexural stiffness parallel to 3–2 when compared with conventionally produced mouldings. This corresponded with a 19% decrease in flexural stiffness in tests carried out parallel to gates 1–4.

## 5.5.3  Multicavity SCORIM for the production of moulded rings

SCORIM may be applied to mouldings produced in multicavity moulds.[21] One example of a multicavity arrangement for SCORIM is illustrated in Fig. 5.24, and was used to produce glass-reinforced polypropylene rings. The mould design also featured the use of a hot runner system, and as a whole the design provided for completely automatic production of rings with a substantially enhanced internal weldline strength. The dependence of the flexural load

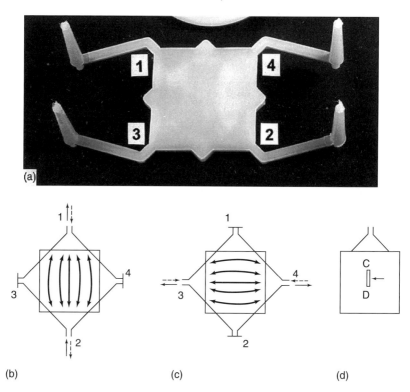

(a)

(b)          (c)          (d)

X EDGE

y

X

y
X

5.22 (a) The 85 mm square moulding showing four independently
controlled gates; (b) operation of the live feeds at gates 1 and 2
tends to cause alignment in the direction indicated; (c) operation of
the live feeds at gates 3 and 4 tends to cause alignment in the
direction indicated; (d) the orientation of X-ray microradiograph in
Part (3); (e) the action of SCORIM four live-feed moulding.[20]

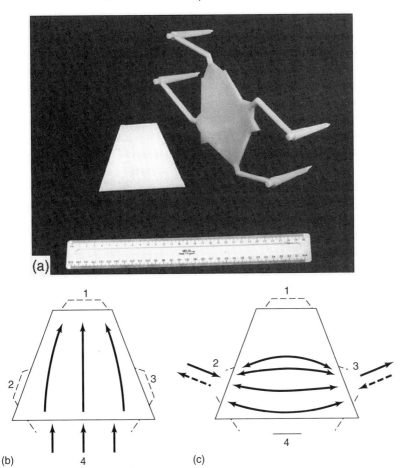

*5.23* (a) Tapered fin mouldings, with and without the runners associated with the four live-feeds; (b) initial mould filling through gate 4; (c) the operation of live feed 3 and 4 to produce transverse orientation of fibres through the thickness of the fin.[20]

recorded at break, the process conditions and cycle times are summarized in Table 5.3. A procedure for using four live-feed SCORIM for the production of moulded rings has been described,[20] and provided for the elimination of internal weldlines and production of circumferential alignment of reinforcing fibres through the section thickness of thick-sectioned mouldings.

## 5.6    Extensions of the shear controlled orientation concept

From the above examples, the one that best illustrates the benefit of control of preferred orientation in a specified region of a moulding, is the tapered fin. More

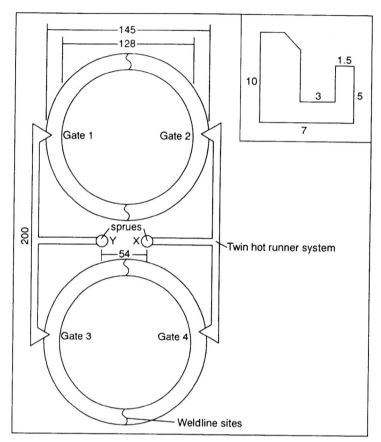

*5.24*  Two cavity hot runner tool used for the production of rings using two live-feed SCORIM. The cross-section is shown in the inset.

generally the removal of flow defects, such as internal weldlines, may be effected by the incorporation of SCORIM pistons within the mould cavity. The schematic diagram of Fig. 5.25 illustrates an arrangement that provides for local modification of microstructure and physical properties, and a mould that is self-contained in terms of SCORIM, allowing its use on any single barrelled moulding machine with sufficient dimensions and clamp capacity to accommodate the mould. In general it may not be necessary to control the orientation of fibres throughout the volume of a moulded part. SCORIM may provide for control in specified regions of a moulding and an in-mould SCORIM offers a cost effective and elegant solution to enhancing locally deficient properties.

The SCORIM concept also provides for the *in situ* formation and management of fibres in polymer matrices, illustrated in Fig. 5.26 by the formation and purposefully controlled orientation of thermotropic liquid crystal polymer fibres in a polypropylene matrix. The copolyester amide liquid crystal polymer (Vectra B -Hoechst Celanese) and the polypropylene copolymer were blended by co-

Table 5.3. Flexural load at break of glass fibre-reinforced polypropylene mouldings produced by conventional and SCORIM moulding

Weldline: flexural load at break

| Moulding | CONVENTIONAL | | | SCORIM | | | |
|---|---|---|---|---|---|---|---|
| Process conditions | Cavity pressure (bar) | | | Total hold time (s) | Piston frequency (Hz) | | Piston cycles per moulding |
| | 55 | 100 | 260 | 8/0.5/4 | 4/0.5/2 | 2/1/2 | 1/1/1 |
| Flexural load at break Cavity 1 (kN) | All mouldings broke on ejection from tool | | | 710 | 693 | 725 | 795 |
| Cavity 2 (kN) | | | | 793 | 813 | 788 | 760 |
| Total cycle time(s): | 21.3 | 21.3 | 21.3 | 25.3 | 21.3 | 19.5 | 18.7 |

Material: 35% 10 mm glass fibre in polypropylene.

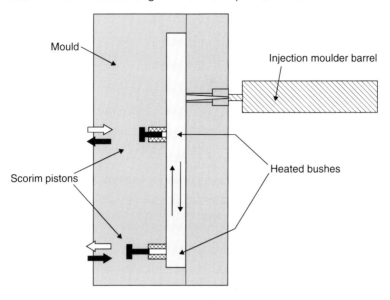

Mould

Injection moulder barrel

Scorim pistons

Heated bushes

*5.25*   One embodiment of an in-mould SCORIM.

rotating twin-screw extrusion, and the blend injected into the mould cavity where SCORIM was applied to form the fibres and determine their orientation.[22] A compounding injection moulding machine[23] according to the schematic diagram in Fig. 5.27, and manufactured by Dassett Process Engineering, UK, is proposed for this purpose. The machine which is also installed in the Wolfson Centre for Materials Processing, was designed to provide for the compounding of metal and ceramic matrix composites, processed with the aid of sacrificial polymer binders. With the addition of SCORIM the compounding injection machine provides for the compounding of inorganic matrix composites and the management of short fibres within a sacrificial polymer binder filled with ceramic or metal powder. The moulding may be subsequently debound and then sintered to give a ceramic- or metal-matrix composite component with management of the orientation of short fibres.[24]

The extension of the SCORIM process to provide for the control of fibres through the thickness of mouldings, from the mould surface to the central core arises from the integration of inductive heating and SCORIM. The induction of very high mould temperatures[25] enables the alignment of fibres by macroscopic shear before solidification can occur. The combination of bright surface moulding (BSM) and SCORIM also allows for the removal of visual defects in mouldings and in particular in transparent polymer matrices containing aluminium flake.

The concept of applying macroscopic shears to solidifying melts may also be utilized in the continuous extrusion of profile. The basic functions of the SCOREX extrusion die have been detailed previously.[6] These are:

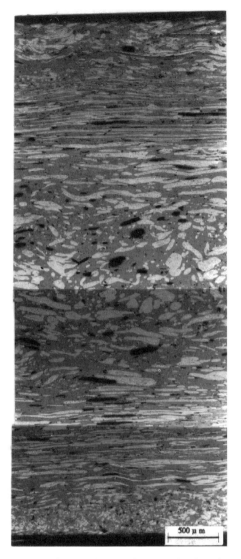

*5.26* Reflected light micrograph showing the copolyesteramide LCP fibril morphology produced through substantial part of the thickness of a moulding by the action of SCORIM.

- to pass molten material from an extruder through a one way valve into the die chamber,
- to solidify the material before exit from the die,
- to apply a macroscopic shearing action on the solidifying melt to influence the alignment of reinforcing fibres in extrusions.

The schematic diagrams in Fig. 5.28 illustrate design features of a die used for the production of the containing circumferentially aligned short fibres. The

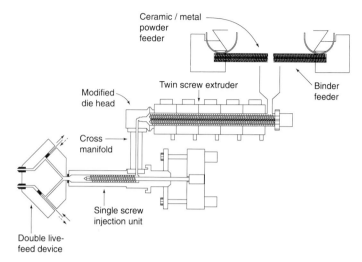

*5.27* Schematic diagram of the compounding injection moulding machine used for the compounding of ceramic and metal matrix composites, and the moulding of the composites into moulded parts with management of fibre orientation using SCORIM.

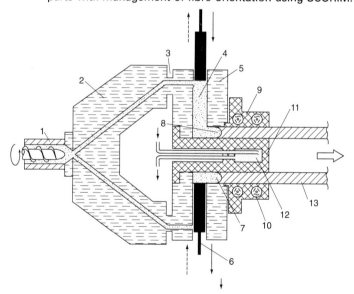

*5.28* The centre section through the SCOREX die assembly used for the production of tube. The features indicated are as follows: (1) extruder (supply of molten material); (2) heated adapter block; (3) connection bushes; (4) piston chamber (one of four); (5) heated piston manifold block; (6) piston (one of four); (7) molten material being subjected to the shearing action of the pistons; (8) melt–solid interface (the position of this defines the cross-sectional areas of (7)); (9) cooling rings (external cooling); (10) circular channels for cooling fluid; (11) cooled die core assembly; (12) cooling chamber for core; (13) solid extrudate.[6]

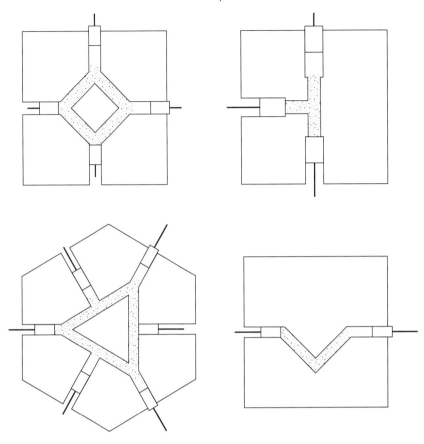

*5.29*  The live-feed arrangements used to produce preferred orientation
       in a selection of profiles.

schematic diagrams in Fig. 5.29 illustrate arrangements for managing the
orientation of short fibres in a selection of profiles, with emphasis on the
production of transverse orientation. The sequencing of the operation of pistons
is selected to produce the desired level of preferred orientation, and may be
continuous or intermittent. When applied to a circular die,[6] SCOREX can
produce a uniform structure in both the hoop and axial directions, resulting in
substantial enhancement of the hoop properties of the pipe arising from the
controlled circumferential orientation of fibres. The process also provides for
excellent control of dimensions and surface finish. As with SCORIM, the
SCOREX process may in principle be applied to a wide range of melt
processable materials, for the purpose of optimizing microstructure and physical
properties, and additionally, for the purpose of producing curvilinear profiles[26]
and for enhancing the visual appearance of moulded parts.

## References

1. R.A. Chivers, D.R. Moore and P.E. Morton, *Plastics Rubber Compos. Process. Applic.*, 1991, **15**, 145.
2. M.J. Folkes, in *Short Fiber Reinforced Thermoplastics*, Research Studies Press, John Wiley and Sons, Chichester, 1982.
3. C.L. Tucker and S.G. Advani, in *Flow and Rheology in Polymer Composites Manufacturing*, ed. S.G. Advani, Elsevier, London, 1994.
4. P.S. Allan and M.J. Bevis, *Plastics Rubber Process. Applic.*, 1987, **7**, 3
5. P.S. Allan and M.J. Bevis, *UK Patent* 2170-140-B.
6. P.S. Allan and M.J. Bevis, *Plastics Rubber Process. Applic.*, 1991, **16**, 133.
7. H. Becker and L.M. Gutjahr, *Kunststoffe*, 1993, **79**, 1107.
8. H. Becker, G. Fischer and U. Muller, *Kunststoffe*, 1993, **83**, 165.
9. M.J. Bevis, P.S. Allan and P.C. Emeanuwa, *Proceedings 2nd Institute Mechanical Engineers Conference on Fibre Reinforced Composites*, Liverpool, 1986, 95–98.
10. H. Hamada, Z. Maekawa, T. Horrino and K. Lee, *Internatl. Polym. Proc.*, 1988, **2**, 131.
11. S. Piccarolo, F. Scargiali, G. Crippa and G. Titomalia, *Plastics Rubber Compos. Process. Applic.*, 1993, **19**, 205.
12. L. Wang, P.S. Allan and M.J. Bevis, *Plastics Rubber Compos. Process Applic.*, 1995, **23**, 139.
13. J.R. Gibson, P.S. Allan and M.J. Bevis, *Compos. Manuf.*, 1990, **1**, 183.
14. P.J. Hine, R.A. Duckett, I.M. Ward, P.S. Allan and M.J. Bevis, *Polym. Compos.*, 1996, **17**, 400.
15. P.J. Hine and I.M. Ward, *J. Mater. Sci.*, 1996, **31**, 371.
16. L.Wang, P.S. Allan and M.J. Bevis, *Plastics Rubber Process. Applic.*, 1996, **25**, 385.
17. P.S. Allan and M.J. Bevis, *Plastics Rubber Process Applic.*, 1983, **3**, 85 and 331.
18. P.S. Allan, M.J. Bevis and A. Zadhoush, *Iran J. Polym. Sci. Technol.*, 1995, **4**, 50.
19. P.S. Allan, M.J. Bevis, S.T. Hardwick, M.W. Murphy and D.J. Sculley, *Plastics Rubber Process. Applic.*, 1990, **13**, 15.
20. P.S. Allan and M.J. Bevis, *Compos. Manuf.*, 1990, **1**, 79.
21. J.R. Gibson, P.S. Allan and M.J. Bevis, *Plastics Rubber Internat.*, 1991, **16**, 12.
22. H. Liang, P.S. Allan and M.J. Bevis (in preparation).
23. P.S. Allan, M.J. Bevis, B.A. McCalla, J.R. Gibson, P.R. Hornsby, W.H. Lee and K. Tarverdi (in preparation).
24. I E Pinwill, F Ahmad, P.S. Allan and M.J. Bevis, *Powder Metallurgy*, 1992, **35**, 107.
25. P.S. Allan, M.J. Bevis and K Yasuda. *UK Patent Application*, 1996, 2299779.
26. P.S. Allan and M.J. Bevis *UK Patent* GB 2237 237B.

# 6

## Theory and simulation of shear flow-induced microstructure in liquid crystalline polymers

ALEJANDRO D REY

## 6.1 Introduction

The synthesis of liquid crystal polymers (LCPs) has enlarged the range of applications of polymers to areas where superior mechanical performance is required, as typified by the diverse uses of Kevlar fibers. Other important non-mechanical applications of LCPs are in electrooptical devices, where side chain LCPs are being used. Further uses of LCPs as membrane and barrier materials are in rapid development. A number of reviews, textbooks and monographs on the theory and applications of liquid crystalline polymers are available in the literature.[1-8]

The mechanical applications of LCPs rely on the precise control of orientation of the rigid anisotropic molecules that form a stable liquid crystalline phase. Therefore a fundamental understanding of flow-induced orientation will enhance the range of processing routes and will allow for the design and control of new structures.

As in other material systems, modeling is an integral component in the development, understanding, optimization and control of processing, fabrication and uses of liquid crystalline materials. The macroscopic modeling of these materials takes into account the internal structure, as defined by partial positional and orientational order of these phases. The consequence of an internal structure in the modeling process is the requirement of adding new balance equations to those that govern structureless fluids. For example, for a uniaxial nematic liquid crystal an internal momentum balance equation is required to describe the average macroscopic orientation of the liquid. Constitutive equations, required to specify relations between forces and fluxes, must reflect the symmetry properties of the phases.[7] These requirements give rise to a variety of theories applicable to different liquid crystalline phases.

This chapter only considers the nematic liquid crystalline phase of rod-like molecules, characterized by positional disorder and uniaxial orientational order. For the case of rigid elongated molecules, the average orientation, described by a unit vector, the director $\mathbf{n}$, is along the long molecular axis. The presence of order admits the possibility of defects, from which the name nematic comes. The imperfect molecular alignment along **n** is described by the scalar order parameter, which is affected by strong flows in LCPs. A successful theory that describes the macroscopic behavior of nematic phases is the Leslie–Ericksen (L–E) vector theory,[1,4,7] applicable in the absence of spatio-temporal variations of the scalar order parameter. For strong flows, tensorial theories seem to be necessary to capture a wide range of unusual rheological behavior, currently believed to be driven by the coupling of molecular alignment and macroscopic orientation.

The objective of this chapter is to present a series of modeling applications of vector and tensor theories whose predictions are in at least good agreement with experimental data. Emphasis is placed on recent results, where theory and simulation are able to reproduce and explain significant features of flow-induced orientation in nematic LCPs.

## 6.2     Shear flow-induced orientation phenomena in rod-like nematic polymers

### 6.2.1   Leslie–Ericksen continuum theory of nematics

The continuum theory of flow phenomena in uniaxial nematic liquids was given its present form by Leslie.[4] The anisotropic structure is described by the director **n**, which represents the average molecular orientation and defines the axis of cylindrical symmetry. The classical theories of fluids describe the kinematics by specifying only the velocity field, while for a fluid with microstructure the additional specification of the director field is required. This explicit account of the microstructure gives rise to the internal angular momentum balance equation. The L–E balance equations, using Cartesian tensor notation, are:[4]

$$v_{i,i} = 0; \quad \rho \dot{v}_i = F_i + \hat{t}_{ji,j} - p_{,j}\delta_{ji}; \quad 0 = \Gamma_i^e + \Gamma_i^v \qquad \text{[1a, b, c]}$$

where $v$ is the velocity vector and $\delta$ is the unit tensor.

The fluid is assumed to be incompressible; $\rho$ is the density and $p$ is the pressure. The superposed dot denotes the material time derivative. The inertia of the director is neglected. The mechanical quantities appearing in the theory are defined as follows: $F_i$ = external body force per unit volume, $\hat{t}_{ji}$ = extra-stress tensor, $\Gamma_i^e$ = elastic torque per unit volume and $\Gamma_i^v$ = viscous torque per unit volume. The constitutive equations are :

$$\hat{t} = \tilde{t} + t^e \qquad \text{[2a]}$$

$$\begin{aligned}\tilde{t} = &\alpha_1(\mathbf{n\,n} : A)\mathbf{n\,n} + \alpha_2\,\mathbf{n\,N} + \alpha_3\,\mathbf{N\,n} \\ &+ \alpha_4\,\mathbf{A} + \alpha_5\mathbf{nn}\cdot\mathbf{A} + \alpha_6\,\mathbf{A}\cdot\mathbf{nn}\end{aligned} \qquad \text{[2b]}$$

$$\mathbf{t}^e = \frac{\partial F}{\partial \mathbf{n}} \cdot (\nabla \mathbf{n})T \tag{2c}$$

$$\Gamma^e = \mathbf{n} \times \mathbf{h} = -\mathbf{n} \times \frac{\partial F}{\partial \mathbf{n}} \tag{2d}$$

$$\Gamma^v = -\mathbf{n} \times (\gamma_1 N + \gamma_2 \, A \cdot \mathbf{n}) \tag{2e}$$

where

$$\gamma_1 = \alpha_3 - \alpha_2 \tag{3a}$$

$$\gamma_2 = \alpha_6 - \alpha_5 = \alpha_3 + \alpha_2. \tag{3b}$$

The $\alpha_i = 1, \ldots, 6$ are the Leslie coefficients. $\mathbf{h}$ is the molecular field and $\delta/\delta\mathbf{n}$ denotes the functional derivative. The kinematic quantities appearing in the constitutive equations are:

$$N_i = \dot{n}_i - \Omega_{ik} n_k \tag{4a}$$

$$\Omega_{ik} = (v_{i,k} - v_{k,i})/2 \tag{4b}$$

$$A_{ik} = (v_{i,k} + v_{k,i})/2 \tag{4c}$$

$N_i$ is the angular velocity of the director with respect to that of the fluid, $\Omega_{ik}$ is the vorticity tensor and $A_{ik}$ is the rate of deformation tensor. The free energy density, $F$, is given by:[4]

$$2F = K_{11}(\nabla \cdot \mathbf{n})^2 + K_{22}(\mathbf{n} \cdot \nabla \times \mathbf{n})^2 + K_{33}|\mathbf{n} \times \nabla \times \mathbf{n}|^2 \tag{5}$$

where $K_{11}$, $K_{22}$, $K_{33}$ are the splay, twist and bend elastic constants.

Table 6.1 gives values of the nine parameters of the L–E theory for an extensively studied nematic rod-like LCP, poly-$\gamma$-benzylglutamate (PBG).[9] An important factor in flow-induced orientation in LCPs is the anisotropies of the viscoelastic material parameters. For PBG the elastic anisotropies are: $K_{11}/K_{22}$ = 15.5, and $K_{11}/K_{33} = 1.58$. Representative measures of viscous anisotropies are given by the Miesowicz shear viscosities, defined by:[4]

$$\eta_a = \tfrac{1}{2}(\alpha_4); \quad \eta_b = \tfrac{1}{2}(\alpha_3 + \alpha_4 + \alpha_6); \quad \eta_c = \tfrac{1}{2}(-\alpha_2 + \alpha_4 + \alpha_5)$$
$$\text{[6a, b, c]}$$

where the director is oriented along the vorticity axis, the velocity gradient and the flow direction, respectively. For PBG the viscous anisotropies are: $\eta_c/\eta_b$ = 187.3 and $\eta_c/\eta_a = 39.83$. It is interesting to note that viscosity reduction is not, in most cases, the operating mechanism in orientation selection. For example, a director orientation along the vorticity axis has the lowest shear viscosity possible but is unstable.

Table 6.1. Physical constants
for PBG

| | |
|---|---|
| $\alpha_1$ (Pa s) | $-3.66$ |
| $\alpha_2$ | $-6.92$ |
| $\alpha_3$ | $0.018$ |
| $\alpha_4$ | $0.348$ |
| $\alpha_5$ | $6.61$ |
| $\alpha_6$ | $-0.292$ |
| $K_{11}$ ($10^{-11}$ N) | $12.1$ |
| $K_{22}$ | $0.78$ |
| $K_{33}$ | $7.63$ |

The main parameters that control flow-induced orientation are the Ericksen number $E$ and the reactive parameter $\lambda$:[4]

$$E(T) = \frac{\gamma_1 h U}{(K_{11} K_{22} K_{33})^{1/3}}; \quad \lambda(T) = -\frac{\gamma_2}{\gamma_1} \qquad \text{[7a, b]}$$

where $h$ is a characteristic sample length, $U$ a characteristic velocity and $T$ the temperature. The temperature dependence arises through the temperature variations of the viscoelastic coefficients. The Ericksen number is the ratio of viscous torques to elastic torques and $\lambda$ is the ratio of strain to vorticity torque effects.

A fundamental rheological property of nematics is whether or not they orient in the direction of flow during shearing deformations.[4,7] It is currently agreed that the majority of nematic polymers do not orient in the flow direction at low shear rates. In the L–E theory the shear flow-orienting behavior is captured by the magnitude of $\lambda$, as defined above. Neglecting surface effects, if $\lambda > 1$, the flow orienting effects of strain dominate over the tumbling effects of vorticity, and the director orients in the shear plane and close to the flow direction. On the other hand, if $0 < \lambda < 1$, the tumbling effects of vorticity dominate over the orienting effects of strain, and the director does not orient in the shear plane or close to the flow direction. In the latter case, there is no stable orientation at which the flow torques vanish and stable steady states appear through complex spatially non-homogeneous orientation patterns.

Since surface conditions affect the orientation through transmission of elastic torques, the Ericksen number is another important parameter that controls the orientation. Surface orientation effects can be controlled by physicochemical treatments. If the fixed director orientation is normal (parallel) to the surfaces the anchoring condition is called homeotropic (planar).[4] At sufficiently low $E$ the orientation dictated by the surface orientation is weakly perturbed, but at higher $E$ the orientation is dictated by a balance between the elastic and viscous torques.[5] Increases in $E$ above critical thresholds often lead to flow-induced orientation instabilities, expressed as subcritical or supercritical bifurcations.[10,11] In the former case multistability between different orientation modes sets in, and in the latter exchange of stability is present. The appearance and loss of stability signals the exchange of orientation modes that control the classical rheological material functions such as apparent viscosity and normal stress differences.[10–15] The next section deals with the main orientation modes in simple shear flow.

## 6.2.2  Orientation modes and multistability

This section presents a summary of flow-induced orientation phenomena in PBG subjected to isothermal rectilinear simple shear start-up flow, as predicted by the L–E theory.[13] The solution vector is: $\mathbf{v} = (v_x, 0, v_z)$ and $\mathbf{n} = (n_x, n_y, n_z)$. The geometry is defined by two parallel plates, with the upper plate located at $y = h$ moving in the $+x$ direction with a velocity $U$ and the lower fixed plate located at $y = 0$. The $x$–$y$ plane is the shear plane and $z$ is collinear with the vorticity axis. The independent variables are $(x,t)$: $\mathbf{v} = \mathbf{v}(y,t)$ and $\mathbf{n} = \mathbf{n}(y,t)$. Details of the numerical methods are given by Han and Rey.[13] The solutions satisfy the following initial conditions:

$$t = 0, 0 \leq y \leq h, \mathbf{v} = 0, \mathbf{n} = (1 + \varepsilon_x, \varepsilon_y, \varepsilon_z)/((1 + \varepsilon_x^2) + \varepsilon_y^2 + \varepsilon_z^2)^{1/2} \quad [8]$$

where the $\{\varepsilon_i\}$ are random perturbations with a maximum magnitude of 0.01 rad. The initial director state is nearly oriented along the flow direction. The boundary conditions are:

$$t > 0, \ y = 0 \text{ and } y = \text{h}, \mathbf{v} = 0, \mathbf{n} = (1, 0, 0) \qquad [9]$$

Figure 6.1 presents a definition of the terminology, symbols and the main orientation characteristics of the seven stationary solutions to the L-E equations. The seven solution branches are classified according to their orientation dimensions (planar or non-planar), their chirality (left twist or right twist), their rotation number $\Lambda$ (number of turns the director performs in going from the bottom plate to the top plate), their orientation with respect to the shear plane (in-plane or out-of-plane) and the magnitude of the out-of-plane twist component (small twist or large twist). The four achiral (i.e. no net twist, $\Lambda = 0$) out-of-plane OPN± and OPS± solutions may have one way left ($-$) or right ($+$) twists. The two chiral out-of-plane OPC± solutions may be left ($-$) or right ($+$) handed. The chiral solutions have rotation numbers $\Lambda = \pm 1$ and the director rotates by $2\pi$ radians when going from the bottom surface to the top surface. Here we only show the $+$ solution branches. A visualization of the four (IP, OPN+, OPS+, OPC+) stationary solutions is shown in the last row of Figure 6.1(a). We wish to emphasize that for the OPN, OPS and OPC branches, the $+$ and $-$ pair are dissipatively equivalent and each is equally likely to arise, thus predicting banded patterns consisting of alternating $+$ and $-$ bands,

| Orientation dimension | In-shear-plane 2-D orientation | 3-D | 3-D | 3-D |
|---|---|---|---|---|
| Chirality | Achiral | Achiral | Achiral | Chiral |
| Rotational number (A) | 0 | 0 | 0 | (+/−)1 |
| Orientation structure | IP, In-plane | OPN(+/−), twistable | OPS(+/−), supertwistable | OPC(+/−), cholesteric |

(a)

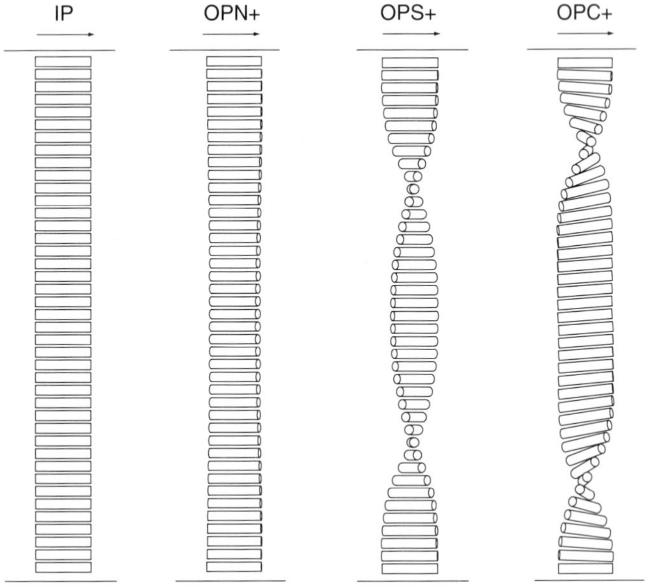

IP        OPN+        OPS+        OPC+

(b)

*6.1* (a) Terminology, symbols and orientation characteristics of solutions to the L–E equations. (b) Three-dimensional visualization of stable steady state orientation profile.

typical of sheared LCPs. Next some representative textural evolutions are presented.

Figure 6.2 shows a three-dimensional visualization of the director profile for the OPS+ solution for several strains (strain $= Ut/h$). The initial director change involves in-plane tilting, while at later times simultaneous tilting and twisting in the centerline region results in a twisted OPS+ orientation state. At this shear rate the timescales for tilting and twisting are similar, so that after an initial transient, escape of the director from the shear plane is concurrent with in-plane tilting. The orientation of the OPS± solutions is characterized by the absence of tilting and the presence of twisting.

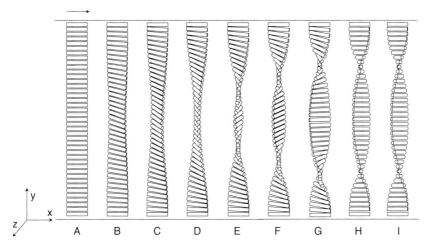

*6.2*  Three-dimensional visualization of the director profile evolution for
$E=350.2$. Each visualization corresponds to the following strains $\gamma$:
$A=0$;  $B=121.8$;  $C=126$;  $D=128$;  $E=130$;  $F=132$;  $G=138.7$;
$H=157.4$; $I=238.5$.

Figure 6.3 shows the evolution of the director as a function of strain $\gamma$ and the
scaled thickness $H=y/H$, for the OPC+ chiral solution. The figure shows that at
the centerline in-plane tilting is faster than out-of-plane twisting, which
eventually leads to a defect-free nucleation of a chiral texture having a full $2\pi$
director rotation. The explanation of the nucleation mechanism for the chiral
orientation OPC mode is as follows. At relatively low $E$, where the achiral

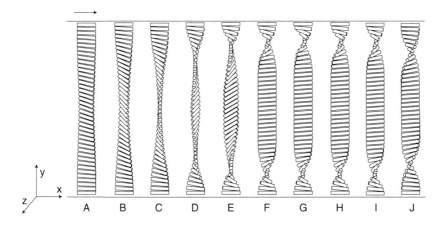

*6.3*  Three-dimensional visualization of the director profile evolution for
$E=466.9$. Each visualization corresponds to the following strains $\gamma$:
$A=72.69$; $B=76.8$; $C=78.4$; $D=80.2$; $E=82.2$; $F=87.4$; $G=92.8$;
$H=106.4$; $I=128.7$; $J=252.9$.

modes are favored, the director tilting speed is slow and any out-of-plane fluctuation has time to grow, thus leading to the one way uniformly twisted out-of-plane modes. On the other hand, at relatively higher $E$, the tilting speed at the centerline is faster than at the bounding surfaces, so that the core tumbles past the compression axis $(-\pi/4)$ without twisting, but at the two wall regions the slow tilting allows for a director escape from the shear plane.

Next we present the parametric dependence of the basins of attractions of the various stable steady state solutions. In other words, we identify which steady state (IP, OPN±, OPS±, OPC± ) is selected when the nematic sample is suddenly driven with a given value of $E$. Figure 6.4 shows the stability diagram, given by $2\Lambda + n_x(y = h/2)$ as a function of the Ericksen number, for PBG in shear flow. The parametric range of the adopted steady states are: (i) $0 < E < E_{ci} = 148$: IP mode, (ii) $E_{c1} < E < E_{c2} = 300$: OPN± modes, (iii) $E_{c2} < E < E_{c3} = 430$: OPS± modes, (iv) $E_{c3} < E$: OPC± modes. The dotted lines represent the seven solutions for values of $E$ that are not selected when starting from the initial condition, Eqn [8].

In partial summary, non-orienting nematic polymer subjected to simple shear flow may exhibit a range of steady state in-plane and out-of-plane orienting

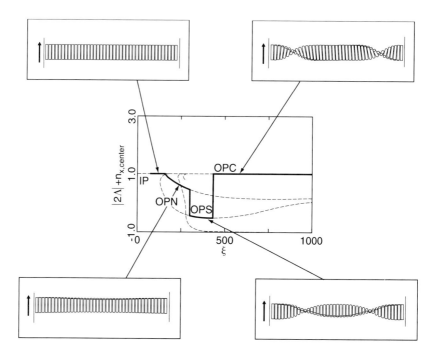

6.4   Orientation mode selection diagram for shearing flow. The bold line segments represent the selected steady state IP, OPN, OPS and OPC modes (see Fig. 6.1), and the inserts are representative visualizations for the four modes.

modes, according to the magnitude of $E$. For PBG, relatively low shear rates promote twisted textures, while relatively large shear rates promote chiral twisted textures. The surprising result of this work is the prediction of the nearly complete absence of tilt deformations, which is expected due to the lack of flow-orienting behavior ($\lambda < 1$) and the large elastic anisotropy ($K_{11} > K_{33} \gg K_{22}$) of nematic LCPs.

### 6.2.3 Pattern formation under forced convection

For over forty years it has been reported that light transmission patterns of shearing liquid crystal polymers under cross polars shows a ubiquitous banded texture[16,17] normal to the direction of flow and of periodicity that scales with the sample thickness. Experimental studies have only established that the banded texture is a result of some spatial orientation modulation, leading to a spatial variation of the effective birefringence. This section presents simulations[17] that are in remarkable agreement with experiments and that explain the origin of this as of yet unexplained ubiquitous phenomenon.[16]

To investigate the representative optical responses of non-homogeneous non-planar three-dimensional orientation during shear flow of polymeric nematics, we solve the L–E equations for simple shear start-up flow, using PBG as the model material and with the following auxiliary conditions:

$$t = 0, 0 \leq y \leq h, \; \mathbf{v} = 0, \mathbf{n} = (\varepsilon_x, \; 1 + \varepsilon_y, \varepsilon_z)/((1 + \varepsilon_y^2) + \varepsilon_x^2 + \varepsilon_z^2))^{1/2}$$

[10]

$$t > 0, \; y = 0 \text{ and } y = h, \mathbf{v} = 0, \; \mathbf{n} = (0, 1, 0)$$

[11]

where the $\{\varepsilon_i\}$ was defined above. The solution vector is given by the director field: $\mathbf{n}(x, y, \gamma) = (n_x(x, y, \gamma), \; n_y(x, y, \gamma), \; n_z(x, y, \gamma))$, where $\gamma$ is the applied strain, $x$ the flow direction, $y$ the velocity gradient direction and $z$ the vorticity direction. The only parameter is $E$. The orientation field is used to compute the light transmission between cross polars, with $y$ being the light propagation direction. When the orientation is uniform along the light propagation direction, the computation of light transmission through a nematic layer is the same as that for a uniaxial single crystal, while when the orientation variation is along the light propagation direction, a more elaborate method is required.[17] Here the light transmission simulation, light with normal incidence first, goes through the polarizer parallel to the flow direction ($x$-axis), then through the nematic polymer sample of non-homogeneous orientation and finally through the analyser that is orthogonal to the polarizer. Details of the computation are given in Han and Rey.[17]

Figure 6.5(a) shows a scientific visualization of the out-of-shear plane component ($n_z$) for a shear strain of $\gamma = 239.7$; the gray scale corresponds to the magnitude and sign of $n_z$: $n_z > 0$ and a large magnitude correspond to light regions, $n_z < 0$ and a large magnitude correspond to dark regions, and $n_z \approx 0$ (planar orientation) is shown as gray. The figure shows the presence of an array of tubular inversion walls[20] immersed in a matrix of planar ($n_z = 0$) orientation, whose axes are orthogonal to the flow direction. As mentioned above, dark and light stripes represent $n_z \approx -1$ and $n_z \approx +1$ director orientations, respectively, while a medium gray scale area represents a planar orientation matrix. The tubular orientation inversion walls divide the continuous planar orientation matrix ($n_x \approx +1$) lying outside the inversion walls, from the planar orientation matrices ($n_z \approx -1$) lying inside the tubular inversion walls.

Figure 6.5(b) shows the light transmission intensity as a function of dimensionless length $L$ along the $x$-axis under crossed polars, corresponding to the three-dimensional orientation structure for a shear strain of $\gamma = 239.7$. According to the experimental observations,[16] as the shear rate increases the wavelength decreases from infinity to a saturation value equal to the half gap thickness. At intermediate shear rates the wavelength of the banded texture is of the order of the gap thickness ($h$). The infinite wavelength occurs at a critical shear rate, or equivalently a critical Ericksen number $E_c$, corresponding to the planar-non-planar transition described above. The critical $E_c$ for periodic pattern formation is found to be very close to the critical Ericksen number for the orientation transition between non-periodic planar textures and non-periodic

## (a)

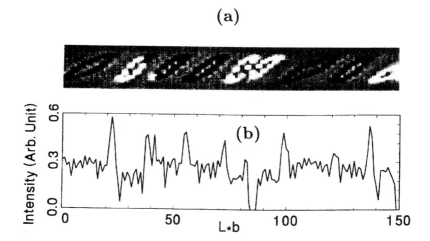

6.5  (a) Scientific visualization of the $n_z$ director component. The long and short axes of the plot represent the flow and thickness directions, respectively. Dark and light stripes represent $n_z \approx -1$ and $n_z \approx +1$ director orientations, repectively, and a mid-gray area represents a planar orientation matrix ($n_z \approx 0$). The Ericksen number is $E = 2918$ and the strain is $\gamma = 239.8$. (b) Light transmission intensity as a function of dimensionless length ($Lb$) corresponding to Fig. 6.5(a). The scale factor $b$ is 9.375.

non-planar textures. The transition from non-periodic non-planar orientation to periodic non-planar orientation arises from the coupling between competing elastic and viscous torques. Elastic effects favor a non-periodic non-planar texture since they contain less elastic distortion, while viscous effects favor the formation of a periodic non-planar texture since the associated governing Miesowicz viscosities are relatively lower.

Figure 6.5(b) shows that there are about eight peaks in the light transmission distribution, which means that the wavelength of the banded texture for the shown Ericksen number is $O(h)$, where $h$ is the gap thickness, in excellent agreement with the results of Yang and Labes.[16]

In partial summary, the present simulations of banded textures in sheared nematic polymers observed between cross polars are in excellent agreement with the experimental data.[16] In addition they provide exact information on the evolution of the underlying orientation texture that is responsible for this generic and universal LCP effect.

## 6.2.4  Non-equilibrium defect structures

This section presents theory and simulation of the effects of shear flow on optical textures of thermotropic liquid crystalline materials, as reported by Alderman and Mackley.[18] Defects are singularities or near-singularities in the director field, characterized by strength and dimension.[19] Dimensionless defects are singular points in the director field. One-dimensional defects are called disclination lines, while two-dimensional structures are called inversion walls. Liquid crystalline textures are characterized by the arrangement, number density, dimensionality, character and strength of defects.[19,20]

In the framework of the elastic continuum theory, disclinations or line defects appear as planar solutions to the minimization of the total free energy of distortion. In the one constant approximation the free energy density is written as:[4]

$$F_d = \frac{K}{2}(\nabla \cdot \mathbf{n} + \nabla \times \mathbf{n})^2 = \frac{K}{2}(\nabla \phi)^2 \qquad [12]$$

if $\mathbf{n} = (\cos \phi, \sin \phi, 0)$ is in the $x$–$y$ plane and $F_d$ is the free energy density and $K$ is the elastic constant. The minimization of the total energy of distortion in a volume $V$ leads to Laplace's equation in two dimensions, $\nabla^2 \phi = 0$ which admits the singular solutions $\phi = S \tan^{-1}(y/x) + C$, $C$ is a constant and $S = m/2$, where $m$ is an integer. The singular line in the $z$-direction is the disclination line and the Schlieren texture[4,19] is a collection of those lines. $S$ is the strength of the singularity and the number of brushes seen under the polarized microscope is given by $4S$. Wedge disclination lines are parallel to the director rotation axis, while twist disclinations lines are parallel to the rotation axis and give rise to threaded textures.[4]

Nematic LCPs exhibit disclination singular lines of strength $S = \pm 1/2$ (thins), coreless disclinations of integral strength $S = \pm 1$ in the form of loops attached to the surfaces or other line segments, singular points and inversion walls.[19] The stress fields generated by these defects produce relative motions and interactions among them, with the result that textures are continuously changing with dynamics that depend on the viscosities involved in the defect motions.[20,21] For example, disclination loops of strength $S = +1/2$, are known to shrink, due to their line tension, into singular points of the same topological charge with a characteristic retraction time that scales with the inverse of the rotational viscosity. Detailed experimental rheo-optical studies of main chain thermotropic nematic LCPs[19] in shear flows between concentric discs show that at a critical shear rate there is a profound textural change characterized by the presence of a high density of disclination loops. For the case of a series of main chain thermotropic nematic LCPs in oscillatory shear flows between circular discs it is reported that at a critical shear rate a massive multiplication of disclination loops occurs. Increasing the frequency (shear rate) refines the texture by shrinkage of the characteristic loop size and at sufficiently large shear rates no loops appear in the field of vision of an optical microscope.

The texture in this section is characterized by a number density of singular disclination loops of strength $S = \pm 1/2$ (thins) that deform and tumble in the flowing material. As reported in the experiments,[19] the loops considered in this section nucleate at the surface at a critical shear rate and as the shear rate increases, its nucleation rate increases. As the loops adopt an orientation parallel to the bounding surfaces they quickly shrink, so that the actual loop density is given by a loop population balance between shrinking deformable loops and continuously created loops. The system is taken to be a suspension of non-interacting elastic loops in a Newtonian isotropic fluid matrix.

Loop emission by pinned disclination segments at the bounding surfaces is a likely mechanism to affect the texture of a liquid crystalline material. A model of loop emission, by pinned disclination segments at the bounding surfaces in a simple shear flow has been presented by Rey.[21] This model, analogous to the Frank–Read model for dislocations in metals, predicts that at a critical shear rate $\gamma_c$, the line tension force of a pinned disclination segment is unable to balance the viscous drag force of a shear flow, and the line will emit loops at regular intervals. An approximate balance of forces along the flow direction gives the critical shear rate as:

$$\dot{\gamma}_c = \frac{2D}{Lb} \qquad [13]$$

where $D$ is the orientation diffusivity, $L$ is the distance between anchoring points of the disclination line segment and $b$ is the maximum distance that the segment rises above the surface. This model neglects elastic anisotropy and image forces. For thermotropic LCP the plate separations used in shear experiments are usually very small and there is scant reported experimental data, so tentatively we take $D = 5 \times 10^{-8} \, \text{cm}^2 \, \text{s}^{-1}$, $L = b = 1 \, \mu\text{m}$ and obtain $\gamma_c = 10 \, \text{s}^{-1}$, approximately the value reported by Kleman.[19] Similar results have been observed with a mainchain thermotropic LCP in rectilinear oscillatory shear: at a frequency of $10 \, \text{sec}^{-1}$ with a plate separation of $10 \, \mu\text{m}$ and an amplitude of $A = 10 \, \mu\text{m}$ a large number of disclination loops suddenly cover the field of vision when viewing the sample with an optical microscope.[22] The frequency of loop emission by a surface source is estimated using Eqn [14]:

$$B' = (\dot{\gamma} - \dot{\gamma}_c) \frac{b}{L} \tag{14}$$

which gives the number of loops emitted by one source per unit time. For simplicity we assume $L = b$. Details of the loop radius and orientation at the instance of emission and dependance of the source operation on the number of loops already emitted are unknown, so we shall assume that if $\dot{\gamma} > \dot{\gamma}_c$ a source consisting of a pinned disclination line segment emits loops of constant radius $r_o$ at a frequency given by Eqn [20], and take the initial loop orientation as a constant or as a simple function of $\dot{\gamma}$. In this section we take $\alpha_o = \pi/2$, $\pi/4$ and $\pi \dot{\gamma}_c/(2\dot{\gamma})$, as some likely initial orientations. Given the present lack of experimental data or observations it seems reasonable at this time to study the predictions of the simplest possible model.

Assume that a nematic liquid crystal is sheared at constant shear rate $\dot{\gamma} > \dot{\gamma}_c$; after transients die out, a constant loop density balance is found between the shrinking loops due to their elasticity and emitted loops due to the surface sources. The effects of increasing shear rate are to modify the retraction dynamics and the frequency of loop emission. The loop retraction is described by convection in the loop size space (a,b) while the emitted loop frequency on a per source basis is described by $B'$. Formally, the stationary population loop balance on a per source basis is expressed as a constant flux $\mathbf{J}$ condition:

$$\nabla \cdot \mathbf{J} = \frac{d}{da} \cdot \left( \frac{da}{dt} n' \right) + \frac{d}{db} \cdot \left( \frac{db}{dt} n' \right) = 0 \tag{15}$$

where the space is defined by (a,b), $\mathbf{J}$ is the flux of loops with semi-axes $\mathbf{a}$ and $\mathbf{b}$, da/dt and db/dt are the convection velocities and $n'$ is the number of loops per unit length having semi-axes length between $a$ and $a + da$ along $\mathbf{a}$ and between $b$ and $b + db$ along $\mathbf{b}$. Assuming that all the loops are emitted with the same angle $\alpha_o$ when they are convected by the flow and have initially the same radius $r_o$, the dimensionless constant flux equation is given by:

$$\frac{da^*}{dt^*} n + \frac{db^*}{dt^*} n = -B[\cos \alpha_o \mathbf{i} + \sin \alpha_o \mathbf{j} + \mathbf{k}] \tag{16}$$

where starred symbols denote dimensionless quantities, with $t^* = tD/r_o^2$, $\mathbf{a}^* = \mathbf{a}/r_o$, $\mathbf{b}^* = \mathbf{b}/r_o$, $\mathbf{n}$ is the number of loops per unit length, $\dot{\gamma}^* = \dot{\gamma} r_o^2/D$,

$B = (\dot{\gamma}^* - \dot{\gamma}_c^*)$ is the number of circular loops emitted per unit time of unit radius (in dimensionless units), $\alpha_o$ defines the initial semi-axis $\mathbf{a}^*$ orientation and $(i, j, k)$ are the unit vectors. Here we assume that $\mathbf{a} = (a_x, a_y, 0)$, $\mathbf{b}^* = (0, 0, b^*)$. The number of loops per unit length is:

$$n(b^*) = b^*(\dot{\gamma}^* - \dot{\gamma}_c^*)[\sin(\alpha_o)]^{2/3} \left[ 1 + \left(\dot{\gamma}^* b^{*2} + \cot(\alpha_o)\right)^2 \right]^{1/3} \qquad [17]$$

Figure 6.6 shows the loop density n as a function of $b^*$ for three initial loop orientations and three increasing shear rates. The theory predicts that at low shear rates the loop density is not affected by the initial orientation and it increases with increasing $b^*$. At higher shear rates loops nucleated at $\pi/2$ exhibit a higher density than those nucleated at $\pi/4$; the increase is less severe if loops nucleate at smaller angles with increasing shear rates.

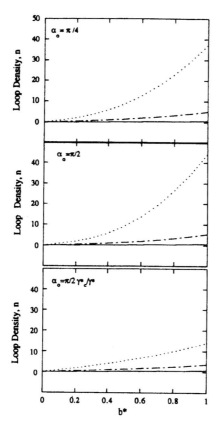

6.6   Loop density *n* as a function of the dimensionless loop semi-axis $b^*$ for three initial orientations and for three dimensionless shear rates $\dot{\gamma}^*$: 1 (full line), 3 (dashed-dot line), and 10 (dashed line).

The total number of loops $N$ per source is found by integrating $n$:

$$N = \int_0^1 n(b^*)db^*.$$ [18]

The average dimensionless loop perimeter $\langle P^* \rangle$ is found from:

$$\langle P^* \rangle = \frac{1}{N} \int_0^1 P^* n \, db^* = \frac{\sqrt{2}\pi}{N} \int_0^1 n\sqrt{a^{*2} + b^{*2}} \, db^*$$ [19]

where $a^*$ is given in terms of $b^*$.

Figure 6.7 shows the corresponding total number of loops per active source $N$ and the average perimeter $\langle P^* \rangle$ as a function of shear rate $\dot{\gamma}^*$, obtained from Eqn (18) and (19), respectively. $N$ increases for the three initial orientations, but the increase is weaker for the case of shear dependent initial orientation. The

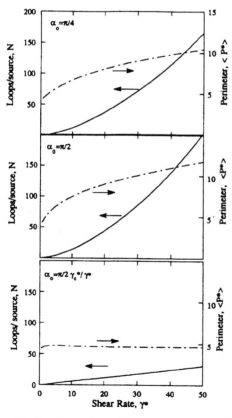

6.7 Total number of loops per source $N$ and average perimeter $\langle P^* \rangle$ as a function of shear rate $\dot{\gamma}^*$, for the same initial orientations as in Fig. 6.6.

average perimeters are monotonically increasing for the constant initial orientations but it exhibits a maximum for the shear dependent initial orientation. This can be explained by the presence of two competing effects: larger shear rates increase the frequency of emission of the source but decrease the stretching effect of shear since loops are nucleated closer to the bounding surfaces.

In partial summary, we have shown that a simple population model of elastic loops nucleated at a shear dependent rate, that deform, tumble and shrink can predict a stationary loop density. The total number of loops is predicted to increase with shear rate, regardless of initial conditions. These results are in agreement with experiments.[19]

## 6.3    Shear flow-induced molecular alignment phenomena in rod-like nematic polymers

### 6.3.1  Tensor continuum theory

This section presents a macroscopic tensor theory for nematic LCPs.[23,24] A second order traceless symmetric tensor, known[4] as the tensor order parameter $\mathbf{Q}$, is used to account for first order variations in the degree of molecular alignment with respect to the macroscopic orientation. A general expression of $\mathbf{Q}$ in the principal frame is:

$$\mathbf{Q} = \mu_n \mathbf{n}\,\mathbf{n} + \mu_m \mathbf{m}\,\mathbf{m} + \mu_l \mathbf{l}\,\mathbf{l} \qquad \text{[20a]}$$

where the eigenvalues expressed in terms of scalar order parameters $(S, P)$:

$$\mu_n = 2S/3, \; \mu_m = (P - S)/3, \; \mu_l = -(P + S)/3 \qquad \text{[20b, c, d]}$$

and where the following restrictions hold:

$$-1/3 \leq \mu_i \leq 2/3; \; -1/2 \leq S \leq 1; \; S - 1 \leq P \leq 1 - S \qquad \text{[20e, f, g]}$$

where $\mathbf{i} = \mathbf{n}, \mathbf{m}, \mathbf{l}$ are the orthogonal unit eigenvectors and the $\{\mu_i\}$ are the corresponding eigenvalues. If the three eigenvalues are equal the orientation state is isotropic, if two are equal the state is uniaxial with respect to the direction corresponding to the distinct eigenvalue, and if all three eigenvalues are distinct the state is biaxial. For equilibrium rod-like nematics the phase is uniaxial $(P = 0)$. The main eigenvector (eigenvalue) is the director $\mathbf{n}$ $(2S/3)$. In this chapter $\mathbf{n}$ denotes orientation and $S$ denotes alignment. For example, a poorly oriented and strongly aligned nematic represents strong spatial variations in $\mathbf{n}$ with high values of $S$.

An approximate expression for the dimensionless free energy density of a uniaxial lyotropic nematic close to the isotropic–nematic transition, inferred from Landau–DeGenne's expansion,[23] is given by:

$$
F = \frac{1}{2}\left(1 - \frac{U}{3}\right)(\mathbf{Q}:\mathbf{Q}) - \frac{U}{3}\mathbf{Q}:(\mathbf{Q}.\mathbf{Q}) + \frac{U}{4}(\mathbf{Q}:\mathbf{Q})^2
$$
$$
+ \frac{L_1}{2cKT}(\nabla\mathbf{Q}:(\nabla\mathbf{Q})^{\mathrm{T}}) + \frac{L_2}{2cKT}(\nabla.\mathbf{Q}).(\nabla.\mathbf{Q})
$$

[21]

where $U$ is the non-dimensional intensity of the nematic potential and $L_1$ and $L_2$ are elastic constants. It is possible to show that the spatially variant terms reduce to the Frank elastic energy in the case that the splay elastic constant $K_{11}$ is considered equal to the bend constant $K_{33}$; this approximation seems realistic for most known cases. The alignment equation is given by the balance between a molecular elastic field $\mathbf{H}^e$ and a molecular viscous filed $\mathbf{H}^v$:

$$
\mathbf{H}^v - \mathbf{H}^e = \mathbf{0}.
$$

[22]

The elastic molecular field $\mathbf{H}^e$ is given by the negative of the variational derivative of the free energy:

$$
\mathbf{H}^s = (cKT)\left(\nabla.\frac{\partial F}{\partial(\nabla\mathbf{Q})} - \frac{\partial F}{\partial \mathbf{Q}}\right)
$$

[23]

and reads:

$$
\mathbf{H}^e = (cKT)\left\{-\left(1 - \frac{U}{3}\right)\mathbf{Q} + U\left(\mathbf{Q}.\mathbf{Q} - \frac{1}{3}(\mathbf{Q}:\mathbf{Q})\delta\right)\right.
$$
$$
- U(\mathbf{Q}:\mathbf{Q})\mathbf{Q} + \frac{L_1}{cKT}(\nabla.\nabla\mathbf{Q}) + \frac{L_2}{2cKT}
$$
$$
\left.+ \left[\nabla(\nabla.\mathbf{Q}) + [\nabla(\nabla.\mathbf{Q})]^{\mathrm{T}} - \frac{2}{3}[\mathrm{tr}\,(\nabla(\nabla.\mathbf{Q})]\delta\right]\right\}
$$

[24]

where the superscript T denotes the transpose and tr the trace. The spatially non-homogeneous terms introduced by $L_1$ and $L_2$ represent macroscopic elasticity, while the spatially homogeneous terms represent molecular elasticity. Neglecting spatial variations eliminates possible coupling between molecular and macroscopic elasticity. The viscous molecular field $\mathbf{H}^v$ is found by expanding it in terms of the rate of deformation tensor $\mathbf{A}$ and the Jaumann derivative of the nematic tensor $\hat{\mathbf{Q}}$:

$$
\mathbf{H}^v = (cKT)\{\sigma_4\mathbf{A} + \tau_4\hat{\mathbf{Q}} + \sigma_6[\mathbf{Q}.\mathbf{A} + \mathbf{A}.\mathbf{Q} - \tfrac{2}{3}(\mathbf{Q}:\mathbf{A})\delta]
$$
$$
+ \tau_6[\mathbf{Q}.\hat{\mathbf{Q}} + \hat{\mathbf{Q}}.\mathbf{Q} - \tfrac{2}{3}(\mathbf{Q}:\hat{\mathbf{Q}})\delta]\}
$$

[25]

$$
\hat{\mathbf{Q}} = \frac{\partial \mathbf{Q}}{\partial t} + (\mathbf{v}.\nabla)\mathbf{Q} - \boldsymbol{\Omega}.\mathbf{Q} + \mathbf{Q}.\boldsymbol{\Omega}
$$

[26]

where $\sigma_i$ $(i = 4,6)$ and $\tau_i$ $(i = 4,6)$ are scalar constants with units of time and must satisfy thermodynamic inequalities resulting from the fact that the entropy production should be positive definite, $c$ is the number of rods per unit volume, $T$ is the absolute temperature and $K$ is the Boltzman constant. For a given flow kinematics, the dynamics of the nematic tensor order parameter $\mathbf{Q}$ is found by integrating the five coupled non-linear parabolic partial differential equations.

The three dimensionless parameters that control the dynamics of $\mathbf{Q}$ are the Ericksen number $E$, the Deborah number or dimensionless deformation rate $D_e = \| \mathbf{A} \| \tau_4$ and the reactive parameter $\lambda$. In comparison to the L–E theory, there is now an additional dimensionless parameter $D_e$, since the present model accounts for molecular elasticity, characterized by a time constant $\tau_4$. When $D_e \ll 1$, a vector theory appears to be sufficient to describe nematic LCPs, but when $D_e \geqslant 1$ a tensor theory is needed, since alignment is now coupled to orientation.[6,25]

In the uniaxial approximation, obtained by setting $\mathbf{Q} = S(\mathbf{nn} - \delta/3)$, the governing parameters for a steady shear flow are:

$$D_e = \dot{\gamma}\tau_4; \quad E = \frac{2cKT\tau_4\left(1 + \tau_6^* S/3\right)h^2\dot{\gamma}}{\left[L_1(L_1 + L_2)^2\right]^{1/3}}; \quad \lambda = -\frac{\left(3\sigma_4^* + \sigma_6^* S\right)}{S\left(3 + \tau_6^* S\right)}$$

$$[27a, b, c]$$

where the star denotes scaling with $\tau_4$. We note that for 'normal' nematics $S$ is non-negative definite: $S \geqslant 0$. The three parameters are functions of the shear rate and temperature. For a given nematic LCP the reactive parameter $\lambda$ is a function of $S$, and since the scalar order parameter $S$ is a function of the shear rate $\dot{\gamma}$, $\lambda$ itself is a function of the shear rate. For many nematic LCPs it is found that they are non-orienting at low shear rates and flow-orienting at high shear rates.[23,24] The reason being that for many materials $\lambda < 1$ for high equilibrium $S$ values and thus slow flows that weakly perturb the equilibrium value of $S$ will not orient the material, since $\lambda$ will remain less than one. On the other hand, the imposition of stronger shear flows tends to decrease $S$ such that $\lambda$ now becomes larger than one and the material orients in shear. The effect of the shear rate on $E$ remains to be investigated.

## 6.3.2 Alignment–Orientation coupling

This section presents a summary of the predictions of the tensor theory for simple shear start-up flow of rod-like nematic LCPs, using the uniaxial approximation $(P = 0)$, neglecting macroscopic elasticity $(L_1 = L_2 = 0)$ and assuming a planar two-dimensional orientation.[23,24] We use the same coordinates and orientation definitions as in Section 6.2.3. The director field is given by $\mathbf{n}(t) = (\cos\theta, \sin\theta, 0)$ and the known velocity field by $\mathbf{v} = (\dot{\gamma}y, 0, 0)$.

With these approximations the dimensionless orientation and alignment equations become:

$$\frac{d\theta}{dt^*} = -\frac{\dot{\gamma}^*}{2}[1 - \lambda \cos(2\theta)]; \quad \frac{dS}{dt^*} = \dot{\gamma}^* \beta_1 \sin(2\theta) + \beta_2 \qquad \text{[28a, b]}$$

where

$$\beta_1 = -\frac{(9\sigma_4^*/4 + 3\sigma_6^* S/2)}{(3 + 2\tau_6^* S)}; \quad \beta_2 = \frac{(-3S + US + US^2 - 2US^3)}{(3 + 2\tau_6^* S)}$$

[29a, b]

where $t^* = t/\tau_4$ is the dimensionless time and $\dot{\gamma}^* = \dot{\gamma}\,\tau_4$ is the dimensionless shear rate $(D_e = \dot{\gamma}^*)$; the starred constants were scaled with $\tau_4$ and $\lambda$ is given in Eqn [27c]. The selected parameter values are: $\sigma_4^* = -0.21$, $\sigma_6^* = -1.33$, $\tau_6^* = -0.98$ and the corresponding $\lambda$ curve is shown in Fig. 6.8, where the dashed (full) line corresponds to $\lambda > 1$ $(0 < \gamma < 1)$.

The main feature of the orienting mode discussed here is the lack of a stable steady state if the shear rate is less than a certain critical value and the nematic potential larger than a certain critical value. In the absence of flow, Eqn [28b] shows that the equilibrium alignment $S_{eq}$ is given by:

$$S_{eq} = \frac{1}{4} + \frac{3}{4}\sqrt{1 - \frac{8}{3U}}; \quad U > 8/3. \qquad \text{[30]}$$

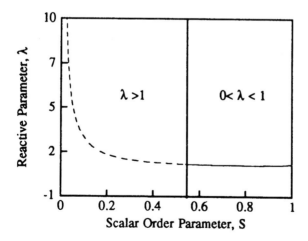

6.8  Reactive parameter $\lambda$ as a function of the scalar order parameter $S$. The vertical line is for $\lambda = 1$, at which the shear flow-orienting properties of nematic LCPs experience a dramatic change.

The particular value of the nematic potential $U^*$ that differentiates these two behaviors of the complex mode is found by inserting $S_{eq}(U)$ into:

$$\lambda = 1 = -\frac{(3\sigma_4^* + \sigma_6^* S_{eq})}{S(3 + \tau_6^* S_{eq})}. \tag{31}$$

For the parameters used in this paper $U^* = 4.49$ and $S_{eq} = 0.72$, which correspond to the vertical line in Fig. 6.8. When $U < U^*$ a stable steady state always exists and to large extent it is similar to the case of flow-orienting nematics. On the other hand when $U > U^*$, a steady state only exists if the shear rate is sufficiently large. For the parameters used here the dimensionless shear rate is $\dot{\gamma}_c^* = \dot{\gamma}/\tau_4 = 1.68$. When $U > U^*$ and the shear rate is less than the critical ($\dot{\gamma}^* < \dot{\gamma}^*_c$), the model predicts stable time periodic solutions. However, it remains to be seen how accounting for macroscopic elasticity modifies this last prediction.

The steady state solutions ($\theta_S$, $S_S$) to the governing Eqn [28] are:

$$\cos(2\theta_s) = \frac{1}{\lambda}; \quad 1 - \lambda^2 + \left(\frac{\beta_2 \lambda}{\beta_1 \dot{\gamma}^*}\right)^2 = 0. \tag{32a, b}$$

The stability of the multiple solutions that satisfy Eqn [32] is found by calculating the real part of the eigenvalues ($\mu_1$, $\mu_2$) of the Jacobian matrix $\mathbf{J}$:

$$\mathbf{J} = \begin{vmatrix} \dfrac{\partial f}{\partial \theta} & \dfrac{\partial f}{\partial S} \\[2mm] \dfrac{\partial g}{\partial \theta} & \dfrac{\partial g}{\partial S} \end{vmatrix}. \tag{33}$$

The pair of eigenvalues of the Jacobian matrix $\mathbf{J}$, written in terms of the determinant of the Jacobian matrix $\det(\mathbf{J})$ and its trace $\operatorname{tr}(\mathbf{J})$, are as follows:

$$\mu_{1,2} = \left[ \operatorname{tr}(\mathbf{J}) \pm \sqrt{[\operatorname{tr}(\mathbf{J})]^2 - 4 \det(\mathbf{J})} \right]/2. \tag{34}$$

The stability of the steady states thus depends on the signs of the determinant and trace of the Jacobian matrix. The left column below indicates the conditions that give rise to the various steady states, shown in the middle column and their stability, which are shown on the right column:

| | | |
|---|---|---|
| $\det(\mathbf{J}) < 0$ | saddles | unstable |
| $\det(\mathbf{J}) > 0$ and $\operatorname{tr}(\mathbf{J}) < 0$ | sinks | stable |
| $\det(\mathbf{J}) > 0$ and $\operatorname{tr}(\mathbf{J}) > 0$ | sources | unstable |

In addition to the stability properties, steady states are further differentiated by their type: real eigenvalues correspond to nodes, while complex eigenvalues correspond to spirals. The two types of sources and sinks, found from the sign of the discriminant in Eqn [34], are as follows:

$$\det(\mathbf{J}) > \text{tr}(\mathbf{J})^2/4 \qquad \text{spirals}$$
$$\det(\mathbf{J}) < \text{tr}(\mathbf{J})^2/4 \qquad \text{nodes}$$

Spiral sources correspond to resonant periodic solutions while spiral sinks denote underdamped periodic motions. Nodes and saddles result from possible combinations of two principal directions and lack of oscillatory behavior (overdamped response). Below we evaluate the nature and stability of the steady states by plotting $\det(\mathbf{J})$ as a function of $-\text{tr}(\mathbf{J})$, using the magnitude of the shear rate as a parameter.

In the immediate vicinity of the steady states, the transient solutions to Eqn [28] are approximated by their linearization around them, and are given by:

$$(\theta, S) = \mathbf{c}_1 e^{\mu_1 t} + \mathbf{c}_2 e^{\mu_2 t} \tag{35}$$

where $\mathbf{c}_1$ and $\mathbf{c}_2$ are two row vector constants and $\mu_1$ and $\mu_2$ are the two eigenvalues given in Eqn [34]. After replacing Eqn [34] and Eqn [35] into Eqn [33] it is possible to show that the trace of the Jacobian matrix is proportional to the shear rate and its determinant to the square of the shear rate, and therefore the solutions represented by Eqn [35] scale with the strain $\gamma = \dot{\gamma}t$. Furthermore, in the presence of spiral sinks the approach to steady state is through underdamped oscillations, whose frequency is proportional to the shear rate. These results are in agreement with previous reports.[26,27]

Below we first present and discuss the existence and nature of the stable steady states, and then the type of approach to steady state after shear start-up; all the results correspond to one representative value of the nematic potential larger than the critical one, i.e. $U = 5 > U^* = 4.49$.

### 6.3.2.1  Stationary regimes

Figure 6.9(a) shows the stable steady director orientation $\theta$ and the scalar order parameter $S$ as a function of the dimensionless shear rate ($\dot{\gamma}^* > \dot{\gamma}_c^*$). The vertical full line denotes the value of the dimensionless shear rate at which spirals become nodes, as explained below. The figure shows that the director orients in the fourth quadrant but close to the flow direction ($\theta < 0$). The figure also shows that the magnitude of the scalar order parameter is less than its equilibrium value ($S_{eq} = 0.76$). This is explained by noting that the fourth quadrant represents the compression sector of shear, and therefore the steady value of the scalar order parameter during shear should be less than in the absence of flow, if the director orients in the fourth quadrant. The curves start at a value of the dimensionless shear rate of 1.68, since for smaller values stability is lost and the stable regime is oscillatory.[6,23]

Figure 6.9(b) shows the corresponding plot (full line) of the determinant of the Jacobian matrix, det($\mathbf{J}$), as a function of the negative of the trace, $-\text{tr}(\mathbf{J})$; the direction of the arrow on the full line indicates increasing magnitudes of the dimensionless shear rate. Each point on the full line corresponds to one specific value of the dimensionless shear rate and was obtained by first computing the Jacobian matrix for the corresponding stable steady state values of the director angle and the scalar order parameter. The figure shows that at sufficiently low values of the dimensionless shear rate, the steady states are spiral sinks and that at higher values of the dimensionless shear rate the spiral sinks are transformed into node sinks. The parabola (dash-dotted line) represents the following special condition:

$$\det(\mathbf{J}) = [\text{tr}(\mathbf{J})]^2/4 \qquad [36]$$

and gives the particular values of the determinant and trace of the Jacobian matrix at which the spiral–node transformation occurs. The crossing of the full line into a region corresponding to positive values of the trace of the Jacobian matrix represents the transformation of stable spiral sinks into unstable spiral sources and physically it represents the loss of stability of the steady director orientation and molecular ordering, in agreement with previous results.[28]

In partial summary, we found that if the shear rate increases above a certain critical value, a stable steady state appears, and its nature is that of a spiral sink; physically it means that at these shear rate values the transient approach to steady state is through damped oscillations. By further increases in the shear rate, the stable spiral sinks are transformed into stable node sinks; physically it means that at these higher shear rates the approach of the director angle and the scalar order parameter to their steady state values is overdamped, lacks oscillations, but possibly contains overshoots and undershoots.

### 6.3.2.2   Start-up dynamics

We next present results corresponding to shear rates for which the following steady state exist and conditions hold:

$$\text{Spiral sinks:} \begin{cases} \det(\mathbf{J}) > 0 \\ \det(\mathbf{J}) > [\text{tr}(\mathbf{J})]^{-2}/4 \end{cases} \qquad [37]$$

Figure 6.10 shows the evolution of the director orientation $\theta$ (top) and the scalar order parameter $S$ (bottom) as a function of strain, $\gamma = \dot{\gamma}t$, for two representative dimensionless shear rates close to the critical one; $\dot{\gamma}^* = 1.71$ (dashed) and $\dot{\gamma}^* = 1.8$ (full). Note that the strain is given by $\gamma = \dot{\gamma}^*t^* = \dot{\gamma}t$. At these flow conditions the stable steady state is a spiral sink, denoting final convergence through underdamped oscillations. These results agree with Fig. 6.9(b), since at both shear rate conditions it can be shown that the defining inequality

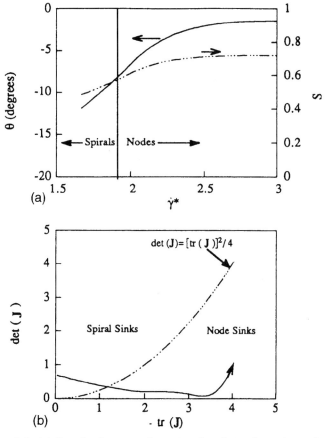

**6.9** (a) Steady director orientation $\theta$ and steady scalar order parameter $S$ as a function of the dimensionless shear rate $\dot{\gamma}^*$, for a nematic potential $U=5$. The vertical line defines the values for the shear rates for which the steady states are nodes or spirals. The director angle is negative and the magnitude of $S$ is less than its equilibrium value $S_{eq}=0.76$. (b) Characterization of the steady states. Determinant of the Jacobian matrix det (**J**) as a function of $-\mathrm{tr}(\mathbf{J})$. The arrow in the full line indicates increasing shear rates. The parabola (dashed triple dot line), defined by det $(\mathbf{J})=(\mathrm{tr}(\mathbf{J}))^2/4$, separates the region of existence of spiral sinks (underdamped mode) from that of node sinks (overdamped mode).

$\det(\mathbf{J}) > [\mathrm{tr}(\mathbf{J})]^2/4$ holds. Figure 6.10 shows that the strength of the damping increases as the shear rate increases, which is also corroborated by Fig. 6.8(b), since as shear increases the corresponding point on the full line moves away from the vertical axis $(\mathrm{tr}(\mathbf{J})=0)$ of the figure and towards the parabola that denotes the transformation of spirals into nodes. Figure 6.10 also shows that after the initial under- and overshoot, the underdamped oscillations scale with

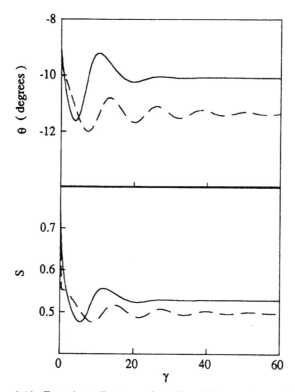

*6.10* Transient director orientation $\theta$(degrees) as a function of strain $\gamma$ (top), and transient scalar order parameter $S$ as a function of strain $\gamma = \dot{\gamma}t$ (bottom), for two characteristic dimensionless shear rates corresponding to spiral nodes. $U = 5$, $\dot{\gamma}^* = 1.7$ (dashed line) and $\dot{\gamma}^* = 1.8$ (full line). For dimensionless shear rates less than 1.68 no steady state exists.

the strain, i.e. there is no significant lag between the responses at the two different shear rates.

It was previously shown by Fig. 6.9(b) that at sufficiently large shear rates the condition given in Eqn [37] is fulfilled, and that a further increase in the shear rate results in the transformation of spiral sinks into node sinks; the value of the shear rate indicated by the vertical divisory line in Fig. 6.9(a) was found numerically to be close to two. Physically it means that at higher shear rates, the approach of the considered rheological functions to their final steady state values is not oscillatory, but possibly exhibits transient over- and/or undershoots.

We next discuss the results corresponding to shear rates for which the following steady states exist and conditions hold:

$$\text{Node sinks:} \quad \begin{cases} \det(\mathbf{J}) > 0 \\ \det(\mathbf{J}) > [\operatorname{tr}(\mathbf{J})]^{-2}/4 \end{cases} \qquad [38]$$

Figure 6.11 shows the evolution of the director orientation $\theta$ and the scalar order parameter $S$ as a function of strain. At these flow conditions the stable steady states are node sinks, denoting final convergence through exponential decay. Both the director angle $\theta$ and the scalar order parameter $S$ exhibit overshoots before the final exponential decay. These overshoots are also present for other initial conditions. The presence of overshoots in the evolution of the director $\theta$ and the scalar order parameter $S$ for this particular initial orientation is explained by noting that the director is initially in the fourth quadrant and thus $S$ drops to smaller values under the compression effects of a shear deformation. The consequence of these two facts is that the quantity $\lambda \cos(2\theta)$, which is the ratio of strain torques to spin torques and governs the director velocity, increases and causes a fast clockwise rotation of $\mathbf{n}$ towards the first quadrant, where at some orientation, defined by $\lambda \cos(2\theta) = 1$, the director velocity is zero. At this point

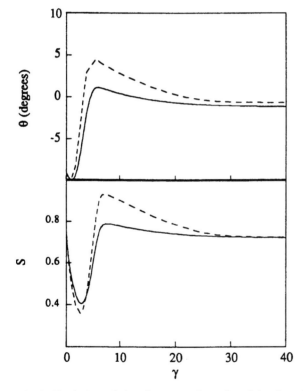

6.11  Evolution of the director orientation $\theta$ (top) and the scalar order parameter $S$ (bottom) as a function of strain $\gamma = \dot{\gamma}t$, for two representative dimensionless shear rates corresponding to the overdamped mode; $\dot{\gamma}^* = 3.6$ (full) and $\dot{\gamma}^* = 6$ (dashed); the nematic potential is $U=5$. At these $\dot{\gamma}^*$ the stable steady states are node sinks, denoting overdamping. Typical overshoots arise at time lags between $S$ and $\theta$.

the director motion ceases but the scalar order parameter keeps increasing, since Eqn [7] is not satisfied. As a result of this lag, the controlling quantity $\lambda\cos(2\theta)$ becomes less than one and the director rotates clockwise with a relatively smaller rate. Finally both the scalar order parameter and the director approach their final steady state values through an exponential decay.

## 6.4    Conclusions

At low deformation rate, flow-induced orientation in nematic LCPs is governed by the viscoelastic anisotropies and non-orienting properties. Distinguishing features are stable steady three-dimensional twisted orientation textures and texture multistability. The optical banded patterns frequently observed during shear start-up are simulated and their origin is elucidated using the well known principles of orientation dynamics of non-orienting nematics. Defect population models are able to simulate and explain typical defect proliferation and deformation phenomena in the presence of shear flows. At sufficiently high shear rates, molecular alignment is coupled to orientation. A tensor theory predicts stable orientation only at high deformation rates and oscillatory transients in agreement with rhological data. Future problems that merit further efforts are the origin of the ubiquitous optical patterns that form after cessation flow, the nature and stability of nematic textures, the role of textures and defects on flow-induced orientation and the role of macroscopic elasticity on orientational instabilities.

## Acknowledgments

This work is supported by a grant from the Natural Sciences and Engineering Research Council of Canada. The results of Sections 6.2.2–6.2.3 were obtained in collaboration with Dr. W.H. Han, and those in Section 6.3 with Ms. Y. Farhoudi.

## References

1. A.N. Beris and B.J. Edwards, *Thermodynamics of Flowing Systems*, Oxford University Press, New York, 1994.
2. A. Ciferri (ed.) *Liquid Crystallinity in Polymers*, VCH, New York, 1991.
3. S. Chandrasekhar, *Liquid Crystals*, Cambridge University, Cambridge, 1991.
4. P.G. de Gennes and J. Prost, *The Physics of Liquid Crystals*, Clarendon Press, Oxford, 1993.
5. A.M. Donald and A.H. Windle, *Liquid Crystalline Polymers*, Cambridge University Press, Cambridge, 1992.
6. G. Marrucci and F. Greco, *Adv. Chem. Phys.*, **LXXXVI**, 331–404, 1993.

7. A.D. Rey, *Macroscopic Modeling of Dynamical Phenomena in Liquid Crystalline Materials, Advances in Transport Processes*, volume IX, eds. A.S. Mujumdar and R.A. Mashelkar, Elsevier, Amsterdam, Chapter 5, pp. 185–229, 1993.

8. M. Srinivasarao, *Internat. J. Mod. Phys. B*, **9**, 2515–2572, 1995.

9. G. Srajer, S. Fraden and R.B. Meyer, *Phys. Rev. A*, **39**, 4828–4834, 1989.

10. W.H. Han and A.D. Rey, *J. Non-Newtonian Fluid Mech.*, **48**, 181–210, 1993.

11. W.H. Han and A.D. Rey, *J. Non-Newtonian Fluid Mech.*, **50**, 1–28, 1993.

12. W.H. Han and A.D. Rey, *Phys. Rev. E*, **49**, 597–614, 1994.

13. W.H. Han and A.D. Rey, *Phys. Rev. E*, **50**, 1688–1691, 1994.

14. W.H. Han and A.D. Rey, *J. Rheol.*, **38**, 1317–1334, 1994.

15. W.H. Han and A.D. Rey, *J. Rheol.*, **39**, 301–322, 1995.

16. N.X. Yang and M.M. Labes, *Macromolecules*, **25**, 7843–7847, 1994.

17. W.H. Han and A.D. Rey, *Macromolecules*, **28**, 8401-8405, 1995.

18. N. J. Alderman and M.R Mackley, *Faraday Discuss. Chem. Soc.*, **79**, 149–160, 1985.

19. M. Kleman, *Points, Lines and Walls*, Wiley, New York, 1983.

20. A.D. Rey, *Liq. Crys.*, **7**, 315–334, 1990.

21. A.D. Rey, Mol. Cryst. Liq. Cryst., **225**, 313–335, 1993.

22. M. Srinivasarao, personal communication.

23. Y. Farhoudi and A.D. Rey, *J. Rheol.*, **37**, 289-313, 1993.

24. Y. Farhoudi and A.D. Rey, *J. Non-Newtonian Fluid Mech.*, **37**, 289–313, 1993.

25. R.G. Larson, in *Spatio-Temporal Patterns in Nonequilibrium Complex Systems*, eds P.E. Cladis and P. Palffy-Muhoray, Addison-Wesley, Massachusetts, pp.219–228, 1995.

26. J. Mewis and P. Moldenaers, *Mol. Cryst. Liq. Cryst.*, **153**, 291–300, 1987.

27. H.L. Doppert and S.J. Picken, *Mol. Cryst. Liq. Cryst.*, **153**, 109–116, 1987.

28. G. Marrucci, in *Liquid Crystallinity in Polymers*, ed. A. Ciferri, VCH, New York, pp.395–422, 1991.

# 7

# Mesostructural characterisation of aligned fibre composites

A R CLARKE, N C DAVIDSON AND G ARCHENHOLD

## 7.1 Introduction

Composite materials are truly heterogeneous systems to characterise and model. The presence of stiff fibres embedded in the brittle matrix with different fibre packing fractions together with voids makes for a finite element modeller's nightmare. The situation has not been helped by the fact that good quality, three-dimensional structural data had been lacking until a few years ago. Physical sectioning of composite samples and observation of the fibre images and void cross-sections under the optical microscope has been possible since the mid 1970s. However, the tedium of manually collecting statistically meaningful fibre orientation data has only recently been superseded by fully automated image analysis. In order to predict composite thermal and mechanical properties, one must have a reasonable structural model and ensure that any simplifying assumptions are permissible. Also, in order to assess the effectiveness of the processing conditions for the manufacture of a complex part, one must be able to check the fibre orientations in the part. The ultimate goal must be to provide a sound theoretical linkage between the processing route and the final global material properties and we do not see how that goal will be achieved without good quality, three-dimensional mesostructural data ('meso-' meaning on a length scale from a few microns through to centimetres, see Piggott[1]).

The aim of this chapter is to outline the progress that has been made in the study of composite mesostructures with two complementary measurement systems: the standard two-dimensional (2D), optical reflection microscope with image analyser and the three-dimensional (3D) confocal laser scanning microscope (CLSM).

### 7.1.1 Definition of the measurement problem

What would make an ideal measurement solution for full characterisation of this heterogeneous composite structure? Clearly, if there were a full set of 3D fibre coordinates for every fibre in the sample and void centre coordinates and sizes, the structure could be completely specified! Unfortunately, the effort required to create this dataset would be prohibitively high. Consider Fig. 7.1(a) in which a single, *ith* fibre is shown characterised by a set of orthogonal axes placed in the material. For convenience, the *XY* plane is defined to be parallel to the surface of the sample (which has been sectioned to show elliptical fibre images). Reinforcing fibres (whether carbon, Kevlar or glass) have diameters $d_i$ typically in the range, 5 μm $\leqslant d_i \leqslant$ 15 μm. The fibres could be classified as either short (length, $l_i \leqslant$ 1 mm), long (1 mm $\leqslant l_i \leqslant$ 1 cm) or continuous through the part. The matrix could be polymeric, ceramic or metallic. The fibres are usually considered to be smooth, straight and of circular cross-section (which is arguably true of glass but is a dubious assumption for certain carbon and Kevlar fibres). Therefore, in short fibre composites, the *ith* fibre might be represented uniquely by the set of parameters $\{x_i, y_i, z_i, l_i, d_i, \theta_i, \Phi_i\}$ as shown in Fig. 7.1(a). The elliptical image of the *ith* fibre on the single 2D section through the sample allows an estimate of $d_i$, $\Phi_i$ and $\theta_i$ to be made. The minor axis of the image $b_i$ is equivalent to $d_i$. The orientation of the major axis, $a_i$, in the *XY* plane defines $\Phi_i$ and the out-of-plane angle, $\theta_i$ is given by

$$\theta_i = \cos^{-1}(b_i/a_i). \qquad [1]$$

Note that $\Phi_i$ is ambiguous, because the same elliptical image would be produced by a fibre whose in-plane angle was $(180° + \Phi_i)$. However, in many theoretical studies this is not an important limitation of the technique. In order to derive the fibre lengths, the sample is usually pyrolised to remove the matrix, and fibre lengths are obtained by studying the residual fibres in a petri dish with an optical reflection (or transmission) microscope and image analyser. Clearly, fibre length distributions may be derived in this way, but all reference to the original fibre spatial coordinates in the sample is lost and hence any attempt to correlate fibre length with fibre orientation will be impossible (however, see Section 7.5.2).

For long fibre and continuous fibre reinforced composites, the fibres will exhibit curvature and waviness, as shown in Fig. 7.1(b), and therefore alternative characterisation strategies are required. The most common type of characterisation for these composites is the 'misalignment angle distribution' but with the advent of the confocal laser scanning microscope, more ambitious characterisations become possible, as discussed in Section 7.5.4 and Section 7.5.5.

Note that submicron accuracy in spatial position and subdegree accuracy in orientation implies that a high magnification is required to identify each elliptical fibre image. Therefore, relatively few fibre images will be found within a typical *XY* image frame and most of the early research with optical reflection microscopy involved random or pseudorandom sampling of a polished section plane to build up a statistically meaningful picture of fibre orientations. However, as discussed in Section 7.2.3, it is now possible to scan systematically over a large area by overlapping *XY* image frames and capturing not only the fibre orientations but also the absolute centre coordinates of each fibre image on

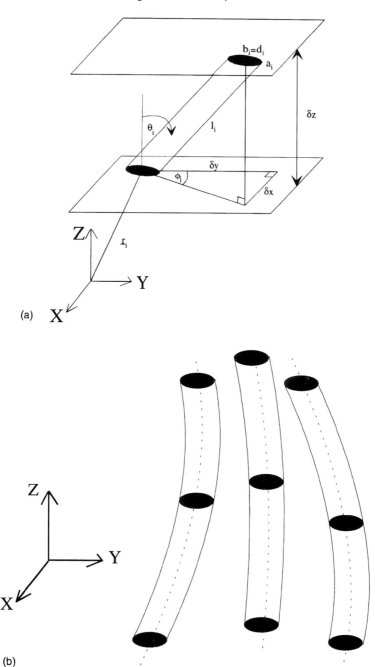

*7.1* (a) Complete characterisation of a short, straight fibre in 3D space. (b) Real fibres or fibre segments (length ⩾ 100 $\mu$m) are curved, making characterisation difficult.

the *XY* section plane. Hence the relationship of each fibre to neighbouring fibres can be assessed (as discussed in Section 7.5.3).

Finally, the cross-sectional areas of fibres and voids within the field of view are used to estimate the fibre packing fraction, $v_f$ and the void packing fraction, $v_v$ respectively, by the equations

$$v_f = 100 \times \Sigma_i(\pi a_i b_i)/\text{total } XY \text{ frame area scanned } \% \qquad [2]$$

$$v_v = 100 \times \Sigma(\text{void areas})/\text{total } XY \text{ frame area scanned } \%. \qquad [3]$$

Care must also be exercised with both of these equations when the reinforcing fibres are short and dispersed within the composite and the voidage is at a low packing fraction of a few percent (see Section 7.6.1). For highly aligned continuous fibre samples (usually at high packing fractions) $v_f$ should be well described by Eqn [2]. However, there could be a problem if polishing damage has occurred and/or if the pixel intensity threshold level has not been chosen carefully for both void and fibre determinations. The reason for this is simple: if there is an intensity gradient at the edge of the image of a void or fibre, a poor choice of threshold would result in extra pixels being assigned all around the image and the fractional error could tend towards $2\pi r/\pi r^2$ (i.e. $2/r$) rather than $1/\pi r^2$ (when only one extra noise pixel is added to a circular void image area).

## 7.1.2 Comparison of measurement techniques

Any measurement technique could be used provided that it effectively generates images of fibres on 2D sections through the material. In a recent publication by Clarke *et al.*,[2] a historical appraisal of the measurement techniques for fibre orientations and fibre misalignments has been presented. These techniques range from contact X-ray microradiography,[3] through to scanning acoustic microscopy,[4] and nuclear magnetic resonance.[5] The typical spatial resolution achievable for the different techniques is shown in Fig. 7.2.

For studies of mesostructural artefacts, the optical technique is preferred despite the fact that it requires (usually) a polished sectioning of the sample. Researchers who have made a significant impact with 2D optical reflection microscopy include Fakirov and Fakirova[6], Yurgartis[7], Toll and Andersson[8], Fischer and Eyerer[9] and Hine *et al.*[10] An excellent, recent review by Guild and Summerscales[11] on the use of image analysis for the characterisation of composites is well worth reading.

Recently, McGrath and Wille[12] have reported an optical technique for determining the 3D fibre orientation distribution in thermoplastic injection mouldings. Their technique depends upon

- refractive index of the matrix being matched to the reinforcing glass fibres,
- the presence of a small percentage of opaque tracer fibres during processing.

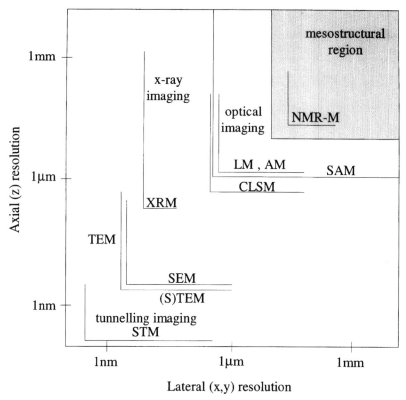

*7.2* Overview of the different measurement techniques showing the ultimate resolution achievable, both laterally (in *XY*) and axially (in *Z*).

Typically, 0.2 wt% of tracer fibres are embedded in composites with 30 wt% glass fibres. The voxel (i.e. minimum volume element) size was approximately 3 $\mu$m $\times$ 2 $\mu$m $\times$ 15 $\mu$m in their *XYZ* coordinate system and they achieved optical sectioning down to a few millimetres below the sample surface. This would appear to be the only other optical technique currently in use which obtains 3D fibre orientation data apart from 2D optical reflection microscopy (see Section 7.2) and the CLSM technique (described in Section 7.3).

### 7.1.3  Characterisation parameters of fibre orientation states

The parameters which have been used to characterise fibre orientation states in composites were proposed to fit in with the state-of-the-art optical measurements which were possible at the time! Hence, as single 2D sections could be analysed by optical reflection microscopy, with or without automated image analysis systems, a number of geometrical coefficients have been popular with experimentalists, for example, one of the first was Herman's orientation parameter, $H_e$ used by, amongst others, Fakirov and Fakirova[6] and O'Donnell and White.[13]

$$H_e = \frac{1}{2}\left(3\frac{\sum_i N(\theta_i).\cos^2\theta_i}{\sum_i N(\theta_i)} - 1\right) = \frac{1}{2}(3\langle\cos^2\theta\rangle - 1) \qquad [4]$$

where $\theta_i$ is the angle of the fibre to the $z$ axis. However, if the fibre orientations are not axisymmetric about the $z$ axis, this is an incomplete characterisation because different orientation states could give the same value for $H_e$.

Fibre orientation states and their links with global elastic moduli have been studied[14–16] using orientation functions, $P_n$, for example, $\langle P_4(\cos\theta)\rangle$ which is related to the orientation averages $\langle\cos^2\theta\rangle$ and $\langle\cos^4\theta\rangle$ by the relationship

$$\langle P_4(\cos\theta)\rangle = (35\langle\cos^4\theta\rangle - 30\langle\cos^2\theta\rangle + 3)/8 \qquad [5]$$

The derivation of tensor coefficients to represent fibre orientation states is also popular (since the paper by Advani and Tucker[17]) and the composites group at IKP, Stuttgart characterise orientation states in this way.[18] The general form of the second order orientation tensor is given in Eqn [6] where $\Psi(\mathbf{p})$ is the fibre probability distribution function and '$\mathbf{p}$' is a vector parallel to the fibre. Higher order orientation tensors can be defined in a similar way.

$$a_{ij} = \int_p p_i p_j \Psi(\mathbf{p})\mathrm{d}p \qquad [6]$$

The nine components of the second rank orientation tensor, $a_{ij}$ have a physical interpretation that is very similar to the components of the stress tensor. For example, if the fibres within the sample volume are randomly oriented in three dimensions, the components reduce to

$$a_{ij} = \begin{bmatrix} \frac{1}{3} & 0 & 0 \\ 0 & \frac{1}{3} & 0 \\ 0 & 0 & \frac{1}{3} \end{bmatrix}. \qquad [7]$$

A Japanese group[19] have suggested a fractal approach to the characterisation of fillers (mainly particulates and short fibres) in composite materials. They have introduced a number of micromorphological parameters, for example $\xi$ the homogeneity distribution parameter. If $A$ is the area of the observation window and $d_p$ is the most frequent of the interparticle distances and $N$ is the total number of particles in the observation window,

$$\xi = \frac{d_p}{\sqrt{A/N}} \qquad [8]$$

Taya $et\ al$[19] argue that the degree of homogeneous distribution of fibres (or particulates) depends upon the window size, $L$ and if the number of fibres $N$

within the window obeys fractal behaviour, $N(L) = \alpha L^D$, where $D$ is the fractal dimension and $\alpha$ is a constant. Both $\alpha$ and $D$ are proposed as potentially useful characterisation parameters.

In the case of continuous, well aligned fibres, Yurgartis[20] has explored other characterisation parameters, for example 'included angles' which are derived from nearest neighbour fibres. Clarke *et al.*[21] have used their 2D, large area scanning capability to investigate nearest neighbour in-plane orientation effects (which are described more fully in Section 7.5.3). Pyrz[22] has also been concerned with the apparently random positioning of fibre centre coordinates seen on physical sections taken perpendicular to the preferred fibre directions. In common with a number of investigators, he has considered a Voronoi cell approach and various clustering parameters, like the second order intensity function, $K(r)$ (see Eqn [9]) in his studies of material strength linked to composite microstructure. The function $K(r)$ characterises different fibre patterns in a 2D section and is defined by

$$K(r) = \frac{A}{N^2} \sum_{k=1}^{N} w_k^{-1} I_k(r) \qquad [9]$$

where $N$ is the number of points in the observation area A, $I_k(r)$ is the number of points within a circle of radius $r$ centred on one of the points and $w_k$ is a weighting factor to allow for image plane edge effects, that is the proportion of the circumference contained within A to the total circumference of radius $r$.

All of these characterisation parameters may be measured with the automated optical systems discussed in Sections 7.2 and 7.3, but the confocal laser scanning microscope allows more novel 3D characterisations to be devised, as mentioned later in Section 7.5.

## 7.2     Optical reflection microscopy

### 7.2.1  Sample preparation

If samples are being prepared for 2D optical reflection microscopy and analysis by image processing techniques, great care must be taken with both sectioning and the final polishing process. Physical sectioning of short fibre reinforcements can easily cause fibre pullout and the sectioning of aligned fibre composites can leave chipped fibres or fibre fragments at the surface.

Polishing is normally undertaken with a set of graded SiC polishing mats starting out with the roughest and gradually introducing the finest. Contrast issues for 2D optical reflection microscopy are no problem for carbon and Kevlar fibres in polymer composites, but glass fibres have very poor contrast with the polymer matrix because of the closely matched refractive indices. Therefore, special treatment of the surface of the polished sample, sputtering for example, is necessary to pick out the glass fibres. Clearly, the fibre orientation characterisations that result from any 2D image analyser depend crucially upon the quality of the surface preparation.

## 7.2.2 Overview of microscope and image analyser

Figure 7.3 shows an overview of the optical reflection microscope and associated 2D image analyser. The Sony CCD camera is parfocally mounted on a standard Olympus BH2 microscope. A monitor displays the video signal from the camera for the benefit of the operator. The field of view displayed depends upon the magnification of both the objective and eyepiece lenses. For fibre orientation work, an objective magnification of $20\times$ and an eyepiece magnification of $6.7\times$ gives about a 150 $\mu$m by 150 $\mu$m field of view.

A fast Frame Grabber fitted inside a standard host microcomputer converts the $XY$ image field into $512 \times 512$ individual image elements, pixels, at videorate frequencies. The pixellated images enable the $XY$ field to be stored in the computer's video random access memory (VRAM) in readiness for digital algorithms to process the image frames.

A Prior $XY$ stage is used to control the systematic scanning in $X$ and $Y$ over large sample areas. A Prior $Z$-drive, through a microcomputer interface, incorporates an autofocusing algorithm (which selects that $Z$ position giving the sharpest contrast along a line of the $XY$ image). This feature is an essential item for large area studies, because of the limited depth of focus of the objective lens and the limits to the flatness of the sample surface. The monitor gives visual confirmation that the system is focusing each $XY$ image frame and is scanning over the required surface area.

There are many commercial image analysers on the market, for example, Kontron (Kontron Elektronik GmbH, 85385 Eching, Munich, Germany) and although they have excellent basic image processing facilities they do not have the specialised features which are needed for high accuracy, large area, fibre orientation studies.

## 7.2.3 An automated large area high spatial resolution design

The Leeds 2D image analyser system has been described in detail elsewhere.[23–26] A network of three transputers from INMOS are used to form a small parallel processing network. The system, as shown in Fig. 7.3, has a number of unique features for determining fibre orientations:

- autofocusing for each new $XY$ frame,
- the image threshold is automatically set for each frame (to account for contrast changes within each $XY$ image frame),

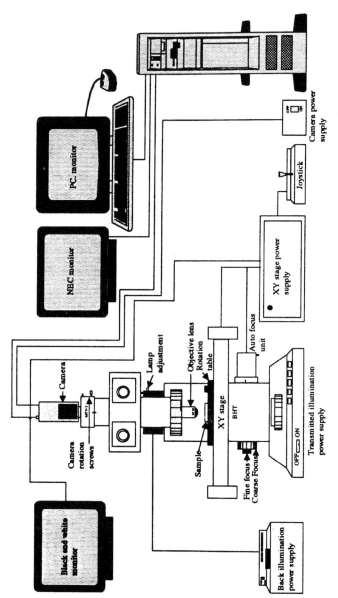

7.3  Overview of the Leeds, large area, 2D image analyser connected to a standard microscope.

- automatic fibre splitting of fibre images within the $XY$ frame,
- fitting of a partial image in one frame to the rest of that image in a neighbouring $XY$ frame,
- a 'fit factor' is assigned to each fibre image, giving a measure of the quality of the elliptical fit to that image.

These features are described in more detail below.

As all transputers are working in parallel, the system is extremely fast in operation. The hardware/software configuration has been designed to process an image frame during the time taken for the $XY$ stage to move from one position to the next. The system can scan sample areas of 50 mm$^2$ or more with submicron resolution in $X$ and $Y$ (containing over 20 000 fibre images) automatically in about 30 min. At the end of the scan, the images have been automatically analysed for their ellipticities and fibre orientation distributions are obtained within seconds. Once the sample has been prepared and has been placed under the microscope on the $XY$ stage, the system has to decide on the best threshold for converting the 256 intensity level $XY$ image into a 'binarised' image, that is where fibre image pixels are white and matrix pixels are black. A typical $XY$ image and the associated histogram of pixel intensities is shown in Fig. 7.4(a) and (b). It can be seen that the intensities corresponding to those pixels within a fibre image are well separated from those intensities corresponding to matrix pixels. In this case, by choosing an intensity threshold of 210, a binarised image which is suitable for further processing will be produced (like the image in Fig. 7.5).

The next task is to determine the ellipticity of each fibre image and fortunately, there are a number of robust algorithms which could be chosen. Some commercial image analysers determine the ellipticity of an object from the longest and shortest chords which can be found within that object. However, Stobie[27] proposed that the first moments, $M_x$ and $M_y$ and the second moments, $M_{xx}$, $M_{xy}$ and $M_{yy}$ of the binarised image would yield a best-fit ellipticity via Eqns [10] to [17]. Each fibre image is actually a set of '$n$' connected pixels. Each pixel assigned to the fibre image is given a weight, $w_i$ of '1' and non-fibre pixels are given a weight '0'. Note that the centre coordinates of each elliptical image $(x_c, y_c)$ are given by $(M_x, M_y)$ and the attraction of the first and second moments technique is that it is a 'single-pass' technique, that is, as the $XY$ image frame is being scanned, the moments of each object in the field are being determined and at the end of the scan, the elliptical parameters are immediately calculable. Using Eqn [1], the out of plane angle, $\theta$, is computed from the values of semi-major axis, $a$, and the semi-minor axis, $b$. The in-plane angle, $\Phi$ is computed directly from the second moments, as shown in Eqn [18].

The only limitation of this technique is that the elliptical image must be complete (because partial images would give incorrect ellipticities). The attraction of the moments technique is its high speed of execution.

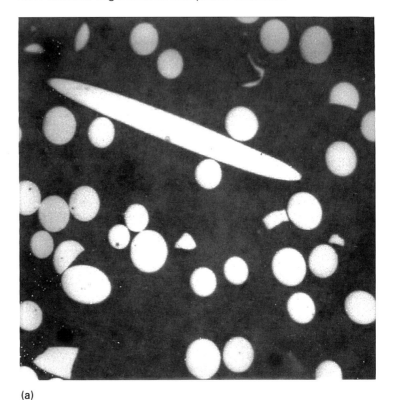

(a)

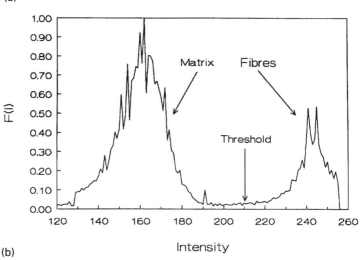

(b)

7.4  (a) Typical *XY* image frame after sputtering and (b) the intensity
distribution for all pixels within this *XY* frame showing those pixels
within the fibre images and a suitable threshold (intensity value,
210) to prepare the frame for further processing.

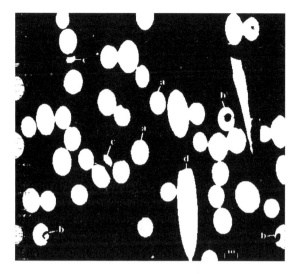

7.5 *XY* frame illustrating the different types of fibre polishing artefacts: (a) perfect fibre images, (b) holes in images, (c) fibre fragments, (d) sectioning through a fibre end.

$$\sum_{i=1}^{n} w_i = n = \text{number of pixels assigned to an image} \qquad [10]$$

$$M_x = \frac{\sum_{i=1}^{n} x_i w_i}{\sum_{i=1}^{n} w_i} = \frac{\sum_{i=1}^{n} x_i}{n} = X \text{ centroid} \qquad [11]$$

$$M_y = \frac{\sum_{i=1}^{n} y_i w_i}{\sum_{i=1}^{n} w_i} = \frac{\sum_{i=1}^{n} y_i}{n} = Y \text{ centroid} \qquad [12]$$

$$M_{xx} = \frac{\sum_{i=1}^{n} x_i^2 w_i}{\sum_{i=1}^{n} w_i} - M_x^2 = \frac{\sum_{i=1}^{n} x_i^2}{n} - M_x^2 \qquad [13]$$

$$M_{yy} = \frac{\sum_{i=1}^{n} y_i^2 w_i}{\sum_{i=1}^{n} w_i} - M_y^2 = \frac{\sum_{i=1}^{n} y_i^2}{n} - M_y^2 \qquad [14]$$

$$M_{xy} = \frac{\sum\limits_{i=1}^{n} x_i y_i w_i}{\sum\limits_{i=1}^{n} w_i} - M_x M_y = \frac{\sum\limits_{i=1}^{n} x_i y_i}{n} - M_x M_y \qquad [15]$$

$$a = \sqrt{2(M_{xx} + M_{yy}) + 2[(M_{xx} - M_{yy})^2 + 4M_{xy}^2]^{\frac{1}{2}}} \qquad [16]$$

$$b = \sqrt{2(M_{xx} + M_{yy}) - 2[(M_{xx} - M_{yy})^2 + 4M_{xy}^2]^{\frac{1}{2}}} \qquad [17]$$

$$\Phi = \tfrac{1}{2}\tan^{-1}\left(\frac{2M_{xy}}{M_{xx} - M_{yy}}\right) \qquad [18]$$

There are many papers discussing alternative elliptical fitting techniques and their relative merits.[28] Recently, an alternative software technique has been explored for fitting elliptical curves to incomplete noisy confocal image edge data, that is, using spokes and least squares fitting.[29] The attraction of this technique is that partial elliptical images can be handled efficiently as well as complete images.

Because of the possibility of chipped and broken fibres (and in short fibre composites, fibre ends intersected by the section plane) it is vital to have a measure of the accuracy of the elliptical fit to the original fibre image data. A simple but useful parameter, the fit factor, $F_f$ which evaluates the quality of the fit is defined by

$$F_f = (\pi ab - \text{actual pixel area})/\text{actual pixel area} \qquad [19]$$

where $\pi ab$ is the area of the fitted ellipse. Clearly, a perfectly elliptical fibre image and a perfectly fitted ellipse to the data would result in $F_f = 0$, a poor fit would correspond to $F_f > 0.01$, that is, a 1% discrepancy. By plotting the semi-minor axis value, $b$, against the fit factor, objects within the field of view could be identified depending upon whether they are touching fibre images, incomplete or damaged fibre ends. An $XY$ image frame which illustrates five different categories of fibre image (i.e. complete fibre, hole in fibre, fibre fragment, incomplete fibre and unsplit fibres) is shown in Fig. 7.5. Typical results are illustrated in Fig. 7.6. Pie charts may be produced which give the operator an indication of both the overall image quality and fibre fragment distributions for each large area scan.

In these high spatial resolution, large area, studies, an $XY$ translation stage is used to scan over large areas by overlapping individual $XY$ frames. The basic problem is that when the computer sends a fixed number of pulses to a stepper motor (to determine the movement in $X$ or $Y$) and the stepper motor shaft rotation is converted into an $X$ or $Y$ linear movement, the actual linear movement produced is variable due to the eccentricity of the lead screw. This effect is only

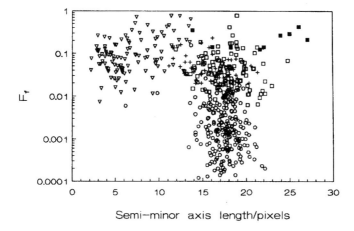

$F_f$

Semi—minor axis length/pixels

7.6 Quality of the elliptical fitting procedure can be quantified by defining a 'fit factor', $F_f$. By correlating this fit factor to the semi-minor axis of each object within the image, different polishing artefacts can be distinguished. ○, good fit; □, holes; ▽, fragments; +, incomplete; ■, unsplit.

noticeable at the high magnifications employed. Hence, the manufacturers' of *XY* stages prefer to quote the repeatability of positioning between two points $(x_a, y_a)$ and $(x_b, y_b)$ rather than absolute positioning accuracy from any initial starting point $(x_i, y_i)$. The cyclic variations in the *X* or *Y* movement error are, to a first approximation, described by a simple sinewave equation of the form $\delta X = a \sin bX$ but over large area scans, there are deviations from this simple equation. Therefore care must be exercised in removing the effect from the raw data.

Fundamentally, there are three possible approaches to the calibration of the absolute positioning in *X* and in *Y* over large scanned areas. The most expensive option is to backup the stepper motor movement with a laser interferometer to check on the absolute positioning of the moving parts! (This technique is used on the large and powerful COSMOS facility at Edinburgh University for analysing large area Schmidt plates).[27] The second method (which is the one preferred at Leeds) is to match fibre images within one high spatial resolution *XY* frame with those same images on the next *XY* frame, that is, after the stepper motors have moved the scanned area to another (overlapping) location. This technique only fails when there are no objects within the field, which is an unlikely event during the analysis of most fibre reinforced composites.

However, there are occasions, for example, scanning near to the edges of a complex part or when the sample has large resin rich regions, when fibre images are lost from the field of view. In this case, a third method is to create a 'lookup table' of linear movements for a fixed set of pulses to the stepper motor, but this assumes that the large area scanning always starts from the same point in *X* and

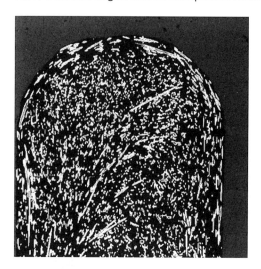

7.7 Example of a complex shaped region (area 4 mm × 7 mm) which has been scanned and reconstructed from 1628 high spatial resolution *XY* frames.

*Y* (and hence always returns to that same point at the end of a large area scan. (If the *XY* stage is accidentally knocked, the operator would have to recalibrate the stage movement and create a new lookup table.) A collage of 1628 image regions has been used to reconstruct a complex automotive sample of glass fibres in epoxy in Fig. 7.7. The actual physical size of the region is 4 mm × 7 mm.

The high spatial resolution large area capability of this design enables the rapid scanning of carbon and glass fibre reinforced composites. The only downside to the 2D optical reflection technique is the ambiguity in the in-plane angle, Φ. Current developments at Leeds, described later in Section 7.6.2, will remove this ambiguity and will enable the deconstruction and reconstruction (in 3D) of fibres in opaque composite materials.

## 7.3    Confocal laser scanning microscopy

The problem with conventional 2D optical reflection microscopy is that the *XY* image, although perfectly acceptable for many applications, is formed from light reflected not only at the focal plane of the objective lens, but also light emitted from a region immediately above and immediately below this focal plane. In other words the *Z* discrimination is poor. However, when focusing on sample surfaces, there is no emission from above the surface and little reflected light from within a typical polished semi-opaque sample. Therefore, the 2D optical reflection microscope is adequate for its main task.

The confocal laser scanning microscope (CLSM) is designed to attenuate this light emission from above and below a thin focal plane by including a small pinhole (or confocal aperture) in the optical path. Also, by using high power laser light, the CLSM can penetrate under the sample surface, that is, not only placing the focal plane at the surface but also being capable of placing this focal plane at different depths, $Z$ within the sample. The depth of penetration depends upon the opacity of the sample at the laser light wavelengths. By choosing suitable filters in the optical path, the CLSM can also be operated in 'fluorescence mode', creating an $XY$ image of the strength of fluorescing artefacts within the sample. Because of this optical depth discrimination, the quality of the surface finish is not too important for the CLSM, especially if fluorescence mode is used.

Over the past eight years, many commercial CLSM systems have appeared, but they have been designed mainly for the biological and physiological research areas and assume that the researcher wants full 'hands-on' control of the confocal instrumental parameters. A number of groups have explored the use of the CLSM for 3D surface topological studies in materials science, for example Gee[30] and Dutch groups have been active in confocal design (e.g. Brakenhoff *et al.*[31]) and in some of the earliest papers which speculated on its potential use in composites research.[32,33] If the reader is interested in the finer points of the CLSM and the systems on the market, there is a superb CLSM reference book edited by Pawley[34] and also a mathematically rigorous account of the CLSM by Wilson.[35]

## 7.3.1  Basic principles

A highly schematic ray diagram and an overview of the optical path in a typical commercial instrument, the Biorad MRC 1000 is shown in Fig. 7.8(a) and (b). The optical path is complicated by the need to insert special filters which select either the laser wavelength (reflection mode of operation) or wavelengths greater than the exciting laser wavelength (fluorescence mode of operation).

In most of the commercial CLSM systems, the $XY$ image frame is built up pixel by pixel by a point scanning process. The detector is usually a sensitive photomultiplier tube located behind the confocal aperture and it receives light from a small point in the sample which is illuminated by the incoming laser light. The focused laser light is directed to the sample by scanning mirrors in the optical path. Hence, the time taken to build up the $XY$ frame is usually a few seconds. (However, videorate scanning has been achieved recently using acousto-optical deflectors in the Noran CLSM.[36])

The intriguing feature of the CLSM is that, as the $Z$ drive is altered, a different focal plane is selected within the sample instead of the surface blur in a conventional 2D optical reflection microscope. Therefore optical sections are effectively thrown into the sample, provided that it is sufficiently transparent at the laser wavelengths.

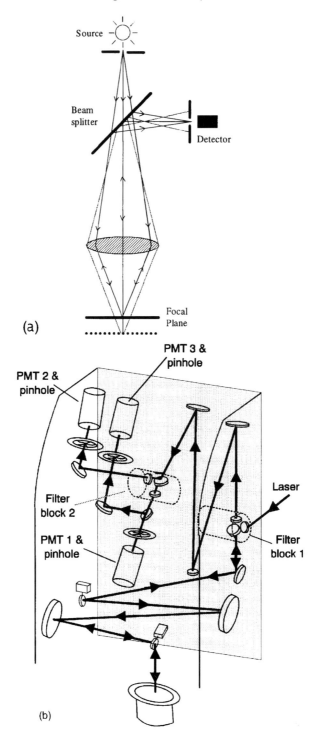

Source

Beam
splitter

Detector

Focal
Plane

(a)

PMT 3 &
pinhole

PMT 2 &
pinhole

Laser

Filter
block 2

PMT 1 &
pinhole

Filter
block 1

(b)

The most common laser subsystem used is the argon-ion laser operating at a wavelength of 514 nm and delivering 25 mW to the sample. The objective lens must be a high numerical aperture (*NA*) lens in order to achieve good optical sectioning and to collect a significant proportion of the returning light from the sample. The significance of the *NA* value and the objective immersion medium refractive index, $n_o$, to the light collection efficiency, $\varepsilon_c$, is shown below

$$\varepsilon_c = \frac{1}{2}\left(1 - \sqrt{1 - \left(\frac{NA}{n_o}\right)^2}\right)$$  [20]

For air objectives, an *NA* of 0.95 is preferred, whereas for oil immersion objectives, an *NA* of 1.4 gives excellent results. The optical sectioning resolution in Z is also improved by a higher *NA* objective, reaching an optimal full width-half maximum response, $\delta z_{fwhm} = \pm 0.25$ $\mu$m.

## 7.3.2 Operational modes for three-dimensional reconstruction

As has been mentioned already, there are two main CLSM operational modes. The reflection mode is the term used to denote that the detector receives light at the same wavelength as the exciting laser radiation. Supersharp images are obtained in reflection mode of the sample surface features and coupled with the off-axis light discrimination, it has led to the CLSM being used for 3D surface topological studies.

Fluorescence mode is the term used to denote that the detector receives the fluorescence radiation which has been excited by the incident laser light but radiates at longer wavelengths ($\lambda > 514$ nm). An example of a typical surface *XY* image seen in reflection and fluorescence modes is shown in Fig. 7.9(a) and (b). In both cases a Nikon, oil immersion objective, *NA* 1.4 and 60× magnification has been used to obtain these images on a Biorad MRC 500 system. Note that the fluorescence signal is weaker than the reflection signal and hence it is necessary to integrate over many image frames to improve the signal-to-noise ratio. This process is called Kalman filtering.

The best mode of operation depends upon the application. It has been found that the fluorescence mode is preferred for the study of fibres in polymeric composites, because the epoxies fluoresce well and the fibres do not fluoresce (therefore giving good contrast).[37] Voidage can be discriminated by a combination of reflection and fluorescence modes (as discussed in Section

*7.8* (a) Highly schematic view of the confocal concept where the light reflected from the sample is directed through a 'confocal aperture' by a beamsplitter (giving better out-of-plane rejection) and (b) a schematic of the folded optics in a Biorad MRC 1000 (modern version of MRC 500).

*7.9* *XY* surface image of a well aligned glass fibre reinforced T800 epoxy taken with an oil objective (60×) in (a) reflection mode and (b) fluorescence mode. The area scanned is about 100 $\mu$m × 175 $\mu$m and the dark diffuse regions are voids.

7.6.1). However, for determining low packing fraction particulates in thin transparent films (e.g. packaging film which only fluoresces weakly), the reflection mode must be used.

This optical sectioning feature of the CLSM suggests two possibilities for the 3D reconstruction of composite samples. These two possibilities are shown in Figs 7.10 and 7.11 and the most appropriate method depends upon the measurements to be made, for example if 3D fibre waviness is to be studied, *XZ* frames at different *Y* are most appropriate (as shown in Fig. 7.10) with the fibres

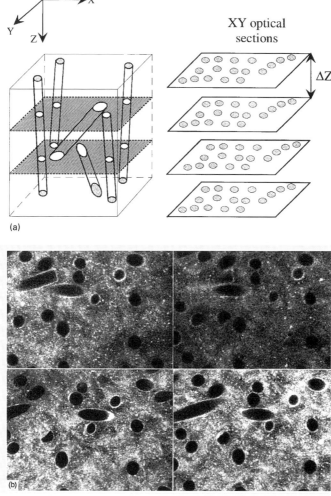

7.10 One method for 3D reconstruction using a CLSM (a) schematic view of a set of *XY* planes taken at different depths *Z* and (b) a set of *XY* frames taken at the surface, 15 $\mu$m, 30 $\mu$m and 45 $\mu$m below the surface of a glass fibre reinforced liquid crystalline polymer (using a 60$\times$, *NA* 1.4, oil immersion objective in fluorescence mode).

lying essentially parallel to the sample surface. However, if a study of the interfibre angular distribution with fibre separation is required, taking *XY* frames at two or three *Z* positions and moving over large areas in *XY* with the sample surface perpendicular to the preferred fibre direction is better (as shown in Fig. 7.11).

The sample volume which can be analysed depends upon the maximum useable depth, $Z_{max}$ and this parameter depends upon many factors, for example

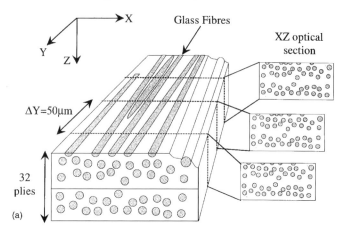

*7.11* Another method for 3D reconstruction using a CLSM (a) schematic view of a set of *XZ* frames taken at different *Y* locations and (b) a typical *XZ* frame for a well-aligned glass fibre reinforced T800 epoxy sample (60×, *NA* 1.4, oil immersion objective and fluorescence mode of operation).

the fibre type (as carbon is opaque, $Z_{max}$ is small), the fibre packing fraction, whether or not the matrix fluoresces and most importantly of all, the sophistication of the image processing routines. Although the image data acquisition may be fully automated, the image data reduction is limited by the considerable difficulties associated with the detection of fibre images within fuzzy, noisy image frames. The design of robust pattern recognition software is a major challenge.

Yet another interesting operational mode is the 'extended focus' facility which allows the user to initiate the acquisition of a sequence of *XY* image frames, each taken at a different depth into the sample. As each *XY* frame is captured, it is added to all of the previous *XY* frames in memory resulting in a composite *XY* frame. The resulting *XY* frame is reminiscent of X-ray transmission

*7.12* (a) Effect of spurious noise pixels around the edge of fibre images on the apparent ellipticity and (b) the variation in the angular error, $\delta\theta$ for a fibre perpendicular to the section plane as a function of different fibre image sizes. The moments technique (—) performs significantly better than the random chords (– – –) technique.

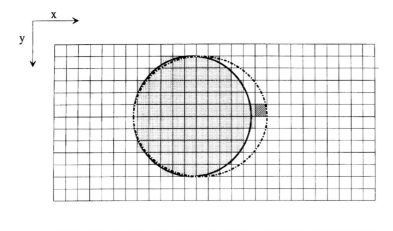

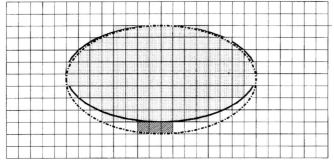

☐ Pixel assigned to fibre image
☐ Pixel assigned as background
▨ Error pixel assigned to fibre image

(a)

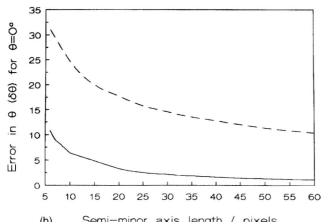

(b)    Semi−minor axis length / pixels

micrographs which show the number density of objects within a sample region, but there is no depth information to separate apparently overlapping objects.

## 7.4     Measurement errors in optical microscopy

Apart from Bay and Tucker,[38] Yurgartis,[7] Hine *et al.*,[10] Fischer and Eyerer[9] and Clarke *et al.*[39] fibre orientation measurement errors have not been discussed in any detail in composites research papers. This is a pity because, without an appreciation of some fundamental instrumental limitations, it is possible to connect a commercial image analyser onto a CCD camera and microscope, use commercial software routines for areas, ellipticities etc. and derive orientation distributions which are at best confusing and at worst totally inaccurate! This section is therefore a cautionary tale to highlight potential pitfalls in the characterisation of fibre orientation states.

### 7.4.1  Noise in pixellated images

Because of the digital nature of the binarised images, it is reasonable to suppose that some spurious pixels might appear on the periphery of the binarised fibre image, as shown in Fig. 7.12(a). The question is 'How does an extra pixel affect the determination of the original fibre direction?' Clearly, using a high magnification, one would expect this effect to be minimised. In Fig. 7.12(b), the effect of a single spurious pixel on the out-of-plane angular error, $\delta\theta$ for a fibre at $\theta = 0°$ and at various image sizes is plotted.

The moral here is that the best estimates of fibre orientation are produced by analysing highly elliptical fibre images. Therefore it is better to section well aligned fibre composites at an angle of 45° to the preferred fibre orientation (and mathematically transform back to that reference plane which is perpendicular to the fibre orientation) as discussed in more detail in Section 7.4.3.

Note that those software packages whose algorithms find image centre coordinates and use random chords through the image centre to deduce the major and the minor axes (and hence the ellipticity of the fibre image) will give even greater orientational errors. The user of commercial software should establish the basis of the algorithm used for image ellipticity determination.

### 7.4.2  Intrinsic fibre ellipticity

One of the key assumptions made by all 2D image analysers of fibre images is that the fibre cross-sections are perfectly circular. If this is not the case, a little thought shows that the observed ellipticity of a fibre's image will be affected by any intrinsic ellipticity or irregularity in that fibre's cross-section. Some carbon fibres have a 'ridged' fibre surface, that is the fibre cross-section is certainly non-circular. Some composites contain carbon or Kevlar fibres which are 'kidney bean shaped' in cross-section!

The two worst case, $\delta\theta$ error curves are plotted in Fig. 7.13(a) in the estimated out-of-plane angle, $\theta$, of a fibre whose intrinsic ellipticity, $e = 0.01$. The schematic diagrams in Fig. 7.13(b) illustrate how an inherently elliptical fibre cross-section can give a misleading $\{\theta, \Phi\}$, as discussed more fully by Davidson.[23] It is extremely difficult to conceive of a sufficiently sensitive experiment to determine the true intrinsic fibre ellipticity but visual inspection of high magnification images indicates that the intrinsic ellipticity of glass fibres is certainly given by $e < 0.01$. Therefore our conclusion is that this may not be a dominant source of error for glass fibre reinforcements, but it certainly could introduce indeterminate orientation errors for carbon and Kevlar reinforced composites.

## 7.4.3  Minimising orientation errors by large angle sectioning

Assuming circular cross-section fibres, the best technique for minimising the orientation errors is to section the sample at an angle of $45°$ or even $60°$ to the preferred fibre orientation. It has been shown,[39] that an apparent frequency distribution of out-of-plane angles, $\theta$ peaking at $15°$ to $20°$ will **always** be seen when the sample is sectioned and polished at right angles to the preferred fibre orientation. This is because of the pixellated nature of the fibre image which produces the same apparent ellipticity as a fibre with an out-of-plane angle of $15°$. However, by sectioning at $45°$, the fibre images become well defined ellipses which are largely unaffected by the pixellation error and which yield more accurate values for the out-of-plane angle, $\theta'$ and in-plane, $\Phi'$ from this section plane. One can mathematically transform back to that section plane perpendicular to the preferred fibre orientation in order to establish the true frequency distributions of out-of-plane angle, $\theta$, and in-plane angle, $\Phi$. The transformation equations are given below:

$$\theta = \cos^{-1}(\sin\alpha.\sin\theta'.\sin\Phi' + \cos\alpha.\cos\theta') \qquad [21]$$

$$\Phi = \tan^{-1}\left(\frac{\cos\alpha.\sin\theta'.\sin\Phi' - \sin\alpha.\cos\theta'}{\sin\theta'.\cos\Phi'}\right) \qquad [22]$$

where $\alpha$ is the angle between the normal to the section plane and the preferred fibre direction, as shown in Fig. 7.14(a).

In Fig. 7.14(b), the frequency distributions for $\theta$ are shown for a sample consisting of well aligned continuous glass fibres in epoxy. The spurious $15°$ to $20°$ peak is seen in the out-of-plane angle distribution from the perpendicular section together with the improved (and consistent) angular distributions obtained after sectioning at $\alpha = 30°$, $45°$ and $60°$ and mathematically transforming back to the perpendicular section. Clearly, the small residual orientation errors from the highly elliptical fibre images apply also to the transformed orientations.

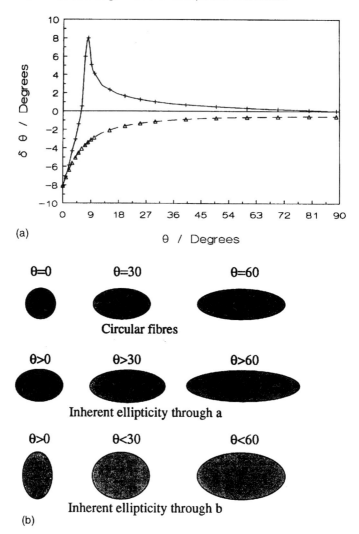

(a)

(b)

7.13  (a) Effect of a non-circular fibre cross-section on the worst case
      angular error $\delta\theta$ as a function of angle, $\theta$. The two curves are a
      result of the section plane alignment with respect to the major and
      minor axes of an intrinsic elliptical fibre cross-section +, through
      a; △, through b. (b) Schematic diagrams of observed elliptical
      footprints due to a circular cross-section fibre and inherently
      elliptical fibres (for selected values of angle, $\theta$).

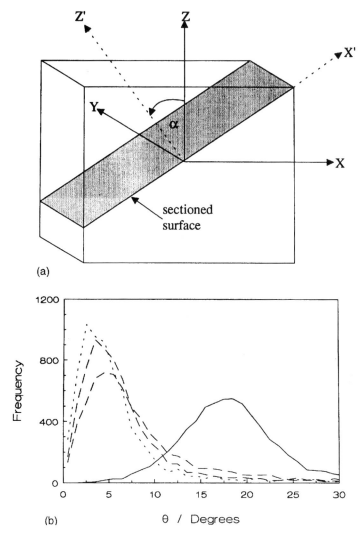

(a)

(b)                    θ / Degrees

*7.14*  (a) Sectioning at an angle, $\alpha$ in order to create elliptical images and hence to improve the estimation of the orientation distribution, $f(\theta)$, of well aligned fibres. (b) A comparison of the orientation distributions of a well aligned sample from the usual section plane perpendicular to the mean fibre direction (giving the spurious peak at 18°) and also from section planes at 30°, 45° and 60° (after mathematical transformation to the perpendicular plane). —, $\alpha = 0$; --, $\alpha = 30°$; ——, $\alpha = 45°$; -----, $\alpha = 60°$.

### 7.4.4  Confocal laser scanning microscope and the apparent depth problem

In the previous section, it has been pointed out that deducing fibre orientations from the ellipticity of the fibre images will only work when each fibre has a circular cross-section. Deducing the orientation of fibres with irregular cross-sections requires a different strategy; namely to locate a fibre image centre on one $XY$ plane and then determine the same fibre centre on a second $XY$ plane, separated from the first by distance $\delta z$, as shown in Fig. 7.1. Hence from the shift in fibre centre coordinates, $\delta x$ and $\delta y$, the angles $\theta$ and $\Phi$ may be deduced unambiguously, using the following equations

$$\theta = \tan^{-1}\left(\sqrt{\frac{\delta x^2 + \delta y^2}{\delta z}}\right) \tag{23}$$

$$\Phi = \tan^{-1}\left(\frac{\delta x}{\delta y}\right). \tag{24}$$

If the CLSM can penetrate below the sample surface by 10 $\mu$m or more, and if a 60× oil immersion objective lens is used to give a high magnification so that the size of each pixel in $X$ and $Y$ represents 0.25 $\mu$m, the relationship between the out-of-plane orientation error, $\delta\theta$, as a function of fibre angle, $\theta$, may be predicted. The curves are shown in Fig. 7.15.

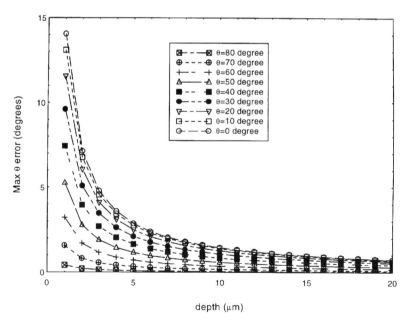

7.15  Error in the out of plane angle, $\theta$ as a function of separation of two registered $XY$ planes for a CLSM system (assuming that the uncertainty in the location of fibre centre coordinates is 1 pixel).

If the composite sample is sufficiently transparent to allow for larger penetration depths to be achieved, the out-of-plane angular error $\delta\theta$ will be even smaller. The only problem is that the distance moved by the Z-drive may not be the actual distance moved by the optical $XY$ plane within the sample, as discussed below!

Referring to Fig. 7.16, it can be seen that if the objective lens immersion medium has a different refractive index to the sample, the effect of refraction at the surface of the sample gives rise to an apparent depth problem. Visser et al.[40] have adopted a geometrical method to estimate the magnitude of this effect and he derives the relationship

$$\Delta f = \frac{\tan[\sin^{-1}(NA/n_o)]}{\tan[\sin^{-1}(NA/n_s)]} . \Delta z \qquad [25]$$

where $\Delta f$ is the effective depth that the focal plane moves for an axial shift of the Z-drive, $\Delta z$. The numerical aperture of the objective lens is $NA$ and the refractive indices of the objective immersion medium and sample are $n_0$ and $n_s$, respectively. Fortunately, many polymer matrices, see Table 7.1, have a refractive index close to 1.518 which is the refractive index of the oil used for oil immersion objectives (at a temperature of 25°C). If the refractive indices are well matched, the apparent depth problem is minimised. However, the sample refractive index is a more complex issue, because it depends upon the refractive index of the reinforcing glass as well as the matrix and another complication is that the polymer matrix may exhibit birefringence.

More research is needed on these issues despite the fact that the topic has been addressed in a number of recent papers.[42,43]

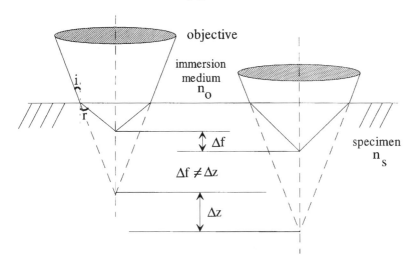

7.16 Schematic diagram of the apparent depth problem for a CLSM where the objective lens immersion medium has a refractive index mismatch with the sample ($\Delta z$ is the movement of the $Z$ drive and $\Delta f$ is the actual movement of the focal plane within the sample).

*Table 7.1* Refractive indices of a number of common polymer matrices[41]

| Matrix type | Refractive index |
| --- | --- |
| PVDF | 1.42 |
| POM | 1.48 |
| PMMA | 1.49 |
| PP | 1.49 |
| PVC | 1.54–1.55 |
| PEEK | 1.67 |

## 7.4.5  Stereological issues

Any estimation of the frequency distribution of fibre orientations from a single 2D section is a biased height-weighted estimate of that distribution.[44-46] There are two main effects:

1  *Height bias* – The probability of a randomly placed section plane intersecting each fibre depends upon the projection of the fibre at right angles to the section plane, as shown in Fig. 7.17(a). If all of the fibres have the same length, the probability that a fibre with out-of-plane angle, $\theta$, intersects the sectioned surface is proportional to $\cos \theta$.

2  *Solid angle bias* – The number of fibres expected within a solid angle, at a mean out-of-plane angle, $\theta$, must be corrected for the area contained within that solid angle, i.e. $\sin\theta.\Delta\theta.\Delta\Phi$ in order that a meaningful frequency distribution of fibre directions can be deduced. The probability of detecting fibres within a small angular range, $\delta\theta$ about $\theta = 0°$, is practically zero because of the small solid angle, but the probability is high for $\theta$ values approaching 90°, as shown in Fig. 7.17(b).

Hence the raw 2D orientation datasets could be corrected for these two effects by using the $\cos^{-1}$ and $\sin^{-1}$ weighting functions. Other, more complicated, expressions have been suggested to take into account the variation in fibre lengths and fibre aspect ratios (ratio of fibre length to fibre diameter).[38,47] Möginger's weighting function, $g(\beta, r, l_n)$ where $\beta = 90° - \theta$, $r = $ mean fibre aspect ratio and $l_n = $ normalised fibre separation (dependent upon the fibre volume fraction) is given as

$$g(\beta, r, l_n)$$

$$= \frac{r}{\sqrt{r^2.\sin^4\beta + \left[\left(\frac{r+d}{l_n+d}\right)^2 + \left(\frac{r(l_n+d)}{r+d}\right)^2\right].\sin^2\beta.\cos^2\beta + \cos^4\beta}}.$$

[26]

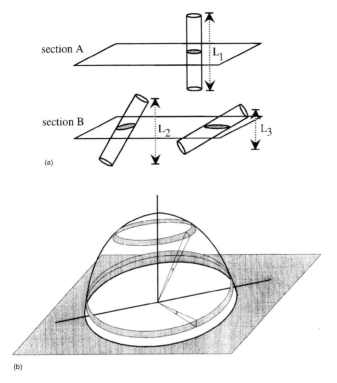

(a)

(b)

*7.17* (a) Height bias effect, i.e. the length of a fibre perpendicular to the sectioning plane determines the probability that that fibre will be detected. (b) Solid angle bias effect, i.e. the number of fibres detectable within a small range of angles $\Delta\theta$ about an angle $\theta$ depends upon that value of $\theta$ (at 0°, the effective solid angle is small, but at 90°, the effective solid angle for detection is large).

Bay and Tucker[38] have been concerned with the orientation states of short fibre reinforced composites and the issue of fibres oriented within 20° of the section plane. If the mean fibre length is $L$ and the mean fibre diameter is $d$, there is a critical angle, $\theta_c$, given by $\theta_c = \cos^{-1}(d/L)$ below which the fibre images are elliptical in shape. (At angles $\theta > \theta_c$, the fibre image is non-elliptical because the section plane cuts through at least one fibre end). Hence they suggested a more appropriate, two part weighting as shown in Eqn 27(a) and (b), where $F_n$ is the weighting function and $F_{90}$ is the weighting function at 90°

$$\frac{F_n}{F_{90}} = \frac{1}{\frac{L}{d}\cdot\cos\theta} \qquad \text{(for } \theta \le \theta_c) \qquad\qquad [27a]$$

$$\frac{F_n}{F_{90}} = 1 \qquad \text{(for } \theta > \theta_c). \qquad\qquad [27b]$$

It is understood that more recent work by Tucker has refined the weighting function above this critical angle, $\theta_c$.

The problem with all of these correction factors is that they are 'model-based' (e.g. requiring assumptions to be made for the length distributions) and therefore it is impossible to evaluate the accuracy of the resulting fibre orientation distributions. As has been argued already, the best 2D data on fibre orientations are produced by large angle sectioning of preferred fibre directions. However, if the sample consists of fibres with a wide range of orientations, this large angle sectioning and transformation procedure breaks down.

The worst case scenario is when the user suspects that the sample contains fibres which are distributed isotropically. When the fibres in a large area 2D section plane are analysed, after correction for the height biasing and solid angle biasing, there will be few fibres close to $\theta = 0°$ (which is due to the pixellation error effect) and few fibres at $\theta = 90°$ (which is due to the fitting of ellipses to rectangular, or partial elliptical, images produced by short fibres lying in the section plane).

If two more 2D sections are taken at roughly orthogonal orientations to the first section, similar frequency distributions will result. The orientation data from these three plots are consistent with the sample being isotropic, but the data do not prove that it is isotropic and do not allow for a measure of deviation from isotropy. Recently, Mattfeldt et al.[48] have suggested a new stereological method for establishing the isotropy of fibre reinforcements using the CLSM technique. Essentially a set of three $XY$ planes, perfectly registered and each separated by 10 $\mu$m are imaged by the CLSM. The fibre image centres are located on all three planes and, from the shift of centres, unambiguous fibre orientations $\{\theta_i, \Phi_i\}$ are derived. This allows for the 3D angular differences, $\beta_{ij}$ between neighbouring fibres to be evaluated according to

$$\cos\beta_{ij} = \sin\theta_i.\sin\theta_j.\cos(\Phi_i - \Phi_j) + \cos\theta_i.\cos\theta_j \qquad [28]$$

The authors derived an expression for the cumulative distribution function, $F(\beta)$, which describes the number of fibres within a certain angular distance of each other on the hemisphere of directions, for the case of completely random fibre orientations. The analytical expression, given in Eqn [29] was confirmed by a Monte Carlo computer simulation of random 3D fibre directions

$$F(\beta) = 2.\sin^2\left(\frac{\beta}{2}\right) \qquad [29]$$

Orientation data was obtained for glass fibres in polyoxymethylene (POM) which had been hydrostatically extruded at different draw ratios, $\Lambda$ (the ratio of extrudate length to the original billet length). A typical set of frames has been shown earlier in Fig. 7.10(b). Note that, although the POM does not fluoresce, the sizing agents coating the glass fibres fluoresce brightly and facilitate the determination of the fibre centre coordinates. The research work sought to check the isotropy of the initial billet before extrusion and also the validity of the 'pseudo-affine' model of deformation[49] during the extrusion process. For related measurements, see also Sections 7.5.1 and 7.5.2 below.

Within the initial POM billet, the fibres on the millimetre scale did not conform to completely random orientations although, as expected, at higher draw ratios the fibres became more and more aligned along the draw axis. The billet was processed from granules (measuring many millimetres by millimetres) in which the fibres may have had preferred orientations. An ultrasonic 'time of flight' technique had been used to determine the 3D elastic moduli of the billet and on the centimetres scale of the ultrasonic beam measurements, the results *were* consistent with isotropy (see Section 7.6.4 for more details of the ultrasonic apparatus).

## 7.5 Characterisation studies of fibre reinforced composites

A series of case studies are presented in this section to illustrate the potential of the two optical techniques: 2D large area optical reflection microscopy and 3D CLSM. As we shall see, in certain cases, full 3D spatial information on fibre orientation states (and voids) has been obtained with the CLSM technique.

### 7.5.1  Geometrical coefficients

The Polymer Physics Group at the University of Leeds has conducted research into hydrostatically extruded glass fibre reinforced POM for over 12 years. Many papers have been written,[50,51] in which the modelling of the extrusion process and the mechanical performance of the extrudates has been explored. Recently, more effort has been expended in the measurement of 3D fibre orientation distributions in the extrudates in order to provide the missing link between process and properties, first, using the 2D large area system and latterly, the CLSM (because in fluorescence mode, the penetration depth of the CLSM in these reasonably low fibre packing fraction samples is about 30 $\mu$m).

The 'pseudo-affine' model, first proposed by Kuhn and Grün[49] in their early work on rubbers, has been investigated for relating fibre reorientation from $\{\theta, \Phi\}$ in the original billet to $\{\theta', \Phi'\}$ in the extrudate where the angle $\theta$ is the angle of the fibre to the draw direction and the angle $\Phi$ is the fibre angle in the plane perpendicular to the draw direction. The model ignores fibre–fibre inter-actions and predicts the reoriented angles after extrusion via the following equations, where $\Lambda$ is the 'draw ratio'.

$$\tan \theta' = \Lambda^{-1.5}.\tan \theta \qquad [30]$$

$$\Phi' = \Phi. \qquad [31]$$

Figure 7.18(a) shows the actual $\theta$ fibre orientation distributions for different draw ratios, as determined by the large area 2D optical reflection microscopy technique. These distributions have been compared to the predicted orientation distributions (assuming that the orientations were randomly distributed in the original billet – which we now know to be an approximation) and the agreement is reasonable, but not perfect. In Fig. 7.18(b), the expected and the actual

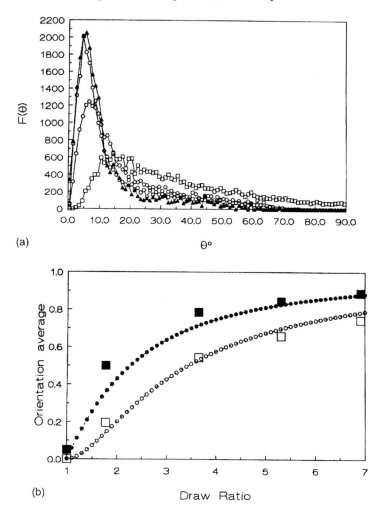

(a)

(b)

7.18 (a) Frequency distributions of out of plane angles, $\theta$ for different draw ratio samples of glass fibre reinforced POM. $\square$, $\Lambda = 1.77$; $\bigcirc$, $\Lambda = 3.65$; $\bigcirc$, $\Lambda = 5.31$; $\blacktriangle$, $\Lambda = 6.91$ (b) A comparison of the predicted geometrical coefficients, $P_2(\cos\theta)$ and $P_4(\cos\theta)$ using the pseudo-affine model and the actual measured values of these coefficients for glass fibre reinforced POM. $\bullet$, $\langle P_2 \rangle$ theory; $\bigcirc$, $\langle P_4 \rangle$ theory; $\blacksquare$, $\langle P_2 \rangle$ measured; $\square$, $\langle P_4 \rangle$ measured.

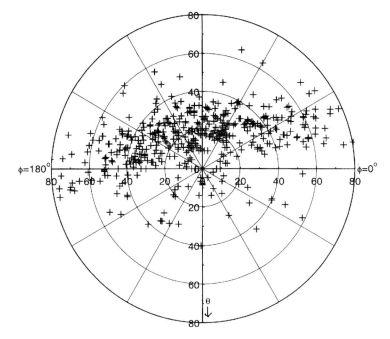

*7.19* Polar plot of unambiguous fibre orientations {θ, Φ} for the initial billet of glass fibre reinforced POM showing the anisotropic fibre orientations on the millimetre scale.

geometrical coefficients, $P_2(\cos\theta)$ and $P_4(\cos\theta)$ are shown plotted as a function of the draw ratio. These plots indicate that, to a first approximation, the pseudo-affine model does indeed give reasonable predictions of fibre orientations at different draw ratios.

In Fig. 7.19, a polar plot of the unambiguous {θ, Φ} data for the original billet are shown. Note that, at angles close to $\theta = 0°$, there are very few fibres and yet there is a broad swathe of fibres from left to right over a wide range of angles. This plot indicates that the fibres are not randomly oriented[48] and the paucity of fibres at angles close to $\theta = 0°$ is not an instrumental effect due to pixellation noise (as in the 2D ellipticity analysis), but a real effect showing preferential fibre orientations on the millimetre scale size.

## 7.5.2    Fibre length distributions

Distributions of fibre lengths are of importance to modellers especially for comparing fibre breakage during the processing cycle and also for investigating the mechanical properties of recycled polymer composites. The standard technique for deducing fibre length distributions is to pyrolise the sample and to analyse the residual fibres in a petri dish. Most of the fibres will be separated, but some will be overlapping others. The 2D image analyser used at Leeds has

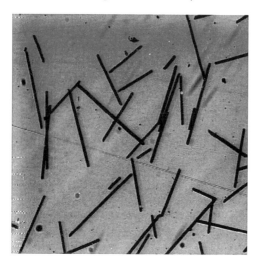

*7.20* Typical *XY* frame of fibres after pyrolysis, ready for fibre length determination.

software routines which take a 'crowded field' of objects and which automatically unscramble crossing fibres in the dataset. It also has the ability to scan over large areas following fibres across many high resolution frames. A typical field of pyrolised fibres is shown in Fig. 7.20 and a derived fibre length distribution for the glass fibres in POM is shown in Fig. 7.22.

The new opportunity for the CLSM is to measure the fibre lengths and fibre orientations at the same time when a processed part is scanned. If the sample is physically sectioned parallel to the preferred fibre orientation (as in Fig. 7.11) and, using the CLSM, a set of optical *XZ* sections is taken into the material, '*in situ*' glass fibre length distributions may be produced. In Fig. 7.21, the reconstruction of a sub-volume of a POM sample at a high draw ratio ($\Lambda = 6.91$) is shown. Note that the true dimensions of this 'reconstructed cube' are $200 \ \mu m \times 1.1 \ mm \times 40 \ \mu m$ in *X*, *Y* and *Z*. The fibres are represented by their central coordinates every $10 \ \mu m$ in the *Y* direction. Clearly, knowing the end coordinates of a fibre, each fibre length may be determined. Hence a frequency distribution of fibre lengths is produced, as shown in Fig. 7.22. This *in situ* fibre length distribution is compared to the fibre length distribution by pyrolysis from the same draw ratio sample and shows excellent agreement.

As expected, there is good agreement between the two techniques; pyrolised fibre lengths and *in situ* fibre lengths. The time taken to generate the 3D dataset with the CLSM does not warrant its use as an alternative to the 2D pyrolised method at the moment. However, the 3D fibre dataset can also be interrogated to search for correlations between fibre lengths and fibre orientations and also for evidence of local fibre curvatures and correlated fibre movement between nearest neighbours.

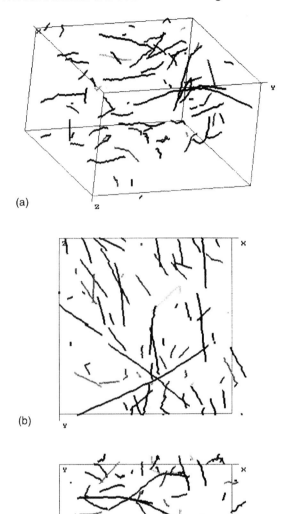

(a)

(b)

(c)

*7.21* (a) Reconstructed 3D 'cube' containing short glass fibres in POM. The actual dimensions were 200 $\mu$m × 1.1 mm × 40 $\mu$m in *XYZ* and (b) a plan view of the fibres in this sample volume and (c) a side elevation view.

*7.22* Comparison between the fibre length distribution of glass fibres in POM deduced from pyrolysis and the fibre length distribution deduced from the fibre lengths in the 3D cube reconstruction of Fig. 7.21(a).

## 7.5.3   Nearest neighbour angular anisotropy

The Instrumentation Group at Leeds has been interested in the nearest neighbour distributions in continuous aligned fibre composites since 1990, that is since the early days of the development of the large area 2D image analyser. Very few papers had been found which discussed the clustering characteristics of fibres or the extent to which real composites mimicked idealised structures (e.g. hexagonal or square arrays of fibres) that were being assumed in theoretical papers on finite element modelling.[52] Two papers have been published on the early work by Clarke and Davidson[53] and Clarke *et al.*[21] The authors are not aware of an analytical solution to the allowed nearest neighbour clustering states at high fibre packing fractions (because it is a classic many–body problem). Therefore it was decided to initiate a programme of 3D computer modelling – using a Monte Carlo type simulation to place fibres in a virtual 3D space in a random way, ensuring that no two fibres occupied the same region of space. In a recent collaborative project, the Instrumentation Group at Leeds and members of the Institute of Mechanical Engineering at Aalborg have approached the 3D simulation problem from different points of view. Aalborg have modified software for modelling low packing fraction spherical inclusions to enable them to simulate short fibres with various packing fractions in 3D. However, they have been restricted to packing fractions below 25% by the available computing power. Using some simplifying assumptions, the Leeds group have simulated higher packing fraction (approaching $v_f = 45\%$) fibres in 3D. This collaborative work is discussed in more detail in an Aalborg PhD thesis.[54]

One parameter which has been explored to characterise the state of order/disorder within high packing fraction, aligned composites is the in-plane, nearest neighbour angle, $\Phi_{nn}$. When the sample contains fibres at a low packing fraction and any fibre orientation is allowed, the frequency distribution of $\Phi_{nn}$ is found to be transversely isotropic, as expected. However, as higher packing fraction parts are processed, the allowable fibre orientation states are strongly affected by the extent of the packing, for example at packing fractions greater than a few percent, fibres interact and tend to become oriented in preferred directions, as shown schematically in Fig. 7.23.

Parts that are processed, for example by compression moulding from prepregs, have layers (i.e. plies) of fibres usually at 40% or higher packing fractions and, in these circumstances, the range of fibre orientations in $\{\theta, \phi\}$ become restricted. The fibres are obviously 'well aligned' and the interesting question is 'can the trend from disordered orientation states to ordered states be characterised?'

A number of papers involving finite element (FE) methods have assumed that the fibres can be represented by hexagonal cells or square cells or by a 2D

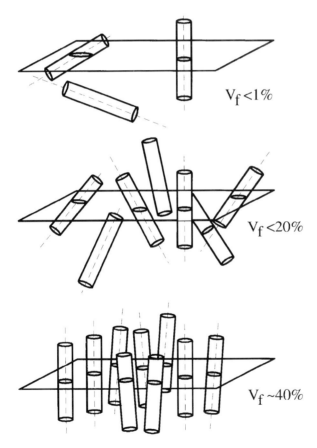

$V_f < 1\%$

$V_f < 20\%$

$V_f \sim 40\%$

*7.23* Schematic view of the increasing influence of neighbouring fibres as the fibre volume fraction changes from 1% through 20% to 40%. The result is a trend towards spatial ordering.

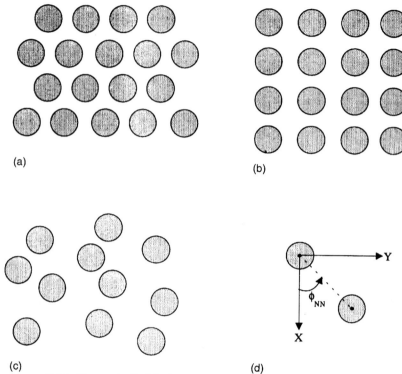

(a)

(b)

(c)

(d)

7.24 Theoretically ideal structures assumed in finite element computations: (a) hexagonal array, (b) square array, (c) Gibbs 'hard core' 2D random and (d) the definition of the nearest neighbour angle, $\Phi_{nn}$.

random Gibbs 'hard core' distribution of fibres, as shown in Fig. 7.24(a), (b) and (c). The Leeds large area 2D image analyser has measured the nearest neighbour distance distributions and also the nearest neighbour angular distributions for a range of composites. None of these archetypal FE models seem to give the true nearest neighbour angular distributions! In Fig. 7.24(d), the nearest neighbour angle is defined as $\Phi_{nn}$ with respect to an arbitrary datum and has a value in the range $0°$ to $360°$. Whereas the frequency distribution of $\Phi_{nn}$ was expected to give an isotropic plot (if the Gibbs hard-core model was valid), when real composites were studied, a clear anisotropy was found, as shown schematically in Fig. 7.25.

7.25 (a) Definitions of the in-phase anisotropy factor, $A_1$, and the out-of-phase anisotropy factor, $A_0$, obtained from the frequency distribution of $\Phi_{nn}$. (b) Plot of $\Phi_{nn}$ for idealised spatial structures. (c) Typical computer simulation where the fibres were given $L/d = 40$, angular ranges $\Delta\Phi = 5°$, $\Delta\theta = 45°$ and a $v_f = 14\%$.

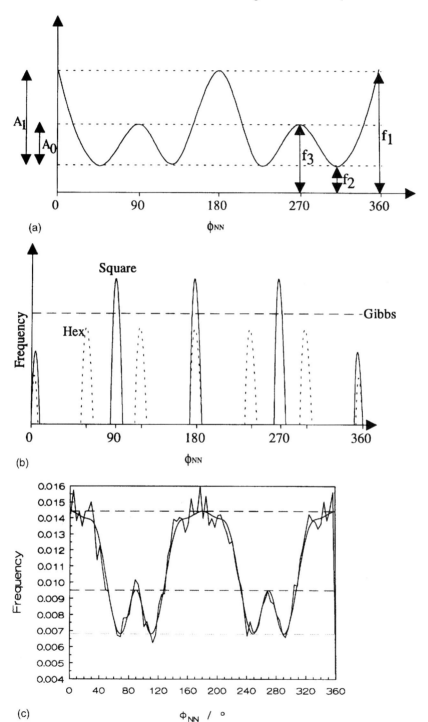

(a)

(b)

(c)

Referring to Fig. 7.25 and Eqns [32] and [33], the anisotropy may be characterised by an 'in-phase' anisotropy factor, $A_1$, and an out-of-phase anisotropy factor, $A_0$.

$$A_1 = (f_1 - f_2)/(f_1 + f_2) \qquad\qquad [32]$$

$$A_0 = (f_3 - f_2)/(f_3 + f_2). \qquad\qquad [33]$$

No analytical function has been found to date which predicts this anisotropy as a function of increasing packing fraction and fibre aspect ratio (i.e. ratio of length to diameter). Therefore Monte Carlo simulations have been made to investigate how the anisotropy increases as the packing fraction increases and how it depends upon the constraints on the fibre orientation angles, $\theta$ and $\Phi$ and also the fibre aspect ratio $L/d$. The computing power required to simulate high fibre packing fractions is considerable (for example, the parallel processing network of transputers used in the 2D image analyser were working overnight to complete our original simulations!).

Figure 7.26 shows the correlation between the in-phase anisotropy factor and the out-of-phase anisotropy factor for a number of simulations at a fibre packing fraction of around 35% to 40%. The simulations imply that if a restricted range of allowable in-plane angles, $\Delta\Phi$, and a range of out-of-plane angles, $\Delta\theta$, is specified and if attempts are made to place the fibres in a 2D spatially random way (to mimic the Gibbs 'hard core' distribution) and ensure that no two fibres occupy the same space, an anisotropy in the nearest neighbour angles is inevitable. As the constraint on the range of angles, $\Delta\theta$, is relaxed, the anisotropy in the frequency distribution of $\Phi_{nn}$ decreases and tends towards the point (0,0) on the plot. A perfect square array of fibres would be represented by the point (1,1) and a perfect hexagonal array would be described by a point at (0,1) on the plot. However, in practice there must be angular misalignments (as discussed in the next section) and so real composites would be expected to give anisotropies in the middle part of the plot. A few real composites have been measured with the large area 2D system and the points are plotted for comparison. Our latest work with Aalborg has investigated the way that the anisotropy factors vary with increasing packing fraction.

## 7.5.4  Fibre misalignments and fibre curvature

Fibre misalignments within the long aligned composites are of interest to users because a significant reduction in compressive strength is to be expected for a degree or so of fibre misalignment. However, the problem for the experimenter is to identify all possible sources of fibre misalignment. There are three different effects which one could attempt to unscramble:

1   In multiple-ply composites, there could be misalignment of the plies.

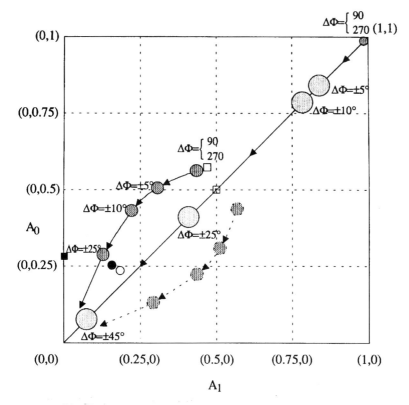

$\Delta\Phi=\begin{cases} 90 \\ 270 \end{cases}$ (1,1)

$\Delta\Phi=\pm5^\circ$

$\Delta\Phi=\pm10^\circ$

$\Delta\Phi=\begin{cases} 90 \\ 270 \end{cases}$

$\Delta\Phi=\pm5^\circ$

$\Delta\Phi=\pm10^\circ$

$\Delta\Phi=\pm25^\circ$

$\Delta\Phi=\pm25^\circ$

$\Delta\Phi=\pm45^\circ$

$A_0$

$A_1$

⊕——▶⊕ 3D hard core model, $\bar{l}=1$mm, packing fraction 33%, $\Delta\theta=5^\circ$.

⊕ ▶⊕ 3D hard core model (with change of $\Phi_{NN}$ origin).

○——▶○ Square array model, $\bar{l}=1$mm, packing fraction 45%, $\Delta\theta=5^\circ$.

7.26  Anisotropy parameters, $A_1$ and $A_0$ for 3D computer simulations of fibres with aspect ratio, $L/d=100$, packing fractions around 33% and restricted ranges of angles $\{\theta, \Phi\}$ are compared with measured anisotropies for a number of aligned composites. ○. continuous glass fibres in epoxy, packing fraction 52% $\Delta\Phi_{FWHM}=50^\circ$; $\Delta\theta=5^\circ$, ●, Cosmos, carbon samples, packing fraction 25%, $\bar{l}=120$ $\mu$m; □, FT14, carbon reinforced, packing fraction 42%, $\bar{l}=1$ mm; ⊡, FT15, carbon reinforced, packing fraction 42%, $\bar{l}=1$ mm; ■, UD913, continuous carbon reinforced, packing fraction 52%.

2  The fibres within a ply will exhibit a spread of misalignments about the mean fibre direction for that ply.

3  Fibres will exhibit some degree of waviness and hence, local fibre curvatures will give local misalignments which might be expected to widen the effective misalignment distributions.

To date, the Yurgartis technique[7] has been the accepted standard technique to specify angular misalignments of well-aligned continuous fibre composites. It has also been argued that the apparent angular misalignment distribution from Yurgartis' method will give an indication of the waviness of the fibres in the sample (because the misalignment angle might be expected to be correlated to the ratio, $A/\lambda$, of the fibre waviness where $A$ is the amplitude of the waves and $\lambda$ is the wavelength of the waviness). The basis of the Yurgartis technique is shown in Fig. 7.27. However, fundamental assumptions are made with this technique which are questionable in most practical composite samples.[55] A section plane is taken within a few degrees to the main fibre direction and therefore very long elliptical-like images will be formed. These fibre images are usually analysed manually with digitiser pads and from the major and minor axis lengths, a frequency distribution of misalignment angles is derived with sub-degree angular errors being quoted. However, perfect elliptical images will be seen only when

- fibres are straight and rod-like over 100 to 200 $\mu$m long fibre segments,
- fibres have circular cross-sections,

and the Yurgartis analysis to derive orthogonal fibre misalignment distributions assumes that there is no correlation between fibre waviness in orthogonal planes (but see below).

Two-dimensional optical techniques have also been used to determine the waviness and/or misalignment of plies in multi-ply materials.[56] However, in the future, we believe that the best way to characterise fibre waviness (when the sample is sufficiently transparent) is using the CLSM technique.[55] Once again, by physically sectioning the sample parallel to the preferred fibre orientation and taking a set of three registered optical $XZ$ sections separated by 50 $\mu$m along the

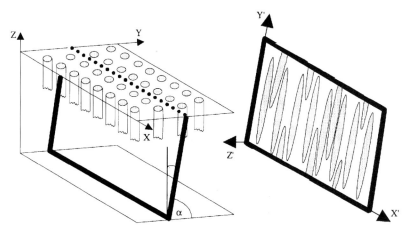

*7.27* Yurgartis technique for estimating small angular misalignments from highly elliptical images on section planes at a small angle, $(90 - \alpha)°$ to the mean fibre direction.

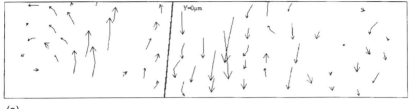

(a)

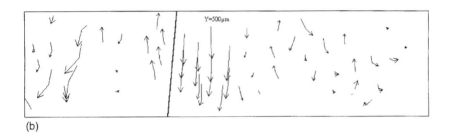

(b)

(c)    7.28    Reconstructed local fibre movements from sets of three XZ planes. The length of the 'arrow' indicates the misalignment of each 100 $\mu$m fibre segment from the Y axis and a 'curved arrow' indicates local fibre curvature. (a) shows the fibre misalignments at an arbitrary start point $Y=0$, (b) shows the fibre misalignments at $Y=500$ $\mu$m, and (c) shows the fibre misalignments at $Y=1000$ $\mu$m. Note the fibre clustering, correlated movement of fibres and the most likely ply boundary.

fibre direction, fibre centre coordinates are obtained and reconstructions of the local fibre curvatures may be made visible, as shown in Fig. 7.28. The dataset shown was part of an investigation into waviness in glass fibre reinforced multi-ply T800 epoxy which had been compression moulded. Note the tendency for correlated movement of neighbouring fibres in the same ply and the identification of the locations of a ply boundary.

Once the loci of each fibre's central axis have been found in *XYZ*, the true 3D radius of curvature of the fibres within the dataset can be investigated. In Fig. 7.29, the cumulative distribution of glass fibre radii of curvature is plotted where

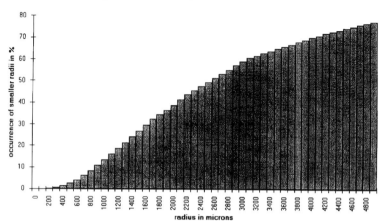

*7.29* Cumulative distribution of radii of curvature of 100 μm fibre segments for a glass fibre reinforced T800 epoxy sample produced by compression moulding. Note the smallest radius of curvature detected was 250 μm and only 25% of the fibre segments had curvatures greater than 5000 μm (i.e. only 25% are essentially straight fibre segments).

the radii have been inferred from 100 μm fibre segments. Note that radii of curvature as small as 250 μm have been seen (for glass fibre diameters of about 10 μm) and that only around 25% of all fibre segments have a radius of curvature in excess of 4000 μm. Therefore, for the particular sample that has been characterised, only 25% of the fibre segments satisfied the condition that they were indistinguishable from 'rigid, straight line segments' over fibre segment lengths of only 100 μm!

In Fig. 7.30(a) and (b) sub-degree resolution angular misalignment distributions are shown in the two orthogonal planes (the *XY* section plane and the *YZ* plane) for fibres within a sample volume of 200 μm × 5 mm × 40 μm. Note that, in the *YZ* plane, the fibres in two neighbouring plies have been recorded and the misalignment of the plies is clearly distinguished. A ply misalignment of about 7° has been found. Also note that the width of the misalignment distribution is determined by both the mean fibre misalignment and also the effect of the local fibre waviness.

## 7.5.5  Power spectral density of fibre waviness

In the previous section, it is clear that the CLSM technique is in a unique position to unscramble 3D waviness in glass fibre reinforced composites and to enable novel characterisation parameters of waviness to be defined. Recent papers by Budianski and Fleck[57] and Slaughter and Fleck [58] have discussed the significance of the power spectral density in their theories of composite strength and failure. In these papers, they had to make intelligent guesses about the form

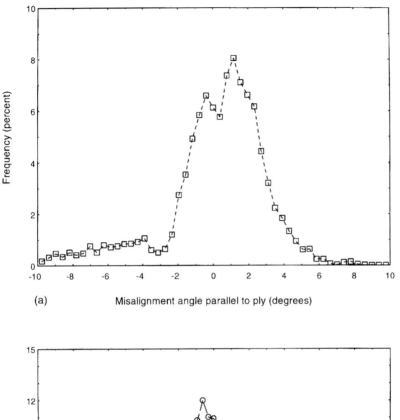

(a)          Misalignment angle parallel to ply (degrees)

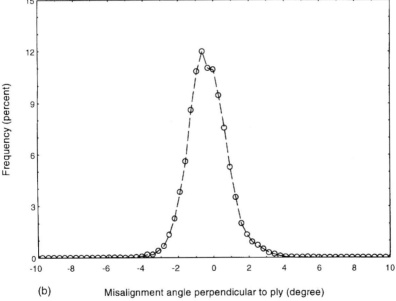

(b)        Misalignment angle perpendicular to ply (degree)

*7.30* CLSM dataset for the glass fibre reinforced T800 epoxy sample may also be used to deduce the misalignment angles for 100 μm fibre segments in (a) a plane parallel to the ply interface and (b) a plane perpendicular to the ply interface.

of the power spectral density because it had been impossible actually to measure this parameter.

The power spectral density, $S_u$ is determined by the Fourier transform of the auto-correlation function, $\mathscr{R}_u$, given by:

$$\mathscr{R}_u(\Delta) = \int_{-\infty}^{\infty} u(y).u(y + \Delta)dy \qquad [34]$$

Therefore, where $u(y)$ represents the deviation in $x$ from the mean fibre path (assumed to be oriented in the $y$-direction), $S_u$ is given by

$$S_u(\omega) = \frac{1}{2\pi} \int_{-\infty}^{\infty} \mathscr{R}_u(\Delta).e^{-i\omega\Delta}d\Delta \qquad [35]$$

Using the CLSM data on glass fibre reinforced epoxy, the power spectral density may now be determined as has been discussed recently by Clarke et al.[59]

In Fig. 7.31, a 3D visualisation of the longest fibres in a sample volume of 200 $\mu$m $\times$ 5 mm $\times$ 40 $\mu$m are shown. These fibres were used to deduce the power spectral density plots presented by Clarke et al.[59] The largest power was found to be in the smallest spatial frequencies, $\omega$. Hence, this confirms that typical fibre 'wavelengths' for aligned continuous fibre reinforcements are expected in the range 1 mm $\leqslant \lambda < 4$ mm.

To our knowledge, these are the first power spectral density distributions which have been derived for any composite material. The potential of the CLSM technique for novel relevant 3D characterisations is obvious. The only problem is the manpower effort to acquire and analyse these data! At Leeds, work is ongoing towards generating fibre waviness data within larger subvolumes (e.g. 400 $\mu$m $\times$ 1 cm $\times$ 40 $\mu$m) using improved image data acquisition and better post-processing of the image data together with pattern matching between $XZ$ sections in order to semi-automate the whole measurement process.

At the time of writing, it is not clear whether the power spectral density approach or the radii of curvature/angular misalignments approach is better to characterise such well aligned composite materials.

## 7.6    Future developments

In this final section, a number of interesting issues are raised which might benefit from the latest developments in the 2D large area and CLSM techniques. First, the problem of low level voidage is discussed, followed by an indication of the software challenge presented by the 3D reconstruction of large sample volumes by both 2D optical reflection microscopy and confocal laser scanning microscopy. Finally, a fully automated ultrasonic 'time of flight' system is

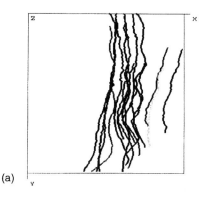

(a)

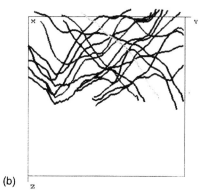

(b)

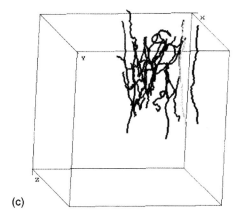

(c)

*7.31* Longest fibres in the glass fibre reinforced T800 epoxy sample are shown in the reconstructed 3D 'cube' (real dimensions are 200 $\mu$m × 5 mm × 40 $\mu$m) (a) in plan view, (b) in side elevation and (c) a 3D cube view.

outlined which enables the determination of 3D elastic moduli for a wide range of samples.

### 7.6.1  Low level voidage determination

In most polymer laboratories, the estimation of void volume fraction, $v_v$ may be undertaken by gravimetric, ultrasonic or optical methods. The 2D image analysis of voids on a section plane is popular because a simple thresholding can identify the void areas and as shown in Eqn [3], the void volume fraction is usually equated to the void area fraction. The question is 'what assumptions are made when 2D image analysis is used to determine voidage?'

It is true that if an infinite number of 2D sections were taken, the mean area fraction would be equal to the mean volume fraction, but in practice only one section plane is taken and a finite number of random or pseudorandom high resolution image frames are analysed for voids. This procedure must assume that:

1    The voids are randomly distributed throughout the sample.
2    The voids are normally distributed about a mean size.
3    No correction is made for height bias (although voids are rarely spherical).
4    Usually, no 'unbiased' counting frame is used – void areas over the whole *XY* image frame are estimated.

Many papers have been written about the estimation of voidage, for example Shi and Winslow[60] present tables for the expected error on an estimate of voidage from a knowledge of the number of voids per *XY* frame and the number of *XY* frames analysed, but their analysis still depends upon the assumptions above. The Leeds 2D large area analyser has the advantage of speed over most commercial systems and also, the mapping of voids (both size and central coordinates) over areas of many square millimetres, will allow assumption (1) to be tested and obviates the need for assumption (4) above. Any tendency for voids to cluster can be established and also whether or not there are systematic void number density gradients throughout the sample. However, the technique can only give an indication of (2) and (3) above. Orthogonal sections could be taken and scanned, but this would be a complicated operation and would increase the analysis time significantly.

The CLSM technique opens up the possibility of estimating the void volume fraction reliably because of its ability to reconstruct a 3D volume. As shown in Fig. 7.32, voids can be imaged at the surface by using the reflection mode and a low *NA* air objective lens. The CLSM could also probe into the void (if a high *NA* air objective were used) and a 3D profile of the surface void could be obtained. However, if good penetration below the sample surface can be achieved, there is the possibility of using both reflection mode and fluorescence mode together to identify sub-surface voidage. Note that there are three

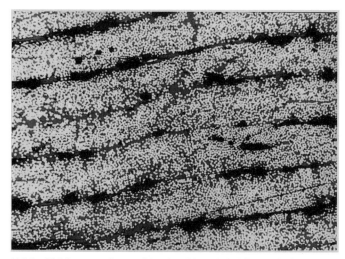

*7.32* *XY* frame taken with the Biorad MRC500 CLSM, using an air objective, 10 × magnification, *NA* 0.3. The sample is a GKN Westland sample, specially prepared to create interply voidage. The area scanned was 1 mm × 0.75 mm.

consequences of using an oil immersion objective and scanning the same region in both reflection mode and fluorescence mode:

- The oil fills in surface voids and they become invisible.
- A very bright signal is produced by subsurface voids in reflection mode (because of the refractive index mismatch between matrix and void).
- The voids do not fluoresce, whereas the matrix most likely does fluoresce.

In Fig. 7.33(a), an oil immersion objective has been used to take an *XZ* section into a glass fibre reinforced liquid crystal polymer matrix. Unfortunately, the voids show more clearly when the overlaid reflection and fluorescence signals are colour coded and are not so obvious in this grey scale image. Also, the gain setting was too high and as a result some flaring of the reflection mode signals from the voids are evident.

In Fig. 7.33(b) the classical Cavalieri approach to estimating the volume, *V*, of an object is shown. The principle assumes that a set of sections separated by '*h*' intersect the object to produce a set of areas, $A_i$. The best estimate of the void volume is then given by $V = h.\Sigma(A_i)$ and with adequate penetration, reasonable estimates of void volumes could be obtained in this manner.

The challenge is to automate the process of CLSM image data acquisition over significant sample volumes, so that it approaches the speed of the 2D system and to develop robust algorithms to determine the total number of voxels (i.e. 3D image cells) which constitute each void.

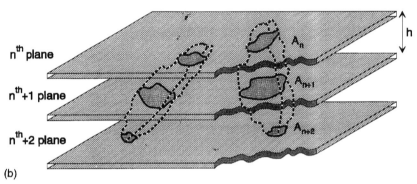

(b)

*7.33* *XZ* scan into a glass fibre reinforced liquid crystalline polymer where the reflection mode image has been superimposed over the fluorescence mode image. Subsurface voids are indicated by bright rellection mode signals which give no corresponding fluorescence signal. (b) A schematic view of the classical Cavalieri approach to volume estimation, *V* from a series of sections separated by '*h*', i.e. if area on the *nth* plane is $A_n$ for one object, then $V = h.\Sigma(A_n)$.

## 7.6.2 Pattern matching between successive two-dimensional sections

Although there are many published papers on pattern recognition and pattern matching between two or more complex images (for example Merickel,[61] Hibbard et al.[62]), the authors are only aware of a few attempts to section a composite sample successfully and then to reconstruct the sample successfully in 3D by pattern matching the fibre images between consecutive sections. However, the only paper that has appeared in a composites conference was by

Paluch[63] and the Paluch technique has been discussed by Clarke *et al.*[55] In essence, Paluch attempted to deduce the waviness of unidirectional carbon fibres by processing a single *XY* image frame at high magnification and then removing another 20 $\mu$m of sample material by sectioning and polishing, before scanning the same region again. Paluch repeated the process 40 times, removing 20 $\mu$m between each scan. The problem was to achieve good registration between these single frames. His method was to minimise the 'random' fibre centre movements between planes, see Fig. 7.34(a). However, as shown in Fig. 7.34(b), if there are any signs of correlated waviness amongst the fibre dataset, his centroid technique tends to straighten out the fibres, that is underestimate the real waviness of the fibres! There is also the problem of knowing, to submicron precision, how much material $\Delta z$ has been removed between sections.

Unlike Paluch who tried to match a whole set of isolated high spatial resolution, small area *XY* frames, the high resolution large area scanning capability of the Leeds 2D image analyser can be used to improve the pattern matching process significantly. The practical problem is to have good quality fibre images and excellent registration between fibre centres. Our aim is to remove the ambiguity in the in-plane angle, $\Phi$, by successful pattern matching between two sections and then to investigate the technique further for the systematic sectioning and 3D reconstruction of opaque fibre composites.

Consider an *XY* area of 5 mm $\times$ 5 mm which is scanned (section A) and the fibre orientation information derived. After removing some material from the sample and polishing, the same region is scanned again (section B). Due to removing and replacing the sample on the microscope stage, there will be a translation ($\Delta x$, $\Delta y$) and an in-plane rotation, $\psi$, between the coordinate systems of section A and section B. The link between corresponding points in the section A $(x, y)$ coordinate system and those points in the section B $(x', y')$ coordinate system is given by

$$x'_i = (x_i.\cos\psi - y_i.\sin\psi) + \Delta x \qquad [36]$$

$$y'_i = (x_i.\sin\psi + y_i.\cos\psi) + \Delta y. \qquad [37]$$

In pattern matching problems, it is important to identify 'control points', that is features on one section that clearly are present on the next section. If *n* control points that appear in both sections can be identified, the best fit set of parameters can be found by minimising the sum of squared errors. The partial derivatives of this with respect to $\Delta x$, $\Delta y$ and $\psi$ are set to zero to derive a linear set of equations which can be solved

$$\cos\psi = \frac{n\sum_i^n (y_i y'_i + x_i x'_i) - \sum_i^n x_i \sum_i^n x'_i - \sum_i^n y_i \sum_i^n y'_i}{C} \qquad [38]$$

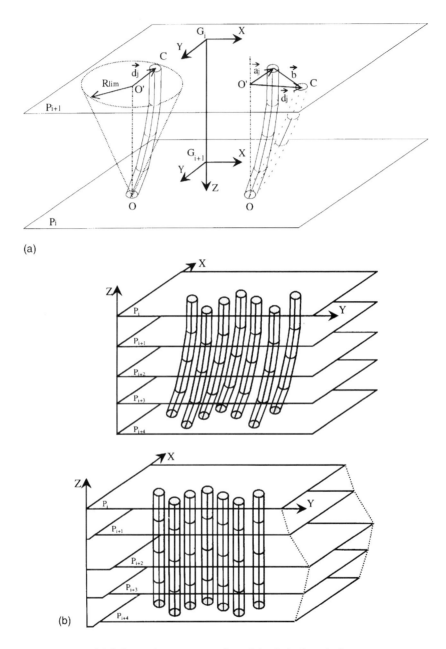

*7.34* (a) Schematic representation of the Paluch technique to reconstruct a sample (which has been physically sectioned) by correlating the fibre image centre coordinates on one section plane to those on a second plane. (b) If fibre movements are correlated, this technique will underestimate the waviness of the fibres in the dataset.

$$\sin\psi = \frac{n\sum_i^n (x_i y_i' - x_i' y_i) + \sum_i^n y_i \sum_i^n x_i' - \sum_i^n x_i \sum_i^n y_i'}{C}$$ [39]

$$\Delta x = \frac{-\sum_i^n x_i \sum_i^n (y_i y_i' + x_i x_i') + \sum_i^n y_i \sum_i^n (x_i y_i' - x_i' y_i) + \sum_i^n (x_i^2 + y_i^2) \sum_i^n x_i'}{C}$$

[40]

$$\Delta y = \frac{-\sum_i^n y_i \sum_i^n (y_i y_i' + x_i x_i') - \sum_i^n x_i \sum_i^n (x_i y_i' - x_i' y_i) + \sum_i^n (x_i^2 + y_i^2) \sum_i^n y_i'}{C}$$

[41]

where

$$C = n\sum_i^n (x_i^2 + y_i^2) - \left[\sum_i^n x_i\right]^2 - \left[\sum_i^n y_i\right]^2.$$ [42]

In aligned fibre composites, sections would be taken normal to the preferred fibre direction and hence, the only objects which can be used as control points are circular fibre images. As the 5 mm $\times$ 5 mm section plane may contain more than 15 000 fibre images, it was decided to define $n$ clusters of six nearest neighbour fibres along the section diagonal. It was assumed that, over a fibre distance of 10 or 20 $\mu$m, the fibres within the cluster would retain their relative positions.

Each cluster selected in section A is characterised by the set of angles $\alpha_n$ and interfibre image distances $l_n$ between members of the cluster. These data produce an orientation independent definition of each cluster which aids the unambiguous identification of cluster control points in section B. In this way, the translational and rotational corrections between sections A and B may be deduced and the centre coordinates of all fibre images on section B are transformed to align with the coordinate system for section A. The first results obtained by this procedure are described by Davidson *et al.*[64]

## 7.6.3  Automation of three-dimensional CLSM for large volume studies

The CLSM gives excellent signal-to-noise in reflection mode when examining the 3D spatial structure of particulates or low packing fraction fibres in thin films or transparent coatings. Therefore the main issues with the further automation of the CLSM in these cases is to have the capability of

- access to the *XY* image frame data as it is scanned,

- detection of the surface of the sample (from the large increase in intensity at the surface for the same system gain) – and a comparison with a previous *XY* image frame to detect the movement of objects within the frame,
- collecting the *XY* data at incrementally increasing *Z* below the surface,
- algorithms tailored for the particular problem to be addressed, e.g. 3D voxel connectivity routines for particle sizing (to achieve data compression wherever possible),
- after acquiring image data over a sufficient depth (i.e. possibly the thickness of the film), repositioning the *Z* drive to just above the sample surface,
- the computer to instruct the automated *XY* stage to move in either *X* or *Y*, with some overlap to the previous sub-volume,

and the cycle is repeated until a large area (and hence sub-volume) is effectively scanned.

If the sample is a thick composite material with a fairly high fibre packing fraction, the best operating mode with most polymeric matrices is the fluorescence mode. Instead of recording and storing all of the bytes in each *XY* frame at each *Z* position, it may only be necessary to save one or two lines of image data from within the *XY* image field, at each *Z* position. Hence, *XZ* frames of data will be produced (as discussed in Section 7.3.2) and an algorithm can be used which will identify each fibre image within this *XZ* field and estimate both fibre centre coordinates and the fibre diameters. A typical raw data *XZ* field is shown in Fig. 7.35(a). Note that the signal-to-noise is good near the surface, but deteriorates with depth into the sample. Hence, the automation of the CLSM for these large volume studies requires pattern recognition of fibres (and voids) in variable intensity image planes. Traditional spatial filter operations on the image data do not improve the situation. At Leeds, some progress has been made towards the automatic recognition of fibres in these *XZ* fields. First, the original image field is corrected for the attenuation of fluorescence as a function of depth, resulting in an improved image, see Fig. 7.35(b). The partial fibre images at the surface and within a fibre diameter below the surface are fitted with a 'spoke and least squares fit' procedure and the result is shown in Fig. 7.35(c). The bottom part of the image field is scanned with a circular spatial filter and from an analysis of minima in 'error space' a best guess is made of the positions of the remaining fibres, see Fig. 7.35(d).

### 7.6.4   Automated ultrasonic testrig for elastic constants determination

Another complementary measurement technique characterising the effect of 3D microstructure on global mechanical properties is the ultrasonic time-of-flight immersion technique, as described by Read and Dean[65] and Lord.[66] Our future plans involve the correlation of ultrasonic measurements to 3D confocal (and

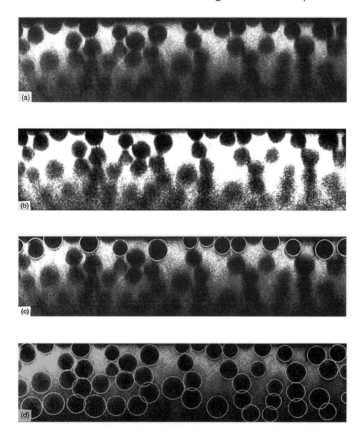

*7.35* Automation of fibre recognition and the estimation of fibre image centres and fibre diameters will become important for CLSM materials research. This shows how fibres can be detected even in noisy images. (a) Raw *XZ* image, (b) after fluorescence normalisation with depth, (c) '*n*-point' least squares fitting of surface fibres, and (d) further processing to pick out fibres deep within the sample.

2D) measurements in order to test models which relate microstructure to material properties. Both 'time-of-flight' and attenuation measurements are made with a fully automated 'immersion technique' as shown in Fig. 7.36.

By measuring the time delay between the transmitter and receiver to nanosecond accuracy, the velocity of a 2.25 MHz ultrasonic pulse through the water can be determined and then through the water and sample. Hence, the time delay through the sample is computed together with the attenuation of the ultrasonic signal. If the sample surface is normal to the ultrasonic beam path, a stepper motor scans the sample through the beam, so that linear scans may be built up of the straight through, tensile velocity component. Because the

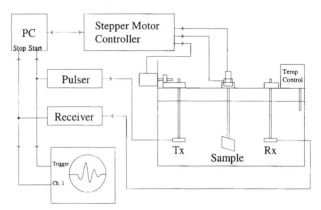

Oscilloscope

*7.36* Overview of the ultrasonic time-of-flight system in which an ultrasonic pulse is generated every millisecond, passed through water (as a coupling medium) and then through the sample, before being detected by a fixed transducer. Ultrasonic wave velocities are deduced from the time delay, Δt between transmitted and received pulses.

computer records a time delay and an amplitude value every millisecond, there are two important consequences:

- The system can integrate over many measurements and, in effect, achieve subnanosecond accuracy in estimating the ultrasonic pulse time delays.
- The system can scan very quickly and after scanning forwards and backwards remove any systematic temperature-dependent effects or short timescale noise (which may be due to external mechanical vibrations).

In this 'fast-scanning mode' of operation, spatial variations in the longitudinal modulus may be measured to within 0.01%, as shown in Fig. 7.37.

However, the system has another powerful mode of operation. The sample may be rotated with respect to the ultrasonic beam, and the variation of delay time through the sample versus angle of incidence yields both tensile and shear wave velocities and hence a number of elastic moduli may be determined. The sample is then rotated by 90° and another angular scan performed in order to yield more elastic constants. If the symmetry of the sample is such that all nine elastic constants are required, the sample is cut into strips, the strips rotated and reassembled before performing a final angular scan.

Once again, the automation of this technique (which was tedious to perform by manual means) decreases the sample analysis time by a factor of 15 to 20 times, increases the sensitivity of the system and permits measurements which would be impracticable with a (slower) manual system. A paper is in preparation, Enderby et al.[67] to describe the system in more detail and to discuss the measurement errors.

*7.37*  Using stepper motors, the sample may be moved in front of the ultrasonic beam so that spatial variations in the delay times can be linked to variations in tensile velocity. The 2D reconstruction shows a plaque area of 4.5 cm × 11 cm and the intensity grey scale covers a range of ± 30 ns in the apparent time delay.

## 7.6.5  Concluding remarks

Our overall goal is to test theoretical assumptions of fibre–fibre interactions, provide finite element modellers with better physical data and relate both process conditions to 3D mesostructure and 3D mesostructure to material mechanical and thermal properties.

The challenge before us is to be able to collect fibre orientation and spatial position data on any composite sample within a reasonable timescale. Computer-assisted microscopy holds the key to high spatial resolution measurements. The full automation of the 2D image analyser goes a long way towards fulfilling our objective especially once the systematic sectioning, polishing and pattern matching of fibre images has been proved to be robust and reliable. Once the CLSM technique can be fully automated, especially for glass fibre reinforcements, even faster 3D reconstruction of voids, fibre orientation states and fibre waviness will be possible. (Probably many carbon and Kevlar fibre reinforced composites will be explored too when fluorescent dopants are used to enhance the signals from greater depths below the surface.) The sensitivity and inherent imaging speed of CLSMs is constantly improving, new techniques for improving spatial resolution are being developed and with the reduction in computing costs, reasonably priced CLSM systems should be available within the near future. No composites laboratory should be without one!

## Acknowledgements

We would like to acknowledge the useful collaborations over eight years with many colleagues: Professor Paul Curtis and Dr Mike Pitkethly at DRA

(Farnborough); Dr Mike Wisnom at the Aerospace Engineering Department at Bristol University; our Brite Euram project partners: Dr Gunther Fischer and Dr Christian Ludwig at IKP Stuttgart, Dr Michel Vincent at Ecole des Mines, Dr Tom Berland and Dr Einar Hinrichsen at Sintef in Oslo; Dr Norman Marks at GKN (Westland Helicopters); Dr Paul Marshall at British Aerospace; Dr Alan Duckett, Dr Peter Hine, Stephen Wire and Professor Ward at the Polymer IRC in Leeds; Dr Rob Bailey, Dr Paul Mills and Dr Simon Allen at ICI Wilton; Dr Paul Smith at Noran Instruments (UK); Dr Andrew Dixon at Biorad Microscience Ltd for the loan of their MRC500 CLSM without which this work would not have been possible; Professor Ryszard Pyrz and Dr Peter From at Aalborg University, Institute for Mechanical Engineering; Dr Garry Burdett at Health and Safety Executive, Sheffield; Dr Norman Fleck at the University of Cambridge and Dr Will Slaughter at the University of Pittsburgh. The ultrasonic system has been created by talented members of the Mechanical and Electronic Workshops within the Department of Physics and Astronomy at Leeds and the programming skills of our PhD student, Mike Enderby. Also, we would like to acknowledge the work of a visiting Erasmus student, Markus Müller (from the University of Karlsruhe) who has helped to unravel the mysteries of fibre curvature and fibre waviness.

Dr Nic Davidson and Geoff Archenhold have been supported by the Brite Euram programme, BE-8081: Push-Pull Processing of Liquid Crystalline Polymers, coordinated by Dr Gunther Fischer at IKP, University of Stuttgart.

## References

1. M. R. Piggott, The effect of fibre waviness on the mechanical properties of unidirectional fibre composites, *Proceedings Conference on Mesostructures and Mesomechanics in Fibre Composites*, held at Niagara-on-the-Lake, ed. M. R. Piggott, University of Toronto, 1994, pp. 145–158.

2. A. R. Clarke, G. Archenhold and N. C. Davidson, 3D confocal microscopy of glass fibre reinforced composites, chapter 3 in *Microstructural Characterisation of Fibre Reinforced Composites*, Woodhead Publishing, 1997, in press.

3. M. W. Darlington, P. L. McGinley and G. R. Smith, Structure and anisotropy of stiffness in glass fibre-reinforced thermoplastics, *J. Mat. Sci.*, 1976, **11** 877–886.

4. F. Lisy, A. Hiltner, E. Baer, J. L. Katz and A. Meunier, Application of scanning acoustic microscopy to polymeric materials, *J. Appl. Polym. Sci.*, 1994, **52**, 329–352.

5. A. L. Segre, D. Capitani, M. Malinconico, E. Martuscelli, D. Gross and V. Leheman, Nuclear magnetic resonance microscopy of multicomponent polymeric materials, *J. Mat. Sci. Letters*, 1993, **12**, 728–731.

6. S. Fakirov and C. Fakirova, Direct determination of the orientation of short glass fibres in an injection moulded polyethylene terephthalate system, *Polym. Compos.*, 1985, **6**(1), 41–46.

7. S. W. Yurgartis, Measurement of small angle fibre misalignments in continuous fibre composites, *Compos. Sci. Tech.*, 1987, **30**, 279–293.

8. S. Toll and P.-O. Andersson, Microstructural characterisation of injection moulded composites using image analysis, *Composites*, 1991, **22**(4), 298–306.

9. G. Fischer and P. Eyerer, Measuring spatial orientation of short fibre reinforced thermoplastics by image analysis, *Polym. Compos.*, 1988, **9**(4), 297–304.

10. P. J. Hine, R. A. Duckett, N. C. Davidson and A. R. Clarke, Modelling of the elastic properties of fibre reinforced composites, 1: orientation measurement, *Compos. Sci. Tech.*, 1993, **47**, 65–73.

11. F. J. Guild and J. Summerscales, Microstructural image analysis applied to fibre composite materials: a review, *Composites*, 1993, **24**, 383–393.

12. J. J. McGrath and J. M. Wille, Determination of 3D fiber orientation distribution in thermoplastic injection moulding, *Compos. Sci. Tech.*, 1995, **53**, 133–143.

13. B. O'Donnell and J. R. White, Young's modulus variation within glass-fibre-filled nylon 6,6 injection mouldings, *Plastics and Rubber Institute, Proceedings 1st Conference on Deformation and Fracture of Composites*, UMIST Manchester, Chameleon Press, Wandsworth, UK, 1991, 24.1–24.6.

14. I. M. Ward, The measurement of molecular orientation in polymers by spectroscopic techniques, *J. Polym. Sci.: Polym. Symp.*, 1977, **58**, 1–21.

15. I. M. Ward, Optical and mechanical anisotropy in crystalline polymers, *Proc. Phys. Soc.*, 1962, **80**, 1176–1188.

16. P. J. Hine, R. A. Duckett and I. M. Ward, Modelling the elastic properties of fibre reinforced composites: II theoretical predictions, *Compos. Sci. Tech.*, 1993, **49**, 13–21.

17. S. G. Advani and C. L. Tucker III, The use of tensors to describe and predict fibre orientation in short fibre composites, *J. Rheol.*, 1987, **31**, 751–784.

18. H.-C. Ludwig, G. Fischer and H. Becker, A quantitative comparison of morphology and fibre orientation in push-pull processed and conventional injection moulded parts, *Proceedings Conference on Mesostructures and Mesomechanics in Fibre Composites*, held at Niagara-on-the Lake, ed. M. R. Piggott, University of Toronto, 1994, 223–232.

19. M. Taya, K. Muramatsu, D. J. Lloyd and R. Watanabe, Determination of distribution patterns of fillers in composites by micromorphological parameters, *JSME Internat. J.*, 1991, **34** (2), 198–206.

20. S. W. Yurgartis, Techniques for the quantification of composite mesostructure, *Process Mesostructures and Mesomechanics in Fibre Composites* held at Niagara-on-the Lake, ed. M. R. Piggott, University of Toronto, 1994, 32–55.

21. A. R. Clarke, N. C. Davidson and G. Archenhold, The measurement and modelling of fibre directions in composites, *IUTAM Symposium on Microstructure-Property Interactions in Composite Materials*, held at Aalborg, Denmark, ed. R. Pyrz, 1995, Kluwer Academic Publishers, 77–88.

22. R. Pyrz, Quantitative description of the microstructure of composites. Part 1: morphology of unidirectional composite systems, *Compos. Sci. Tech.*, 1994, **50**, 197–208.

23. N. C. Davidson, Image analysis for fibre orientation measurement, *PhD Thesis*, Department of Physics and Astronomy, University of Leeds, 1993.

24. A. R. Clarke, N. C. Davidson and G. Archenhold, A multitransputer image analyser for 3D fibre orientation studies in composites, *Transactions of the Roy. Microscopical Society*, ed. H. Y. Elder, Adam Hilger, Bristol, 1990, Volume 1, 305–309.

25. A. R. Clarke, N. C. Davidson and G. Archenhold, A large area, high resolution, image analyser for polymer research, *Proceedings International Conference Transputing '91*, Volume 1, held at Sunnyvale, CA, (eds Welch and Bakkers, IOS Press, 1991, pp. 31–47.

26. A. R. Clarke, N. C. Davidson and G. Archenhold, Image analyser of carbon fibre orientations in composite materials, *Proceedings 2nd International Conference on Applications of Transputers, TA90*, held at Southampton, UK, IOS Press, 1991, pp. 248–255.

27. R. S. Stobie, Analysis of astronomical images using moments, *J. Br. Interplanetary Soc.*, 1980, **33**, 323–326.

28. W.-Y. Wu and M.-J. J. Wang, Elliptical object detection by using its geometric properties, *Pattern Recognition*, 1993, **26** (no. 10), 1499–1509.

29. G. Archenhold, internal report, MPI-95/1, University of Leeds.

30. M. Gee, *Parallel Confocal Microscopy* National Physical Laboratory, Teddington, annual report, 1994.

31. G. J. Brakenhoff, H. van der Voort, M. W. Baarslag, J. L. Oud, R. Zwart and R. van Driel, Visualisation and analysis techniques for 3D information acquired by confocal microscopy, *Scanning Microscopy*, 1988, **2**(4), 1831–1838.

32. A. Knoester and G. J. Brakenhoff, Applications of confocal microscopy in industrial solid materials: some examples and a first evaluation, *J. Microscopy*, 1990, **157**(1), 105–113.

33. J. L. Thomason and A. Knoester, Application of confocal scanning optical microscopy to the study of fibre reinforced composites, *J. Mat. Sci. Letters*, 1990, **9**, 258–262.

34. J. B. Pawley, *Handbook of Biological Confocal Microscopy*, New York and London, Plenum Press, 1995, 2nd edn.

35. T. Wilson, *Confocal Microscopy*, London, Academic Press, 1993.

36. P. Smith, The Odyssey CLSM, *Noran publicity brochure*, 1996.

37. G. Archenhold, A. R. Clarke and N. C. Davidson, 3D microstructure of fibre reinforced composites, *SPIE Biomedical Image Processing and 3D Microscopy*, 1992, **1660**, 199–210.

38. R. S. Bay and C. L. Tucker, Stereological measurement and error estimates for three dimensional fibre orientation, *Polym. Eng. and Sci.*, 1992, **32**(4), 240–253.

39. A. R. Clarke, N. C. Davidson and G. Archenhold, Measurements of fibre directions in reinforced polymer composites, *J. Microscopy*, 1993, **171**, 69–79.

40. T. D. Visser, J. L. Oud and G. J. Brakenhoff, Refractive index and axial distance measurements in 3D microscopy, *Optik*, 1992, **90**(1), 17–19.

41. C. Hall, *Polymer Materials: An Introduction for Technologists and Scientists*, London, Macmillan Education, 1989.

42. S. Hell, G. Reiner, C. Cremer and E. H. K. Stelzer, Aberrations in confocal fluorescence microscopy induced by mismatches in refractive index, *J. Microscopy*, 1993, **169**(3), 391–405.

43. K. Carlsson, The influence of specimen refractive index, detector signal integration and non-uniform scan speed on the imaging properties in confocal microscopy, *J. Microscopy*, 1991, **163**(2), 167–178.

44. H. J. G. Gundersen, P. Bagger, T. F. Bendtsen, S. M. Evans, L. Korbo, N. Marcussen, A. Moller, K. Nielsen, J. R. Nyengaard, P. Pakkenberg, F. B. Sorensen, A.

Vesterby and M. J. West, The new stereological tools: disector, fractionator, nucleator and point sampled intercepts and their use in pathological research and diagnosis, *APMIS 96*, 1988, 857–881.

45. L. M. Cruz-Orive and E. R. Weibel, Recent stereological methods for cell biology: brief survey, *Amer. J. Physiol.*, 1990, **258**, L148–L156.

46. V. Howard, Stereological techniques in biological electron microscopy, *Biophysical Electron Microscopy*, chapter 13, Academic Press, 1990.

47. B. Möginger and P. Eyerer, Determination of the weighting function $g(\beta, r, v)$ for the fibre orientation analysis of the short fibre reinforced composites, *Composites*, 1991, **22**, 394–398.

48. T. Mattfeldt, A. R. Clarke and G. Archenhold, Estimation of the directional distribution of spatial fibre processes using stereology and confocal scanning laser microscopy, *J. Microscopy*, 1994, **173**(2), 87–101.

49. W. Kuhn and F. Grün, The optical anisotropy of rubbers, *Kolloidzeitschrift*, **101**, 248–258.

50. H. Brody and I. M. Ward, Modulus of short carbon and glass fibre reinforced composites, *Polym. Eng. Sci.*, 1971, **11**(2), 139–151.

51. P. J. Hine, N. C. Davidson, R. A. Duckett, A. R. Clarke and I. M. Ward, Orientation measurement in hydrostatically extruded, glass fibre reinforced POM, *Proceedings of Deformation and Fracture of Composites II*, PPS8, held at Institute of Materials, Manchester, Chameleon Press, Wandsworth, UK, 1993.

52. P. J. Davey and F. J. Guild, The distribution of interparticle distance and its application in finite element modelling of composites, *Proc. R. Soc. London A*, **418**, 95–112.

53. A. R. Clarke and N. C. Davidson, Determining the spatial distributions of fibres in composites, *Proceedings of Symposium A4 on Composite Materials, International Conference on Advanced Materials-ICAM91*, held at Strasbourg 1991, Elsevier, North Holland, pp. 55–60.

54. P. From, Stereological characterisation of microstructure morphology for composites, *PhD Thesis*, Aalborg University 1996, chapter 8.

55. A. R. Clarke, G. Archenhold and N. C. Davidson, A novel technique for determining the 3D spatial distribution of glass fibres in polymer composites, *Compos. Sci. Tech.*, 1995, **55**, 75–91.

56. E. T. Camponeschi, Lamina waviness levels in thick composites and its effect on their compression strength, *Proceedings International Conference on Composite Materials*, ICCM8, Honolulu, 1991, 30-E-1 to 30-E-13.

57. B. Budianski and N. A. Fleck, Compressive failure of fiber composites, *J. Mech. Phys. Solids*, 1993, **41**, 183–211.

58. W. S. Slaughter and N. A. Fleck, Microbuckling of fiber composites with random initial fiber waviness, *J. Mech. Phys. Solids*, 1994, **42**(11), 1743–1766.

59. A. R. Clarke, G. Archenhold, N. C. Davidson, W. S. Slaughter and N. A. Fleck, Determining the power spectral density of the waviness of unidirectional glass fibres in polymer composites, *Appl. Compos. Mater.*, 1995, **2**, 233–243.

60. D. Shi and D. Winslow, Accuracy of a volume fraction measurement using areal image analysis, *Journal of Testing and Evaluation*, JTEVA, 1991, **19**(3), 210–213.

61. M. Merickel, 3D reconstruction: the registration problem, *Computer Vision, Graphics and Image Processing*, 1987, **42**, 206–219.

62. L. S. Hibbard, R. A. Grothe, T. L. Amicar-Sulze, B. J. Dovey-Hartman and R. B. Pages, Computed 3D reconstruction of median-eminence capillary modules: - image alignment and correlation, *J. Microscopy*, 1993, **171**(1), 39–56.
63. B. Paluch, Analysis of geometric imperfections in unidirectionally reinforced composites, *Proceedings of European Conference on Composite Materials*, ECCM6, held at Bordeaux, 1993, Woodhead, Cambridge, pp. 305–310.
64. N. C. Davidson, A. R. Clarke and G. Archenhold, Large area, high resolution image analysis of composite materials, *J. Microscopy*, 1997, **185**(2), 233–242.
65. B. E. Read and G. D. Dean, *The Determination of the Dynamic Properties of Polymers and Composites*, Adam Hilger, Bristol, 1978, pp. 162–179.
66. D. Lord, The determination of the elastic constants of fibre reinforced composites by an ultrasonic method, *PhD Thesis*, Department of Physics, University of Leeds, 1989.
67. M. D. Enderby, A. R. Clarke, P. Ogden and A. A. Johnson, An automated, ultrasonic immersion technique for the determination of 3D elastic constants of polymer composites, *Proceedings of 17th International Conference on Ultrasonics*, UI '97, held at Delft, 1997, Elsevier Science Ltd, Oxford, in press.

# 8

# Materials property modelling and design of short fibre composites

R B R O O K S

## 8.1 Introduction

Alignment in composite materials has a significant effect on the mechanical and thermal properties of the material. The complex orientations of reinforcing particles, induced by flow conditions, present the designer with particular difficulties. Materials modelling is an important element in the design process which predicts material properties and gives the designer a tool with which to tailor the material to a specific application.

This chapter concerns one particular aspect of materials modelling, namely, the modelling of *stiffness* in short fibre-reinforced composites. The focus is on polymeric composites reinforced with short glass fibre, materials which are generally converted into components by the injection moulding process. Such materials, by virtue of their structure and processing route, show all the complex features arising from flow-induced alignment. Because of their improved properties over the base polymer and their low cost processing, these materials have wide application and their use is growing in a number of areas. This is particularly so in the automotive industry where they are making significant inroads into more structural engineering applications. Typical examples include clutch and throttle pedals, intake manifolds and radiator endcaps, all demanding applications. Although these materials are the focus of the discussion, the methods described are equally applicable to other particulate and reinforced composites including metal and ceramic matrix materials. The focus on stiffness is deliberate for a number of reasons. First, the modelling of stiffness is in a relatively advanced state of development with a number of reliable models and procedures now available. Second, the methods used for stiffness modelling are applicable to a number of other properties such as thermal expansion, thermal conductivity and dielectric properties. Finally, for polymer matrix composites, with their relatively low moduli, stiffness prediction is particularly important as it is often the main factor controlling component design. It should be added that there are particular difficulties encountered when modelling the strength of short fibre materials as this derives from complex failure mechanisms and local stress concentrations. This aspect of materials modelling is therefore not included in this chapter.

There are two main parts to this chapter. The first is a description of the main models developed for predicting stiffness in fully aligned material which can be regarded as the building block for more complex systems with flow-induced alignment. Only models giving closed form equations for the elastic moduli are described, as these are most useful to the designer. For clarity, the detailed equations are listed separately in Boxes 1 to 7. Wholly numerical solutions are not considered. The effects of reinforcement aspect ratio and constituent properties are discussed and the various models are compared over a range of volume fractions. The second part of the chapter concerns the important aspect of fibre orientation and how the models for the fully aligned material can be combined with orientation information to predict the stiffness of more complex systems. The method is illustrated using finite element analysis to predict the tensile and flexural modulus of a flat plaque component containing complex orientations both spatially and through the thickness. Accuracy is assessed against available experimental data for this system. The chapter concludes with a brief discussion of the limitations of the method and the important areas for future work.

## 8.2    Modelling the stiffness of fully aligned materials

### 8.2.1  Bounds on moduli

Variational principles have been used to obtain upper and lower bounds on moduli for both particulate and continuous fibre-reinforced composites. The simplest form of these bounds is derived by assuming constant stress or constant strain conditions, leading to the well known Reuss (constant stress) lower bound, and Voigt (constant strain) upper bound expressions for stiffness. These are commonly called the 'rule of mixtures' equations and although in some cases they give adequate estimates for specific moduli, in general they give values too far apart to be of practical use.

Closer bounds have been obtained for the five independent elastic moduli of a transversely isotropic continuous fibre-reinforced composite using variational methods and minimum energy theorems (Hill, 1964; Hashin, 1965). Although not applicable directly to short fibre composites, these bounds do provide a benchmark against which results from other theories can be measured. In particular, predictions for a short fibre material with high aspect ratio ($l/d > 100$), which approximates to a continuous fibre-reinforced composite, should lie within these bounds. The detailed expressions are given in Boxes 1 and 2. It is interesting to note that the lower bound expressions for $E_{11}$, $v_{12}$, $G_{12}$ and $k$ correspond to the exact expressions obtained from the composite cylinders

## Box 1

Bounds model (based on Hill, 1964 and Hashin, 1965)

Upper bounds

$$\frac{E_{11}}{E_m} = \frac{E_f}{E_m} V_f + (1 - V_f) + \frac{4V_f(1 - V_f)(v_f - v_m)^2}{E_m\left(\dfrac{V_f}{k_m} + \dfrac{(1 - V_f)}{k_f} + \dfrac{1}{G_f}\right)}$$

$$\frac{G_{12}}{G_m} = \frac{G_f}{G_m} + \frac{(1 - V_f)}{G_m\left(\dfrac{1}{(G_m - G_f)} + \dfrac{V_f}{2G_f}\right)}$$

$$\frac{v_{12}}{v_m} = \frac{v_f}{v_m} V_f + (1 - V_f) + \frac{V_f(1 - V_f)(v_f - v_m)\left(\dfrac{1}{k_m} - \dfrac{1}{k_f}\right)}{v_m\left(\dfrac{V_f}{k_m} + \dfrac{(1 - V_f)}{k_f} + \dfrac{1}{G_f}\right)}$$

$$\frac{k}{k_m} = \frac{k_f}{k_m} + \frac{(1 - V_f)}{k_m\left(\dfrac{1}{(k_m - k_f)} + \dfrac{V_f}{(k_f + G_f)}\right)}$$

$$\frac{G_{23}}{G_m} = \frac{G_f}{G_m} + \frac{(1 - V_f)}{G_m\left(\dfrac{1}{(G_m - G_f)} + \dfrac{V_f(k_f + 2G_f)}{2G_f(k_f + G_f)}\right)}$$

assemblage (CCA) model (Hashin and Rosen, 1964). Although the CCA model is exact, its structure, which is formed of varying cylinder sizes, is unrealistic. It does, however, possess a necessary randomness in fibre distribution and appears to give accurate results. It has been suggested (Hashin, 1983) that in the absence of geometric information the Hill and Hashin bounds are therefore the best that can be achieved. There is no such exact solution for $G_{23}$, and the same conclusions cannot be drawn for this modulus. The closeness of the Hill and Hashin bounds depends very much on the relative stiffnesses of the components of the composite. Closely matching stiffnesses give close bounds whilst very different stiffnesses, as is the case with glass fibres and a polymer matrix, give wide bounds. The usefulness of bounds expressions is therefore limited.

### 8.2.2  Self-consistent models

The self-consistent method for modelling materials properties has been developed by a number of workers (Hill, 1965a,b; Budianski, 1965; Laws and McLaughlin, 1979; Chou et al., 1980). The basic approach assumes that an

**Box 2**

Bounds model (based on Hill, 1964 and Hashin, 1965)
   Lower Bounds

$$\frac{E_{11}}{E_m} = \frac{E_f}{E_m} V_f + (1 - V_f) + \frac{4V_f(1 - V_f)(v_f - v_m)^2}{E_m\left(\dfrac{V_f}{k_m} + \dfrac{(1 - V_f)}{k_f} + \dfrac{1}{G_m}\right)}$$

$$\frac{G_{12}}{G_m} = 1 + \frac{V_f}{G_m\left(\dfrac{1}{(G_f - G_m)} + \dfrac{(1 - V_f)}{2G_m}\right)}$$

$$\frac{v_{12}}{v_m} = \frac{v_f}{v_m} V_f + (1 - V_f) + \frac{V_f(1 - V_f)(v_f - v_m)\left(\dfrac{1}{k_m} - \dfrac{1}{k_f}\right)}{v_m\left(\dfrac{V_f}{k_m} + \dfrac{(1 - V_f)}{k_f} + \dfrac{1}{G_m}\right)}$$

$$\frac{k}{k_m} = 1 + \frac{V_f}{k_m\left(\dfrac{1}{(k_f - k_m)} + \dfrac{(1 - V_f)}{(k_m + G_m)}\right)}$$

$$\frac{G_{23}}{G_m} = 1 + \frac{V_f}{G_m\left(\dfrac{1}{(G_f - G_m)} + \dfrac{(1 - V_f)(k_m + 2G_m)}{2G_m(k_m + G_m)}\right)}$$

inclusion, for example fibre, is embedded in an infinite medium having the same properties as the composite as shown in Fig. 8.1(a). Using established analyses which relate the uniform strain at infinity to the strain within the inclusion (Eshelby, 1957) and assuming that the inclusion strain can be averaged over all inclusions, self-consistent relationships between the moduli can be obtained. The method has been formulated for isotropic composites modelled as spherical inclusions (Hill, 1965a; Budiansky, 1965), for continuous aligned circular fibres (Hill, 1965b) and for aligned short fibres of varying aspect ratio modelled as ellipsoidal inclusions ( Chou et al., 1980). For short fibres, iterative methods are used to obtain a solution. Initial moduli for the composite are estimated, for example using rule of mixtures expressions, and the self-consistent equations solved for the unknown moduli, which are themselves fed back in until convergence occurs. The self-consistent method does not take into account local interaction between fibres and for that reason becomes unreliable at high volume fraction where it has been shown to give greater values than other models

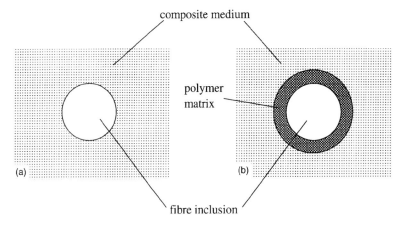

*8.1* Forms of the self-consistent model.

(Hashin, 1983). The discrepancies are also greater where there is significant mismatch between fibre and matrix properties (Hill, 1965b).

To overcome these difficulties, the method has been extended by assuming the fibre to be embedded in a concentric cylindrical annulus of matrix material, which itself is embedded in an infinite composite medium as shown in Fig. 8.1(b). This extended method has been termed the generalised self-consistent (GSC) method. As boundary conditions have to be matched across each of the three layers, the method is more complex and consequently has only been formulated for spherical inclusions and aligned continuous fibre composites in two dimensions (Christensen and Lo, 1979). In the latter case the composite medium is transversely isotropic and has five independent elastic moduli. As with the lower bound expressions, results show that four of these moduli, $E_{11}$, $v_{12}$, $G_{12}$ and $k$ also coincide with the exact expressions obtained from the composite cylinders assemblage (CCA) model (Hashin and Rosen, 1964), giving support to the general approach. In fact it has been demonstrated (Christensen, 1990), by studying the limiting analytical forms for the variation of moduli with volume fraction for several models, that the GSC method is the only currently available model which gives physically realistic expressions at full packing. Furthermore, the GSC model provides results in simple and convenient analytical form. However, it has yet to be formulated for short fibres where fibre aspect ratio is a key parameter. The full form of the GSC expressions for continuous fibre composites are shown in Boxes 3 and 4.

## 8.2.3 Mori–Tanaka model

The limitations of the bounds approach and the self-consistent schemes have led to a more rigorous approach to modelling which takes into account the local

**Box 3**

Generalised self-consistent model (after Christensen, 1990)

$$\frac{E_{11}}{E_m} = \frac{E_f}{E_m} V_f + (1 - V_f) + \frac{4 V_f (1 - V_f)(v_f - v_m)^2}{E_m \left( \frac{V_f}{k_m} + \frac{(1 - V_f)}{k_f} + \frac{1}{G_m} \right)}$$

$$\frac{G_{12}}{G_m} = \frac{G_f(1 + V_f) + G_m(1 - V_f)}{G_f(1 - V_f) + G_m(1 + V_f)}$$

$$\frac{v_{12}}{v_m} = \frac{v_f}{v_m} V_f + (1 - V_f) + \frac{V_f(1 - V_f)(v_f - v_m)\left( \frac{1}{k_m} - \frac{1}{k_f} \right)}{v_m \left( \frac{V_f}{k_m} + \frac{(1 - V_f)}{k_f} + \frac{1}{G_m} \right)}$$

$$\frac{k}{k_m} = 1 + \frac{V_f}{k_m \left( \frac{1}{(k_f - k_m)} + \frac{(1 - V_f)}{(k_m + G_m)} \right)}$$

$$\frac{G_{23}}{G_m} = \frac{-B \pm \sqrt{B^2 - AC}}{A}$$

$A$, $B$ and $C$ are given in Box 4

stress and strain field around and within inclusions and in particular the perturbation of this field due to interaction between fibres. The general method is based on a combination of previous analyses for the average uniform stress inside an ellipsoidal inclusion (Eshelby, 1957) and the concept of average matrix stress (Mori and Tanaka, 1973), the latter work giving rise to the name associated with the method. Using the method, useful closed form expressions for the elastic moduli in an aligned short fibre composite are given by Tandon and Weng (Tandon and Weng, 1984; Zhao et al., 1989), however, a simpler direct description of the method has been given by Benveniste (1987) and further elucidated by Christensen (1990). For the insight it gives into the origin of the method, it is the approach and explanation of the latter two authors which will be outlined here.

Considering a two phase composite comprising matrix and particles of similar ellipsoidal shapes but differing sizes, the average strain in the composite, $\bar{\varepsilon}$, is given by,

$$\bar{\varepsilon} = V_1 \bar{\varepsilon}_1 + V_2 \bar{\varepsilon}_2 \qquad [1]$$

## Box 4

Generalised self-consistent model (after Christensen, 1990)
(see Box 3 for main equations)

$$A = 3V_f(1 - V_f)^2 \left(\frac{G_f}{G_m} - 1\right)\left(\frac{G_f}{G_m} + \eta_f\right)$$

$$+ \left[\frac{G_f}{G_m}\eta_m + \eta_f\eta_m - \left(\frac{G_f}{G_m}\eta_m - \eta_f\right)V_f^3\right]$$

$$\cdot \left[V_f\eta_m\left(\frac{G_f}{G_m} - 1\right) - \left(\frac{G_f}{G_m}\eta_m + 1\right)\right]$$

$$B = -3V_f(1 - V_f)^2 \left(\frac{G_f}{G_m} - 1\right)\left(\frac{G_f}{G_m} + \eta_f\right)$$

$$+ \frac{1}{2}\left[\frac{G_f}{G_m}\eta_m + \left(\frac{G_f}{G_m} - 1\right)V_f + 1\right]$$

$$\cdot \left[(\eta_m - 1)\left(\frac{G_f}{G_m} + \eta_f\right) - 2\left(\frac{G_f}{G_m}\eta_m - \eta_f\right)V_f^3\right]$$

$$+ \frac{V_f}{2}(\eta_m + 1)\left(\frac{G_f}{G_m} - 1\right)\left[\frac{G_f}{G_m} + \eta_f + \left(\frac{G_f}{G_m}\eta_m - \eta_f\right)V_f^3\right]$$

$$C = 3V_f(1 - V_f)^2 \left(\frac{G_f}{G_m} - 1\right)\left(\frac{G_f}{G_m} + \eta_f\right)$$

$$+ \left[\frac{G_f}{G_m}\eta_m + \left(\frac{G_f}{G_m} - 1\right)Vf + 1\right] \cdot \left[\frac{G_f}{G_m} + \eta_f + \left(\frac{G_f}{G_m}\eta_m - \eta_f\right)V_f^3\right]$$

$$\eta_m = 3 - 4v_m$$

$$\eta_f = 3 - 4v_f$$

where $\bar{\varepsilon}_1$ is the average strain in matrix, $\bar{\varepsilon}_2$ is the average strain in inclusions, $V_1$ is the volume fraction of matrix and $V_2$ is the volume fraction of inclusions.

The effective stiffness tensor, $\mathbf{L}^*$, of the composite relates the average strain to the average stress, $\bar{\sigma}$, as follows,

$$\bar{\sigma} = \mathbf{L}^*\bar{\varepsilon}. \tag{2}$$

It has been shown (Hill, 1963) that,

$$\mathbf{L}^* = L_1 + V_2.(L_2 - L_1).A \tag{3}$$

where $L_1, L_2$ are stiffness tensors for each phase.

$A$ is defined as a concentration factor in the literature and relates the far field boundary strain, $\varepsilon_0$, to the average strain in the inclusion, thus,

$$\bar{\varepsilon}_2 = A\varepsilon_0. \tag{4}$$

In general, the determination of the tensor $A$ is a difficult problem, but a solution has been obtained for the dilute case (Eshelby, 1957). For a single ellipsoidal inclusion embedded in an infinite matrix the strain field within the particle is homogeneous and $A$ is denoted by $T$, the dilute solution strain concentration tensor, given by the following,

$$A = T = [I + SL_1^{-1}(L_2 - L_1)]^{-1} \tag{5}$$

where $I$ is the fourth order unit tensor and $S$ is the fourth order Eshelby tensor (Eshelby, 1957).

Solutions have been obtained for $S$ for both isotropic and anisotropic matrices (Zhao, 1989), however, Eqn [5] is only applicable to small concentrations where there is no particle interaction.

The Mori–Tanaka method generalises the above to the non-dilute case. First, a new tensor $G$ is introduced which relates the average matrix strain, $\bar{\varepsilon}_1$, to the average strain in the inclusions, $\bar{\varepsilon}_2$, as follows,

$$\bar{\varepsilon}_2 = G\bar{\varepsilon}_1 \tag{6}$$

Using Eqns [1], [4] and [6] it can be shown that,

$$A = [V_1 I + V_2 G]^{-1} G. \tag{7}$$

Thus if $G$ can be established under non-dilute conditions then $A$ follows and from Eqn [3] the effective composite stiffness can be found. Although, in reality $G$ is concentration dependent, the Mori–Tanaka method assumes that $G$ can be approximated by the concentration independent tensor $T$, that is the dilute solution strain concentration tensor. Thus Eqn [7] becomes

$$A = [V_1 I + V_2 T]^{-1} T. \tag{8}$$

At the two extremes of concentration, $A$ becomes

low inclusion concentration   $(V_1 = 1, V_2 = 0)$   $A = T$
high inclusion concentration   $(V_1 = 0, V_2 = 1)$   $A = I$

The above assumption means that $A$ satisfies the necessary conditions at either end of the concentration scale. It gives the correct dilute solution and also gives the correct inclusion phase property at high concentration. Because of this assumption, the Mori–Tanaka method is an approximation and Eqn [8] may be in error at intermediate concentrations. It is essentially a mathematical fitting method and as stated earlier (Christensen, 1990) physically realistic expressions do not result even at high concentrations. However, in spite of this, simple to use closed form expressions do result which are applicable to short fibre aligned systems. The full form of the Mori–Tanaka expressions (Zhao, 1989) are given in Box 5 and expressions for the Eshelby tensor in Box 6.

**Box 5**

Mori–Tanaka model (after Zhao et al., 1989)

$$\frac{E_{11}}{E_m} = \frac{1}{1 + [V_f(A_1 + 2v_m A_2)]/A} \qquad \frac{G_{12}}{G_m} = 1 + \frac{V_f}{\frac{G_m}{G_f - G_m} + 2V_m S_{1212}}$$

$$\frac{G_{23}}{G_m} = 1 + \frac{V_f}{\frac{G_m}{G_f - G_m} + 2V_m S_{2323}}$$

$$\frac{v_{12}}{v_m} = 1 - V_f \frac{v_m(A_1 + 2v_m A_2) + (A_3 - v_m A_4)}{v_m[A + V_f(A_1 + 2v_m A_2)]}$$

$$\frac{k}{k_m} = \frac{(1 + v_m)(1 - 2v_m)}{1 - v_m(1 + 2v_{12}) + [V_f[2(v_{12} - v_m)A_3 + (1 - v_m(1 + 2v_{12}))A_4]]/A}$$

$A_1 = D_1(B_4 + B_5) - 2B_2 \qquad B_1 = V_f D_1 + D_2 + V_m(D_1 S_{1111} + 2S_{2211})$

$A_2 = (1 + D_1)B_2 - (B_4 + B_5) \quad B_2 = V_f + D_3 + V_m(D_1 S_{1122} + S_{2222} + S_{2233})$

$A_3 = B_1 - D_1 B_3 \qquad B_3 = V_f + D_3 + V_m(S_{1111} + (1 + D_1)S_{2211})$

$A_4 = (1 + D_1)B_1 - 2B_3 \qquad B_4 = V_f D_1 + D_2 + V_m(S_{1122} + D_1 S_{2222} + S_{2233})$

$A_5 = (1 - D_1)/(B_4 - B_5) \qquad B_5 = V_f + D_3 + V_m(S_{1122} + S_{2222} + D_1 S_{2233})$

$A = 2B_2 B_3 - B_1(B_4 + B_5) \qquad D_1 = 1 + 2(\mu_f - \mu_m)/(\lambda_f - \lambda_m)$

$$D_2 = (\lambda_m + 2\mu_m)/(\lambda_f - \lambda_m)$$

$$D_3 = \lambda_m/(\lambda_f - \lambda_m)$$

$\mu, \lambda$ are the Lame constants

$\mu = G = E/2(1 + v)$

$\lambda = (Ev)/(1 + v)(1 - 2v)$

Eshelby's tensor components, $S_{ijkl}$, are given in Box 6

## 8.2.4  Halpin–Tsai model

The most widely used materials property model, particularly in engineering design applications, is the Halpin–Tsai model (Halpin and Kardos, 1976) or Halpin–Tsai equations as they are often termed. Although the model has limitations with respect to its rigour and accuracy, its main advantage is the simple universal form of expression in its formulation and its applicability to a

## Box 6

Eshelby's tensor components (after Zhao *et al.*, 1989)

$$S_{1111} = \frac{1}{2(1-v_m)}\left[1-2v_m+\frac{3a^2-1}{a^2-1}-\left(1-2v_m+\frac{3a^2}{a^2-1}\right)\cdot g\right]$$

$$S_{2222} = \frac{3a^2}{8(1-v_m)(a^2-1)}+\frac{1}{4(1-v_m)}\cdot\left[1-2v_m-\frac{9}{4(a^2-1)}\right]\cdot g$$

$$S_{2233} = \frac{1}{4(1-v_m)}\cdot\left[\frac{a^2}{2(a^2-1)}-\left(1-2v_m+\frac{3}{4(a^2-1)}\right)\cdot g\right]$$

$$S_{2211} = -\frac{a^2}{2(1-v_m)(a^2-1)}+\frac{1}{4(1-v_m)}\left[\frac{3a^2}{a^2-1}-(1-2v_m)\right]\cdot g$$

$$S_{1122} = -\frac{1}{2(1-v_m)}\left[1-2v_m+\frac{1}{a^2-1}\right]+\frac{1}{2(1-v_m)}\left[1-2v_m+\frac{3}{2(a^2-1)}\right]\cdot g$$

$$S_{2323} = \frac{1}{4(1-v_m)}\cdot\left[\frac{a^2}{2(a^2-1)}+\left(1-2v_m-\frac{3}{4(a^2-1)}\right)\cdot g\right]$$

$$S_{1212} = \frac{1}{4(1-v_m)}\cdot\left[1-2v_m-\frac{a^2+1}{a^2-1}-\frac{1}{2}\left(1-2v_m-\frac{3(a^2+1)}{a^2-1}\right)\cdot g\right]$$

$$S_{3333} = S_{2222} \quad g = \frac{a}{(a^2-1)^{\frac{3}{2}}}[a(a^2-1)^{\frac{1}{2}}-\cosh^{-1}a] \quad \textit{for } a>1$$

$$S_{3322} = S_{2233} \quad g = \frac{a}{(1-a^2)^{\frac{3}{2}}}[\cos^{-1}a-a(a^2-1)^{\frac{1}{2}}] \quad \textit{for } a<1$$

$$S_{3311} = S_{2211} \quad a=l/d=\textit{fibre aspect ratio}$$

$$S_{1133} = S_{1122}$$

$$S_{1313} = S_{1212}$$

number of different material forms and their moduli. It is this simplicity which makes it attractive in engineering. The Halpin–Tsai equations were originally formulated for aligned continuous fibre composites from two main sources. Hermans' generalised self-consistent model (Hermans, 1967) provides the equations for plane strain bulk modulus, $k$, longitudinal shear modulus, $G_{12}$, and transverse shear modulus, $G_{23}$, and Hill's universal relations (Hill, 1963), simplified to 'rule of mixtures' forms, that is neglecting the interaction between phases due to the difference in Poisson's ratios, are used for longitudinal modulus, $E_{11}$, and longitudinal Poisson's ratio, $v_{12}$. The Hermans equations are written in a generalised form (Halpin, 1976) as follows.

$$\frac{M_c}{M_m} = \frac{1 + ABV_f}{1 - BV_f} \qquad [9]$$

and

$$B = \frac{\dfrac{M_f}{M_m} - 1}{\dfrac{M_f}{M_m} + A} \qquad [10]$$

where $V_f$ is the volume fraction of fibre, $M_c$ is the composite modulus ($k$, $G_{12}$ or $G_{23}$), $M_f$ is the corresponding fibre modulus and $M_m$ is the corresponding matrix modulus. $A$ is a measure of the reinforcement geometry and loading.

By studying numerical solutions consistent with the equations of elasticity, the Halpin–Tsai equations have also been shown to be applicable to short fibre composites. Using $A$ as a fitting parameter, whose value differs for each modulus, the generalised equations above can be used to evaluate $E_{11}$, $E_{22}$, $G_{12}$ and $G_{23}$ in short fibre materials. Values for $A$ are given in Box 7 where it can be seen that for $E_{11}$ the parameter $A$ is a function of aspect ratio of the reinforcement phase. This is not the case for the other moduli where it takes on a constant value. These values for $A$ are the original Halpin–Tsai values (Halpin and Kardos, 1976). The values for $E_{11}$ and $G_{12}$ are generally agreed whilst alternative values for the transverse moduli have been proposed (Nielsen and Landel, 1994). Although the origin of these alternative values derives from the relationship between $A$ and the generalised Einstein coefficient, $k_e$, in the viscosity of suspensions (Nielsen and Landel, 1994), the actual values suggested indicate very low values for $k_e$. The values of $A$ used in the equations for $E_{22}$ and $G_{23}$ are therefore those proposed in the original Halpin–Tsai equations. As for continuous fibre composites, $v_{12}$ is evaluated using the simple rule of mixtures expression whilst $k$ can be derived from the other moduli or by using the GSC formula given in Box 7.

An important modification to the original form of the Halpin–Tsai equations, that is Eqns [9] and [10] has been proposed. The original equations have been found to be inaccurate at higher volume fractions. This is a consequence of the Herman's solution not taking into account the limit in maximum packing fraction in a real system. The problem was addressed by Nielsen (1970) who produced the amended form of the generalised equation as follows

$$\frac{M_c}{M_m} = \frac{1 + ABV_f}{1 - B\psi V_f} \qquad [11]$$

---

## Box 7

Modified Halpin–Tsai model (after Halpin and Kardos, 1976 and Nielsen and Landel, 1994)

$$M_c = \text{composite modulus}$$

$$\frac{M_c}{M_m} = \frac{1 + ABV_f}{1 - B\psi V_f}$$

$$M_f = \text{fibre modulus}$$

$$M_m = \text{polymer modulus}$$

$$V_f = \text{fibre volume fraction}$$

| $M$ (modulus) | $A$ |
|---|---|
| $E_{11}$ | $2l/d$ |
| $E_{22}$ | 2.0 |
| $G_{12}$ | 1.0 |
| $G_{23}$ | 1.0 |

$$B = \frac{\dfrac{M_f}{M_m} - 1}{\dfrac{M_f}{M_m} + A}$$

$$\psi = 1 + \frac{1 - \phi_m}{\phi_m^2} \cdot V_f \qquad \phi_m = \text{maximum packing fraction}$$

Other equations

$$v_{12} = v_f V_f + v_m (1 - V_f) \qquad \text{(rule of mixtures)}$$

$$\frac{k}{k_m} = 1 + \frac{V_f}{k_m \left( \dfrac{1}{(k_f - k_m)} + \dfrac{(1 - V_f)}{(k_m + G_m)} \right)} \text{(GSC model)}$$

---

where

$$\psi = \frac{1 - \phi_m}{\phi_m^2} \cdot V_f.$$

The introduction of the parameter $\psi$ takes into account the maximum packing fraction $\phi_m$. For cubic packing $\phi_m = 0.785$, whilst for hexagonal packing $\phi_m = 0.907$ (Nielsen and Landel, 1994). However, these high values are only typical of regularly aligned materials. For short fibre materials, where fibres are generally more random in their arrangement, $\phi_m$ can take significantly lower values. For three-dimensional random alignment, for instance, $\phi_m$ depends on aspect ratio and can range from 0.4 ($l/d = 10$) to 0.1 ($l/d = 50$) (Milewski, 1982). In real short fibre systems flow induced alignment occurs and values for $\phi_m$ will lie typically in the range 0.4 to 0.6. Because of the uncertainty of the maximum packing fraction and its dependence on both alignment and aspect ratio, $\phi_m$ is used as an empirical fitting factor in the materials modelling process.

The full form of the modified Halpin–Tsai equations, as used for aligned short fibre composites, and incorporating the modification for maximum packing fraction, are given in Box 7.

## 8.2.5 Comparison between models

The four models discussed in the previous sections are compared by considering the prediction of three moduli, $E_{11}$, $E_{22}$ and $G_{12}$ as shown in Fig. 8.2, 8.3 and 8.4. As only two of the models have the facility to vary fibre aspect ratio, proper comparison can only be made by looking at the case of very large aspect ratio, that is approximating to continuous fibres, results for which can be evaluated for all models. Although not applicable directly to short fibre systems the comparison should illuminate the main differences between the models. The results are presented as a measure of the variation in the ratio of the composite modulus to the matrix modulus, over a range of volume fraction from 0 to 0.45. This range more than covers volume fractions for short fibre polymer composites (typically $V_f < 0.3$) which are the main focus in this discussion. Numerical values are calculated assuming a glass fibre/polypropylene system with properties given in Table 8.1.

Predictions for $E_{11}/E_m$, as shown in Fig. 8.2, agree closely over the full range of volume fraction for all models including the coincidence of the upper and lower bound values. The variation with volume fraction is linear following the

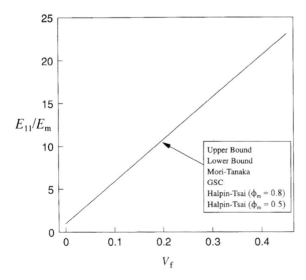

8.2  Comparison of $E_{11}/E_m$ values for all models (GFPP, $l/d = 1 \times 10^6$).

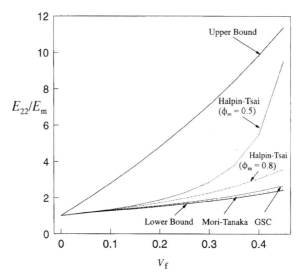

8.3  Comparison of $E_{22}/E_m$ values for all models (GFPP, $l/d = 1 \times 10^6$).

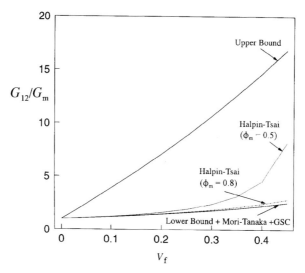

8.4  Comparison of $G_{12}/G_m$ values for all models (GFPP, $l/d = 1 \times 10^6$).

rule of mixtures. It is clear that any terms causing deviation from the rule of mixtures (see full form of equations) are negligible. This is a consequence of these terms, as can be seen for the GSC model (Box 3), depending on the square of the difference between the fibre and matrix Poisson's ratios which is small, and will generally be the case for all composites regardless of reinforcement. For

*Table 8.1.* Property data for short glass fibre-reinforced polypropylene

| Parameter | Value | Units |
|---|---|---|
| *Polypropylene matrix* | | |
| Young's modulus | 1.5 | GPa |
| Shear modulus | 0.55 | GPa |
| Poisson's ratio | 0.36 | — |
| | | |
| *Glass fibre* | | |
| Young's modulus | 75 | GPa |
| | 13 | |
| Shear modulus | 29 | GPa |
| Poisson's ratio | 0.24 | — |
| Average fibre length | 580 | μm |
| Fibre diameter | 11 | μm |

the transverse and shear moduli, $E_{22}$ and $G_{12}$, the models give varying predictions as shown in Fig. 8.3 and 8.4. For both these moduli, the GSC and Mori–Tanaka models are both close to or coincident with the lower bound which itself gives the same result as the exact CCA model (Hashin and Rosen, 1964) as previously discussed. This supports the use of these models in predictions for long and continuous fibre composites and the accuracy of these predictions. The modulus ratios, $E_{22}/E_m$ and $G_{12}/G_m$, as predicted by these models, are relatively low even at high volume fractions, showing the values for these moduli to be matrix dominated as expected. The upper bound predictions are high and the discrepancy from the more accurate lower bound predictions increases with volume fraction. This clearly illustrates the limitations of the bounds approach to modelling. Figures 8.3 and 8.4 also show that the Halpin–Tsai predictions differ from the other models, again generally giving higher values for $E_{22}$ and $G_{12}$, getting worse at high volume fraction. The accuracy of the Halpin–Tsai predictions depends very much on the assumption for maximum packing fraction. Of the two values used for illustration, 0.5 and 0.8, the latter results in better agreement with the other models. This is to be expected, as we are dealing here with continuous fully aligned reinforcement which will pack to a high degree. This, of course, may not be the case for more random arrangements of fibres as might be expected in short fibre materials.

To summarise, for *continuous* fibre materials, the GSC and Mori–Tanaka models give acceptable modulus predictions over the full range of volume fraction, both agreeing closely with the lower bound calculations. Acceptable results can also be obtained using the Halpin–Tsai model on condition a suitable value for the maximum fibre packing fraction is used. The inaccuracy of the upper bound predictions makes the bounds approach to modelling unacceptable.

## 8.3    Effect of fibre aspect ratio on stiffness

Fibre aspect ratio has a significant effect on the longitudinal Young's modulus of an aligned short fibre composite as is clearly illustrated in Fig. 8.5 which shows the predictions of the Mori–Tanaka model over a broad range of aspect ratio and volume fraction. The greatest effect occurs at low aspect ratios, where for example at a volume fraction of 0.4, an aspect ratio change from 2 to 20 results in a modulus increase more than sixfold. However, when the aspect ratio exceeds about 100 the composite modulus reaches a limit as it approaches that of a continuous fibre composite. For real short fibre composites, of course, there is a spectrum of fibre lengths, with typical aspect ratios lying in the range 10 to 100. For modelling purposes, it is therefore necessary to have an accurate measure of the fibre length distribution or at the very least a value for the mean fibre aspect ratio. These measurements can be obtained either by burn off tests or by careful analysis of polished surfaces.

Figures 8.6, 8.7 and 8.8 show the effects of fibre aspect ratio on the three moduli $E_{11}$, $E_{22}$ and $G_{12}$ , for both the Mori–Tanaka and Halpin–Tsai models, the latter assuming a maximum packing fraction of 0.8. For $E_{11}$ the Halpin–Tsai results agree well with the Mori–Tanaka results at high and low aspect ratio, particularly at low volume fractions ($< 0.2$). At intermediate aspect ratio ($\sim 20$ to 50) there are differences of the order of 10% to 20%. In general the Halpin–Tsai model underestimates $E_{11}$ compared to the Mori–Tanaka model. For both models, fibre aspect ratio has very little effect on $E_{22}$ and $G_{12}$ as shown in Fig. 8.7 and 8.8. In fact the Halpin–Tsai model for these moduli has no aspect ratio term. Again there are differences between the two models at high volume

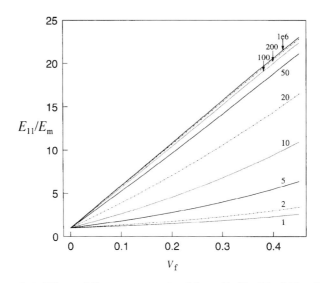

*8.5* Effect of fibre aspect ratio, *l/d*, on $E_{11}/E_m$ (Mori–Tanaka model).

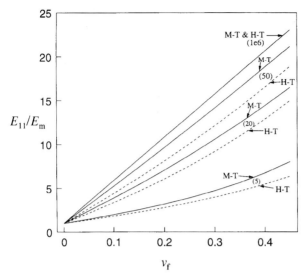

*8.6* Comparison of $E_{11}/E_m$ predictions by the Mori–Tanaka (M-T) model and Halpin-Tsai (H–T) model ($\phi_m = 0.8$) at various fibre aspect ratios. Numbers are fibre aspect ratios, $l/d$.

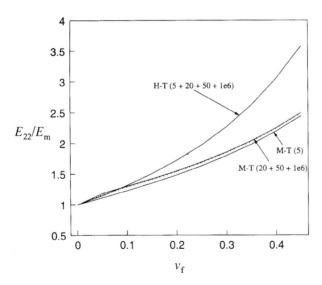

*8.7* Comparison of $E_{22}/E_m$ predictions by the Mori–Tanaka (M-T) model and Halpin-Tsai (H–T) model ($\phi_m = 0.8$) at various fibre aspect ratios. Numbers are fibre aspect ratios, $l/d$.

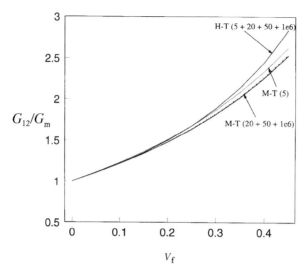

8.8   Comparison of $G_{12}/G_m$ predictions by the Mori–Tanaka (M-T) model and Halpin-Tsai (H–T) model $(\phi_m = 0.8)$ at various fibre aspect ratios. Numbers are fibre aspect ratios, $l/d$.

fraction, particularly for $E_{22}$. For both $E_{22}$ and $G_{12}$ the Halpin–Tsai model overestimates values. At low volume fractions, that is below 0.2, the two models agree well.

## 8.4   Effect of constituent properties on stiffness

Figure 8.9 shows the effect of varying the relative fibre to matrix stiffness, $E_f/E_m$, on the composite modulus ratios $E_{11}/E_m$ and $G_{12}/G_m$, as predicted by the Mori–Tanaka and Halpin–Tsai models. The curves are calculated by assuming a base material of short glass fibre-reinforced polypropylene $(E_f/E_m = 50)$ with a volume fraction of 0.4, fibre aspect ratio of 50 and varying the value of the fibre Young's modulus $E_f$. All other constituent properties remain constant except the shear modulus of the fibre, $G_f$, which is related to the changing Young's modulus by $G_f = E_f/2\,(1 + v_f)$. Basic properties for the constituent materials are given in Table 8.1.

$G_{12}$ is matrix dominated as indicated in Fig. 8.9 by the low modulus ratio $(G_{12}/G_m < 3)$ and is largely unaffected by changes in the constituent properties. Although not shown here, this is also the case for the transverse modulus, $E_{22}$. The predictions for shear modulus agree closely for both models over the full range of fibre stiffness, which is not unexpected in view of the domination of matrix properties. The situation regarding longitudinal modulus, $E_{11}$ is very different. The modulus ratio, $E_{11}/E_m$, is very much higher, indicating a greater influence of the fibre properties. The magnitude of $E_f/E_m$ clearly has a

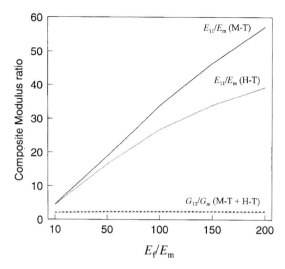

*8.9* Effect of constituent properties ($E_f/E_m$) on composite modulus ratios $E_{11}/E_m$ and $G_{12}/G_m$ (Mori–Tanaka, M-T, and Halpin-Tsai, H–T, models, $\phi_m = 0.8$). Fibre aspect ratio $l/d = 50$.

significant effect on the composite modulus. The predictions of the two models show a discrepancy (Halpin–Tsai giving lower values), which increases at higher modulus ratios, that is when the constituent properties differ significantly. The Mori–Tanaka predictions are influenced more by the consequential changes in shear modulus ratio, $G_f/G_m$, rather than direct changes in $E_f/E_m$. This is related to the stress transfer between matrix and inclusions which is shear controlled. The Halpin–Tsai model, on the other hand, is an empirical model which uses curve fitting to Young's modulus ratio data and thus is directly influenced by changes in $E_f/E_m$. The range of $E_f/E_m$ illustrated in Fig. 8.9 is relevant to a broad range of polymer matrix composites, as shown in Table 8.2.

*Table 8.2.* Young's modulus values and ratios for some composites

| Material | Fibre modulus $E_f$ (GPA) | Matrix modulus $E_m$ (GPa) | Ratio $E_f/E_m$ |
|---|---|---|---|
| Glass fibre/epoxy | 75 | 2.76 | 27 |
| Glass fibre/nylon | 75 | 2.5 | 30 |
| Glass fibre/polyester | 75 | 1.72 | 44 |
| Glass fibre/polypropylene | 75 | 1.5 | 50 |
| Carbon fibre (HS)/epoxy | 200 | 2.76 | 72 |
| Carbon fibre (HS)/nylon | 200 | 2.5 | 80 |
| Carbon fibre (HM)/epoxy | 375 | 2.76 | 136 |
| Carbon fibre (HM)/nylon | 375 | 2.5 | 150 |

All glass–polymer systems have values less than 50 where the discrepancy between the two models is relatively small. It is only in high modulus carbon systems, which have values of $E_f/E_m$ above 100, where the choice of model could be critical.

## 8.5    Modelling the effect of fibre orientation

### 8.5.1    Introduction

It is well known that fibre orientation has a significant effect on the mechanical properties of continuous fibre composites. The theory of classical laminate analysis (Tsai and Hahn, 1980) has been developed to deal with the difficulties of off-axis orientation in these materials and the state of the art is such that designers can now have confidence in their predictions of stiffness for multilayered laminates. The accuracy of these predictions does depend on the micromechanical model adopted for predicting the properties of the individual laminae within the laminate structure. It should be clear from Section 8.2 that suitable models do exist, particularly as much of the early work in this field focused on continuous fibre materials anyway. An alternative approach uses experimental data for the properties of the individual laminae and for reasons of confidence this approach is often adopted.

From a background of confidence in laminate analysis, a number of workers have extended the methods to short fibre materials. The basic principles, detailed in the following sections, involve taking the moduli of the fully aligned short fibre material, referenced to a local set of axes, and transforming them to the composite axes. As the local orientation varies throughout the component, the transformation involves an averaging process over a large number of aligned subunits and consequently the method has been termed the 'aggregate' model (Ward, 1962; Brody and Ward, 1971; Berthelot, 1982; Curtis *et al.*, 1982). The transformation process can be simplified by assuming that the subunits are aligned in the plane or that the composite itself is transversely isotropic. Also by assuming that the composite is subject to in-plane stresses only, the number of compliance and/or stiffness constants required is significantly reduced. The method has also been extended to materials with more general alignment and loading (Hine *et al.*, 1995), where a full tensor transformation is required.

Other approaches have been adopted such as combining the elastic stiffness tensor for aligned material with an expanded form of the orientation distribution function given in terms of generalised Legendre functions (Sayers, 1992) and a similar extension of the Mori–Tanaka method using Wu's fourth order fibre orientation tensor (Biolzi *et al.*, 1994). These latter approaches, however, require the evaluation of complex integral expressions.

In the following sections, the aggregate model, as applied to transversely isotropic short fibre composites, will be described in some detail. The method follows that described by Brody and Ward (1971) but is also used by a number of other authors. Some results of the prediction of tensile and flexural moduli of flat plaque specimens compared with experimental testing will be discussed in order to illustrate the accuracy and limitations of the method.

## 8.5.2  Transversely isotropic subunit

Consider a subunit comprising fully aligned short fibres as shown in Fig. 8.10. This subunit is isotropic in all planes perpendicular to the fibre direction and the fibre axis, that is axis one, is the axis of symmetry. It is said to be 'transversely isotropic'. The Generalised Hooke's Law for this type of material (Kelly and Macmillan, 1986) is

$$\varepsilon_i = S_{ij} \cdot \sigma_j \qquad\qquad [12]$$

where the compliance matrix, $S_{ij}$, is given in Box 8. Writing the compliance matrix in this form, the 'one' direction is the fibre direction (see Fig. 8.10) and there are only five independent compliances as expected for a transversely isotropic material.

Alternatively, the generalised Hooke's law can be written in terms of stiffnesses as follows

$$\sigma_i = Q_{ij} \cdot \varepsilon_j \qquad\qquad [13]$$

where the stiffness matrix, $Q_{ij}$ , is also given in Box 8.

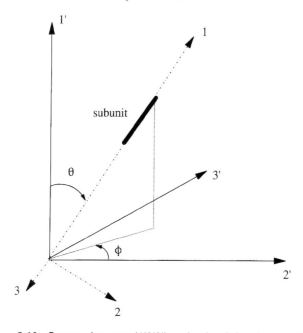

8.10  Composite axes (1'2'3') and subunit local axes (123).

## Box 8

Orientation averaging
  Transversely isotropic subunit

$$\varepsilon_i = S_{ij} \cdot \sigma_j$$

$$S_{ij} = \begin{bmatrix} S_{11} & S_{12} & S_{12} & 0 & 0 & 0 \\ S_{12} & S_{22} & S_{23} & 0 & 0 & 0 \\ S_{12} & S_{23} & S_{22} & 0 & 0 & 0 \\ 0 & 0 & 0 & 2(S_{22} - S_{23}) & 0 & 0 \\ 0 & 0 & 0 & 0 & S_{66} & 0 \\ 0 & 0 & 0 & 0 & 0 & S_{66} \end{bmatrix}$$

$$\sigma_i = Q_{ij} \cdot \varepsilon_j$$

$$Q_{ij} = \begin{bmatrix} Q_{11} & Q_{12} & Q_{12} & 0 & 0 & 0 \\ Q_{12} & Q_{22} & Q_{23} & 0 & 0 & 0 \\ Q_{12} & Q_{23} & Q_{22} & 0 & 0 & 0 \\ 0 & 0 & 0 & \dfrac{(Q_{22} - Q_{23})}{2} & 0 & 0 \\ 0 & 0 & 0 & 0 & Q_{66} & 0 \\ 0 & 0 & 0 & 0 & 0 & Q_{66} \end{bmatrix}$$

$$S_{11} = \frac{1}{E_{11}} \qquad S_{23} = -\frac{v_{21}}{E_{22}} \qquad\qquad Q_{11} = E_{11} + 4v_{12}^2 k$$

$$S_{12} = -\frac{v_{12}}{E_{11}} \qquad S_{66} = \frac{1}{G_{12}} \qquad\qquad Q_{12} = 2kv_{12}$$

$$S_{22} = \frac{1}{E_{22}} \qquad\qquad\qquad\qquad Q_{22} = G_{23} + k$$

$$Q_{23} = -G_{23} + k$$

$$Q_{66} = G_{12}$$

Transformation equations (in-plane subunit alignment and composite loading)

$$S'_{11} = S_{11} \langle \cos^4 \theta \rangle + S_{22} \langle \sin^4 \theta \rangle + (2S_{12} + S_{66}) \langle \cos^2 \theta \, \sin^2 \theta \rangle$$

$$S'_{22} = S_{11} \langle \sin^4 \theta \rangle + S_{22} \langle \cos^4 \theta \rangle + (2S_{12} + S_{66}) \langle \sin^2 \theta \, \cos^2 \theta \rangle$$

$$S'_{12} = (S_{11} + S_{22} - S_{66}) \langle \cos^2 \theta \, \sin^2 \theta \rangle + S_{12} \langle \cos^4 \theta + \sin^4 \theta \rangle$$

$$S'_{66} = 2(2S_{11} + 2S_{22} - 4S_{12} - S_{66}) \langle \sin^2 \theta \cos^2 \theta \rangle + S_{66} (\langle \sin^4 \theta \rangle + \langle \cos^4 \theta \rangle)$$

$$S'_{16} = (2S_{11} - 2S_{12} - S_{66}) \langle \sin \theta \cos^3 \theta \rangle + (2S_{12} - 2S_{22} + S_{66}) \langle \cos \theta \sin^3 \theta \rangle$$

$$S'_{26} = (2S_{11} - 2S_{12} - S_{66}) \langle \cos \theta \sin^3 \theta \rangle + (2S_{12} - 2S_{22} + S_{66}) \langle \sin \theta \cos^3 \theta \rangle$$

$$Q'_{11} = Q_{11}\langle\cos^4\theta\rangle + Q_{22}\langle\sin^4\theta\rangle + 2(Q_{12} + 2Q_{66})\langle\cos^2\theta\sin^2\theta\rangle$$

$$Q'_{22} = Q_{11}\langle\sin^4\theta\rangle + Q_{22}\langle\cos^4\theta\rangle + 2(Q_{12} + 2Q_{66})\langle\sin^2\theta\cos^2\theta\rangle$$

$$Q'_{12} = (Q_{11} + Q_{22} - 4Q_{66})\langle\cos^2\theta\sin^2\theta\rangle + Q_{12}\langle\cos^4\theta + \sin^4\theta\rangle$$

$$Q'_{66} = (Q_{11} + Q_{22} - 2Q_{12} - 2S_{66})\langle\sin^2\theta\cos^2\theta\rangle + Q_{66}(\langle\sin^4\theta\rangle + \langle\cos^4\theta\rangle)$$

$$Q'_{16} = (Q_{11} - Q_{12} - 2Q_{66})\langle\sin\theta\cos^3\theta\rangle + (Q_{12} - Q_{22} + 2Q_{66})\langle\cos\theta\sin^3\theta\rangle$$

$$Q'_{26} = (Q_{11} - Q_{12} - 2Q_{66})\langle\cos\theta\sin^3\theta\rangle + (Q_{12} - Q_{22} - 2Q_{66})\langle\sin\theta\cos^3\theta\rangle$$

The components of the compliance matrix and stiffness matrix are related to the elastic constants of the material as shown in Box 8. Thus knowledge of the elastic constants obtained from materials modelling methods described in Section 8.2 allows the compliance (or stiffness) matrix of the fully aligned subunit to be determined.

## 8.5.3  Transformation of properties and the aggregate model

Referring again to Fig. 8.10, the subunit local axes are the 123 set and the composite axes are the 1'2'3' set. In general the orientation of the subunit with regard to the composite is defined by the two angles $\theta$ and $\phi$ and there will be a distribution of these angles reflecting the distribution of orientations of a large number of subunits comprising the composite. If we now make the assumption that the subunits all lie within a plane, then the angle $\phi$ is constant and can be disregarded in the transformation procedure. In this case it is the distribution of the angle $\theta$ for the subunits which affects the properties of the composite.

The properties of the composite are found by transforming and averaging the properties of all subunits onto the composite axes set, that is the aggregate model. This averaging can be done in one of two ways. First, if the compliances are averaged it involves an assumption of constant stress and will provide a lower bound to the composite stiffness. This is termed the 'Reuss' average. Second, if stiffnesses are averaged, the assumption is one of constant strain and an upper bound to stiffness results. This is called the 'Voigt' average. Which of the averages is correct is open to question as it will depend on such factors as fibre aspect ratio and volume fraction in addition to relative constituent properties. The correct results may, of course, lie between the two bounds.

The general transformation laws for fourth order stiffness and compliance tensors (Ward and Hadley, 1993) are as follows

$$S'_{ijkl} = \alpha_{im}\alpha_{jn}\alpha_{kp}\alpha_{lq}S_{mnpq}$$

$$Q'_{ijkl} = \alpha_{im}\alpha_{jn}\alpha_{kp}\alpha_{lq}Q_{mnpq}.$$     [14]

The components of the transformation matrix, that is $\alpha_{im}$ etc., are the cosines of the angle between the two relevant axes. Where the transformation between the subunit axes and the composite axis is a simple rotation, $\theta$, as in the case of in-plane subunit alignment, the transformation matrix is given by

$$[\alpha] = \begin{bmatrix} \cos\theta & -\sin\theta & 0 \\ \sin\theta & \cos\theta & 0 \\ 0 & 0 & 1 \end{bmatrix}. \qquad [15]$$

Box 8 gives the specific forms of the transformation equations, that is the compliance, $S'_{ij}$, and stiffness, $Q'_{ij}$ (both in contracted form) components relevant to in-plane subunit alignment and composite in-plane loading. If it is also assumed that the composite is transversely isotropic then the '16' and '26' terms can be disregarded. To evaluate the expressions the following orientation functions are required

$$\langle \cos^4 \theta \rangle$$
$$\langle \sin^4 \theta \rangle \qquad [16]$$
$$\langle \sin^2 \theta \cos^2 \theta \rangle$$

where $\langle \ \rangle$ indicates averaging across all subunits, for example,

$$\langle \cos^4 \theta \rangle = \Sigma n_i(\theta). \cos_i^4 \theta / \Sigma n_i(\theta) \qquad [17]$$

where $n_i(\theta)$ is the number of fibres at angle $\theta$.

The orientation functions can be evaluated by assuming a specific fibre orientation distribution (FOD) or directly from FOD measurements obtained from the composite. A number of techniques have been developed to obtain these measurements, the most common of which involves sectioning, optical microscopy and image analysis (Bay and Tucker, 1992; Hine et al., 1993; Hsu and Brooks, 1996). Thus, knowledge of the fibre orientation distribution in the composite and the properties of the fully aligned subunit allow the properties of the composite to be evaluated and the effect of orientation to be determined.

## 8.5.4  Orientation averaging using tensors

The use of orientation tensors for describing fibre orientation distributions has been shown to be a particularly convenient and concise approach (Advani and Tucker, 1987). Their detailed formulation is described elsewhere in this book, however, a brief outline will be given here on their specific use in predicting the effects of orientation on properties. Tensor formulation has become important in recent years, as its conciseness, compared with the orientation distribution function for instance, has made it particularly useful in the computational prediction of orientation. This recent development now makes computer-aided design and property prediction for components made from short fibre materials feasible.

Orientation tensors are derived from the diadic products of the fibre unit vectors, **p**, multiplied by the distribution function and integrated over all directions. Although there are an infinite number of these tensors, they are of even order and the second order, $a_{ij}$, and fourth order, $a_{ijkl}$, tensors are sufficient for accurate orientation description. The procedure for calculating composite properties described in the previous section which involves averaging the subunit properties over all directions by essentially weighting with the orientation distribution, has been termed 'orientation averaging'. It can also be formulated in terms of the orientation tensor description and the basic expressions are now included for completeness (Advani and Tucker, 1987).

The fourth order stiffness tensor, $Q_{ijkl}$, of a transversely isotropic subunit can be averaged to give the composite stiffness, $Q'_{ijkl}$, as follows

$$Q'_{ijkl} = \langle Q_{ijkl} \rangle = B_1(a_{ijkl}) + B_2(a_{ij}\delta_{kl} + a_{kl}\delta_{ij})$$
$$+ B_3(a_{ik}\delta_{jl} + a_{il}\delta_{jk} + a_{jl}\delta_{ik} + a_{jk}\delta_{il}) + B_4(\delta_{ij}\delta_{kl})$$
$$+ B_5(\delta_{ik}\delta_{jl} + \delta_{il}\delta_{jk}). \qquad [18]$$

The constants, $B$, derive from the symmetry and form of the subunit stiffness tensor and are related to the components of that tensor as follows

$$B_1 = Q_{11} + Q_{22} - 2Q_{12} - 4Q_{66}$$
$$B_2 = Q_{12} - Q_{23}$$
$$B_3 = Q_{66} + (Q_{23} - Q_{22})/2 \qquad [19]$$
$$B_4 = Q_{23}$$
$$B_5 = (Q_{22} - Q_{23})/2.$$

The composite stiffness depends on both the second order, $a_{ij}$, and fourth order, $a_{ijkl}$, orientation tensors, though of course the former are incorporated in the latter. In general, to evaluate the orientation average of a fourth order tensor, such as stiffness (or compliance), it is only necessary to know the fourth order orientation tensor (Advani and Tucker, 1987). Full knowledge of the orientation distribution function is not required.

## 8.6    Application in design and some results

To illustrate the application of the above methods of orientation averaging, a study has been undertaken on the prediction of the tensile and flexural moduli of a flat plaque. As shown in Fig. 8.11 the basic geometry comprises a pin-gated rectangular plaque, 150 mm × 50 mm × 4 mm, manufactured by injection moulding from 20 wt% glass-reinforced polypropylene. To assess the accuracy of the predictions, results are compared with experimental tests on injection

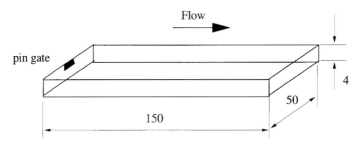

*8.11*   Glass fibre/polypropylene flat plaque (20 wt%).

moulded specimens. Although experimental data is available for orientation distribution in the plaque we are concerned here mainly with prediction and therefore orientation data obtained from mould flow analysis and orientation prediction software (Brooks and Fan, 1996; Fan, 1997) is used. It should be pointed out that the orientation results obtained this way have been compared with measured values and show good agreement (Brooks, *et al.* 1995).

To facilitate flow analysis and orientation prediction, as well as stiffness prediction, the plaque is modelled in two different ways using the finite element method as follows:

- As a coarse two-dimensional triangular mesh (60 elements) for predicting in-plane flow and orientation using the mould flow package FILLCALC, and for predicting in-plane tensile properties using PAFEC-FE (orthotropic elements).
- As a detailed three-dimensional rectangular mesh ($10 \times 5 \times 9$) for predicting three-dimensional flow and orientation using the computational fluid dynamics package FLOW3D, and for predicting both tensile and flexural properties using PAFEC-FE (orthotropic elements).

The orientation data, calculated from the flow analyses, in both cases is derived in tensor form using methods developed by Advani (Advani and Tucker, 1987) and the use of a suitable closure approximation results in only second order components. For the two-dimensional analysis only two independent components result, namely $a_{11}$ and $a_{12}$ (note that $a_{11} + a_{22} = 1$). For the three-dimensional analysis five independent components result, $a_{11}$, $a_{22}$, $a_{12}$, $a_{13}$ and $a_{23}$, however, observations from experimental results show that for the larger part of the plaque the fibres align in the plane of the plaque. The components $a_{13}$ and $a_{23}$ are therefore small and can be disregarded and as $a_{11} + a_{22} \approx 1$, the three-dimensional case has only the same two independent components, $a_{11}$ and $a_{12}$. The key difference between the two cases, however, is the through-thickness distribution present in the three-dimensional case which is accommodated by treating the structure as a laminate with a number of through-thickness layers, nine in this particular case. Thus, whereas for the two-dimensional case properties vary with the spatial distribution of orientation, for the three-dimensional case there is also a through-thickness distribution. However, each individual element in both models has only in-plane orientation and stiffness properties and is subject to in-plane loading.

The elastic constants for the fully aligned material are calculated using the modified Halpin–Tsai equations given in Box 7 with constituent property values for 20% glass fibre-reinforced polypropylene given in Table 8.1. An average value of 50 for fibre aspect ratio for this material has been confirmed by measurements from the polished upper surface of the plaque and a maximum packing fraction of 0.8 is chosen. The stiffness matrix, $Q_{ij}$, for the fully aligned material is calculated from the elastic constants, again using the expressions in Box 7. For each element in the plaque (both two- and three-dimensional) the stiffness matrix of the fully aligned material is combined with the local orientation tensor components, using Eqn [18], to give the local stiffness matrix for the composite, $Q'_{ij}$. In the equation, for simplicity, the fourth order orientation tensor is approximated in terms of second order components using the quadratic closure approximation (Advani and Tucker, 1990). As stated earlier, this method of orientation averaging gives a stiffness average (Voigt) which will result in an upper bound calculation of the plaque stiffness.

Local element stiffness matrices are then inverted to give their equivalent compliance matrices which are needed as input to the finite element structural analysis. Relevant boundary and loading conditions are set to simulate tensile and flexural testing of the plaque and the finite element calculations conducted. The output deflections allow the average tensile modulus and flexural modulus of the plaque to be determined.

The predicted results are given in Table 8.3, along with some measured values (Hsu, 1997) for comparison. Also shown in the table are the mean and range of variation of the orientation tensor component $a_{11}$ in the plaque which gives an indication of alignment in the flow direction. For the tensile modulus there is excellent agreement ( < 2% difference) for both models and likewise the flexural modulus predictions are accurate. As expected the flexural modulus is higher than the tensile modulus ( $\sim 8\%$ ) due to the through-thickness fibre orientation distribution which shows a typical skin–core structure, with greater alignment in the high shear skin regions. From a design point of view it is encouraging to see that the orientation averaging method gives satisfactory results even for a coarse two-dimensional (2D) mesh. This will be particularly important for more complex shaped components where computing power may limit the mesh size. The range of $a_{11}$ values is large, varying from almost fully transverse (0) near the gate to fully aligned (1) in some skin regions. In spite of this variation, the mean $a_{11}$ value ( $\sim 0.55$ ) is close to that of a random-in-plane distribution ( $a_{11} = a_{22} = 0.5$ ). This indicates that although there are areas within the plaque with high and low fibre alignment, overall the plaque behaves close to a random system and this assumption would be valid in design. This is unlikely to be the

*Table 8.3.* Comparison of predicted and measured average stiffness of a 20wt% glass fibre-reinforced polypropylene flat rectangular injection moulded plaque.

|  | 2D finite element model (61 elements) | 3D finite element model (450 elements) | Measured (after Hsu, 1997) |
|---|---|---|---|
| *Modulus* | | | |
| Tensile (GPa) | 3.73 | 3.65 | 3.65 (s.d. 0.04) |
| Flexural (GPa) | — | 3.95 | 3.96 (s.d. 0.04) |
| $a_{11}$ *tensor component* | | | |
| Range | 0.01–0.98 | 0.18–0.8 | — |
| Mean | 0.55 | 0.56 | — |

case, however, where particular flow fields result in greater or lesser fibre alignment, for example convergent or divergent flow. From the range of orientation obtained in the plaque, the Halpin–Tsai equations show that the modulus of any element can vary between about 2 GPa and 6 GPa depending on alignment. If the component geometry were such that significant alignment did occur, then stiffness prediction could show large errors if simplified assumptions were made. Furthermore, where flexural properties are important, such as in thick sections and bending applications, simplified design assumptions are likely to be limited. In these cases a full three-dimensional (3D) or at the very least a through-thickness prediction of orientation and stiffness properties is necessary.

## 8.7    Conclusions

In this chapter materials property modelling has focused on the elastic stiffness of short fibre reinforced composites. A number of models for predicting the stiffness of fully aligned materials have been described and compared. The Mori–Tanaka model is perhaps the most mathematically complete and gives acceptable results. The Halpin–Tsai model has a universal form of equation which is simple to use and consequently attractive in engineering design. It does, however, require empirical fitting parameters, such as a maximum packing fraction, if it is to be used reliably. At low volume fractions, for example $< 0.3$, typical of short fibre materials, both models agree well. Only at higher volume fractions and greater mismatch of constituent properties do the predictions deviate. It should be said that comparison with both experimental data and indeed reliable numerical models is needed if more accurate models are to be developed. It is of course difficult, if not impossible, to manufacture real short fibre materials with the necessary full alignment for comparison, and computer models are limited by both the assumptions involved and by computing power available. Of equal importance, however, is the difficult prediction of the effects of orientation on the stiffness. It has been shown that the combination of the fully aligned material model and orientation data, averaged through a simple component (orientation averaging), such as a flat plaque, does result in reliable predictions of the overall component stiffness. The results given here also confirm that reasonably accurate predictions for in-plane stiffness can be obtained from a 2D flow analysis and orientation prediction using tensors, coupled to relatively coarse mesh finite element calculations. This will be particularly important for complex shapes where the size of the computer model is limited by the computing power available. Already, a number of commercial mould flow analysis packages have incorporated orientation prediction and mechanical property modelling on this basis. Where components are subject to flexural loading, a simplified 3D analysis, based on a laminate model, as described in this chapter, should be used and will give accurate results on condition sufficient elements are used in the model. There is still much work to be done on the accurate prediction of material properties in fully 3D situations, where the computing resources required are not readily available. This applies in many regions such as gate entry positions, ribs and thick sections. The methods described in this chapter are also inapplicable at weld lines and the flow front where complex fountain flow and consequent orientation patterns develop. However, in spite of these difficulties a basic framework for computer-aided design of short fibre materials does now exist, from mould flow analysis, through orientation prediction, to materials property modelling.

# References

Advani, S. G. and Tucker, C. L. (1987), The use of tensors to describe and predict fiber orientation in short fiber composites, *J. Rheology* **31**(8), 751–784.

Advani, S. G. and Tucker, C. L. (1990), Closure approximations for three-dimensional structure tensors, *J. Rheology* **34**(3), 367–386.

Bay, R. S. and Tucker, C. L. (1992), Stereological measurement and error estimates for three-dimensional fiber orientation, *Polym. Eng. Sci.* **32**, 240–253.

Benveniste, Y. (1987), A new approach to the application of Mori–Tanaka's theory in composite materials, *Mech. Mater.* **6**, 147–157.

Berthelot, J. M. (1982), Effect of misalignment on the elastic properties of oriented discontinuous fibre composites, *Fibre Sci. Tech.* **17**, 25–39.

Biolzi, L., Castellani, L. and Pitacco, I. (1994), On the mechanical response of short fibre reinforced polymer composites, *J. Mater. Sci.* **29**, 2507–2512.

Brody, H. and Ward, I. M. (1971), Modulus of short carbon and glass fiber reinforced composites, *Polym. Eng. Sci.* **11**, 139–151.

Brooks, R. and Fan, Y. H. (1996), Quantitative prediction of orientation in short fibre-reinforced thermoplastics and its use in computer-integrated design, *7th European Conference on Composite Materials (ECCM-7)*, London, Institute of Materials.

Brooks, R., Fan, Y. H. and Hsu, C. Y. (1995), Computer-integrated design for short fibre-reinforced thermoplastics, *10th International Conference on Composite Materials (ICCM-10)*, Whistler, Canada, Woodhead.

Budiansky, B. (1965),On the elastic moduli of some heterogeneous materials, *J. Mech. Phys. Solids* **13**, 223–227.

Chou, T-W, Nomura, S. and Taya, M. (1980), A self-consistent approach to the elastic stiffness of short-fiber composites, *J. Compos. Mater.* **14**, 178–188.

Christensen, R. M. (1990), A critical evaluation for a class of micro-mechanics models, *J. Mech. Phys. Solids* **48**, 379–404.

Christensen, R. M. and Lo, K. H. (1979), Solutions for effective shear properties in three phase sphere and cylinder models, *J. Mech. Phys. Solids* **27**, 315–330.

Curtis, A. C., Hope, P. S. and Ward, I. M. (1982), Modulus development in oriented short-glass-fiber-reinforced polymer composites, *Polym. Compos.* **3**, 138–145.

Eshelby, J. D. (1957), The determination of the elastic field of an ellipsoidal inclusion, and related problems, *Proc. R. Soc. Lond.* A**241**, 376–396.

Fan, Y. H. (1997), Fibre Orientation and stiffness prediction in short fibre-reinforced thermoplastics, *PhD Thesis*, University of Nottingham.

Halpin, J. C. and Kardos, J. L. (1976), The Halpin–Tsai equations: a review, *Polym. Eng. Sci.* **16**(5), 344–352.

Hashin, Z. (1965), On elastic behaviour of fibre reinforced materials of arbitrary transverse phase geometry, *J. Mech. Phys. Solids* **13**, 119–134.

Hashin, Z. (1983), Analysis of composite materials – a survey, *ASME J. Appl. Mech.* **50**, 481–505.

Hashin, Z. and Rosen, B.W. (1964), The elastic moduli of fiber-reinforced materials, *ASME J. Appl. Mech.* **31**, 223–232.

Hermans, J. J. (1967), The elastic properties of fiber reinforced materials when the fibres are aligned, *Proc. Koninklijke Nederlandse Akademie van Wetenschappen* Ser. B **70**, 1–9.

Hill, R. (1963), Elastic properties of reinforced solids: some theoretical principles, *J. Mech. Phys. Solids* **11**, 357–372.

Hill, R. (1964), Theory of mechanical properties of fibre-strengthened materials: I Elastic behaviour, *J. Mech. Phys., Solids* **12**, 199–212.

Hill, R. (1965a), A self-consistent mechanics of composite materials, *J. Mech. Phys. Solids* **13**, 213–222.

Hill, R. (1965b), Theory of mechanical properties of fibre-strengthened materials: III Self-consistent model, *J. Mech. Phys. Solids* **13**, 189–198.

Hine, P. J., Duckett, R. A., Davidson, N. and Clarke, A. R. (1993), Modelling the elastic properties of fibre reinforced composites. I : Orientation measurement, *Compos. Sci. Tech.* **47**, 65–73.

Hine, P. J., Davidson, N., Duckett, R. A. and Ward, I. M. (1995), Measuring the fibre orientation and modelling the elastic properties of injection-moulded long-glass-fibre-reinforced Nylon, *Compos. Sci. Tech.* **53**, 125–131.

Hsu, C. Y. (1997), *PhD Thesis*, University of Nottingham (to be published).

Hsu, C. Y. and Brooks, R. (1996), Experimental determination of orientation in short fibre-reinforced thermoplastics using image processing and analysis, *7th European Conference on Composite Materials (ECCM-7)*, London, Institute of Materials.

Kelly, A. and Macmillan, N. H. (1986), *Strong Solids*, 3rd edn, Oxford, Clarendon Press.

Laws, N. and McLaughlin, R. (1979), The effect of fibre length on the overall moduli of composite materials, *J. Mech. Phys. Solids* **27**, 1–13.

Milewski, J. V. (1982), Packing concepts in the use of filler and reinforcement combinations, in *Handbook of Reinforcements for Plastics*, Von Nostrand, New York.

Mori, T. and Tanaka, K. (1973), Average stress in matrix and average elastic energy of materials with misfitting inclusions, *Acta Metall.* **21**, 571–574.

Nielsen, L. E. (1970), Generalized equation for the elastic moduli of composite materials, *J. Appl. Phys.* **41**(11), 4626–4627.

Nielsen, L.E. and Landel, R.F. (1994), *Mechanical Properties of Polymers and Composites*, 2nd edn, New York, Marcel Dekker.

Sayers, C. M. (1992), Elastic anisotropy of short-fibre reinforced composites, *Internat. J. Solids Struct.* **29**, 2933–2944.

Tandon, G. P. and Weng, G. J. (1984), The effect of aspect ratio of inclusions on the elastic properties of unidirectionally aligned composites, *Polym. Compos.* **5**(4), 327–333.

Tsai, S. W. and Hahn, H. T. (1980), *Introduction to Composite Materials*, Westport, Technomic.

Ward, I. M. (1962), Optical and mechanical anisotropy in crystalline polymers, *Proc. Phys. Soc.* **80**, 1176.

Ward, I. M. and Hadley, D. W. (1993), *An Introduction to the Mechanical Properties of Solid Polymers*, New York, John Wiley.

Zhao, Y. H., Tandon, G. P. and Weng, G. J. (1989), Elastic moduli for a class of porous materials, *Acta Mech.* **76** 105–130.

# 9
## Micromechanical modeling in aligned-fiber composites: prediction of stiffness and permeability using the boundary element method

T D PAPATHANASIOU AND M S INGBER

## 9.1 Introduction

The increased use of fiber-reinforced composite materials has sustained a vigorous scientific and technological interest in the area of the prediction of their effective properties. This is hardly surprising since the technical problem is quite a complex one and fiber-reinforced composites are being increasingly used in a number of demanding structural applications where accurate prediction of on-site behavior is a definite requirement for efficient design. At the heart of this problem is the ability to solve the equations which describe the (solid- or fluid-) mechanical response of model systems which are realistic representations of fiber-reinforced composites. Both analytical and numerical approaches can be employed for this purpose. A review of analytical methods for the prediction of the effective stiffness of aligned, short-fiber-reinforced (SFR) composites has been presented earlier in this book. Analytical methods have also been used in the development of models for the prediction of the effective permeability of arrays of fibers (Section 9.3.2 below). This is an area of current interest owing to the wider usage of resin transfer molding and other melt-infiltration techniques for the manufacturing of structural composites in which high fiber concentrations and well controlled orientations are desirable. Analytical approaches, such as those based on variational bounding, the self-consistent method or exact/ asymptotic analyses for the prediction of stiffness and on the lubrication approximation for the prediction of the effective permeability are constrained by a number of assumptions which are introduced, as a matter of necessity, in order to render the problem analytically tractable. The most common assumption is that the reinforcement is in the form of identical particles distributed uniformly throughout the composite. The influence of microstructure (in terms of a non-uniform spatial distribution of the dispersed phase) and micromechanics (in terms of localized particle–matrix and particle–particle interactions) on composite properties is therefore lost or greatly simplified in this approach. Consequently, a number of problems of current technological interest such as prediction of the properties of composites reinforced with non-ideal fibers (such as bent fibers, fibers with rough surfaces or anisotropic properties), the effect of variable material properties (dependent on strain, pressure, temperature, moisture, etc.) or the effect of localized non-homogeneities (e.g. on fracture behavior, Cairns *et al.*, 1995) cannot be realistically addressed in the context of analytical methods.

Numerical modelling is a serious alternative to analytical approaches and has been used extensively to both investigate the limits of validity of analytical models as well as study the influence of factors such as those outlined previously on composite properties. Among numerical methods, the finite difference (FDM) and finite element (FEM) methods (usually called domain methods because they require the generation of a computational mesh on the entire domain of interest) have been used extensively in the field of composites. However, the need to create a suitable computational mesh along with the complexity of the microstructures encountered in real fiber-reinforced composites, make their application to realistic three-dimensional microstructures cumbersome. For this reason and in a manner similar to analytical approaches, these methods have in the past employed representations of a composite body based on a 'unit cell', that is, a structural sub-unit which contains usually one reinforcing particle and which is repeated infinitely in space. As in the case of analytical approaches, the effect of important microstructural aspects such as clustering, localized fiber misorientation or particle size distribution is lost or simplified in this approach.

This disadvantage notwithstanding, when the geometry of interest can be rendered two dimensional, as is the case in cross-sections of composites reinforced with long unidirectional fibers, or when a periodic arrangement of aligned short fibers can be assumed and a representative three-dimensional unit cell identified, the mesh generation problem is significantly simplified and can be tackled by existing commercial or in-house mesh generators. Domain methods have been used with success in numerous such cases.

Brockenbrough *et al.* (1991) used the FEM to investigate the effect of the shape of the fiber cross-section (circular, hexagonal or square) as well as the effect of fiber topology (periodic or random placement) on the transverse mechanical deformation of metal–matrix composites (MMCs) reinforced with long, unidirectional fibers. The two-dimensional geometries considered included cross-sections of up to 30 (elastic) fibers dispersed in various ways in an elastic–plastic matrix. Fiber distribution was found to have a bigger influence on mechanical behavior than fiber shape. Similar two-dimensional simulations have been reported by Shen *et al.* (1995), Bohm *et al.* (1994), Finot *et al.* (1994) and Nakamura and Suresh (1993). Hom (1992) and Weissenbek and Rammerstorfer (1993) have used a three-dimensional unit cell approach to model composites described by infinite arrays of unidirectional identical short fibers, arranged in staggered or unstaggered configurations (that is, with complete or partial overlap between adjacent fiber rows). In the latter work, the fibers were considered to behave thermoelastically with temperature-dependent properties, while the matrix was considered thermo-elastic–plastic with isotropic hardening. In the work of Hom (1992) the fibers were elastic and the matrix elasto–plastic. In these geometrically simple but physically instructive arrangements, domain methods allow the incorporation of complex physics in the analysis, such as the development and relaxation of thermal stresses, the effect of plasticity and strain hardening, etc.

Micromechanical information (such as the evolution of a 'damage' indicator (Weissenbek and Rammerstorfer, 1993) or the development of localized stress fields around the fibers in the various arrangements (Hom, 1992; Brockenbrough et al., 1991) have also been obtained with this approach. Similar three-dimensional fiber arrangements have been modeled by Levy and Papazian (1991) with an emphasis on the effect of thermal treatment on the mechanical behavior of SiC/Al composites, as well as by Sorensen et al. (1992), who studied the effect of small deviations in the arrangement of fibers on the high temperature tensile response of aligned SFR SiC/Al alloys.

Further applications of the FEM in the area of MMCs and phase-transforming metals have been reported by Rammerstorfer et al. (1992). Prediction of crack propagation and ductile fracture through complex microstructures of Al/SiC composites has recently been carried out by Wulf et al. (1996), using the 'multiphase' elements, developed by Steinkopff and Sautter (1995) in order to tackle the difficulties associated with the use of the 'traditional' FEM in multiphase systems of complex microstructure. The effect of fiber breakages on strain distribution in unidirectional, long fiber composites has been treated by Pandey and Sundaresan (1996).

Since the main advantage of domain methods is the ability to include complex physics in the analysis (with the concomitant compromise in microstructural complexity), it is not surprising that most of the above applications are in the area of MMCs, where plasticity, cavitation and thermoelasticity are important. It is unrealistic as well as beyond our objectives to present in this chapter a more detailed coverage of the large amount of FEM simulations in the field of composite mechanics. The use of the FEM in modelling plasticity in composites (an issue of fundamental interest in MMCs) has been further discussed by Dawson et al. (1994). The application of the FEM in modeling unidirectional fiber-reinforced composites is also discussed by Li and Wisnom (1994), while recent applications in the field of biomechanical stress analysis are discussed by Reiter et al. (1994).

Domain methods have also been used in the analysis of fluid flow through arrays of unidirectional obstructions, the principal objective being the prediction of the effective permeability of fibrous porous media. Some recent works in this

*Table 9.1.* Selected works on the application of domain methods for the numerical solution of flow problems across arrays of obstructions (fibers, dendrites, fiber tows)

| Author | Flow | Geometry |
|---|---|---|
| Ranganathan *et al.* (1996) | Stokes'/Darcy's | Arrays of permeable fiber tows of elliptical cross-section. One tow per unit cell. |
| Phelan and Wise (1996) | Stokes'/Darcy's | Arrays of permeable fiber tows of elliptical cross-section. One tow per unit cell. |
| Bhat *et al.* (1995) | Navier–Stokes | Arrays of star-shaped, impermeable dendrites. Consideration of cross-sections of realistic and arbitrary shape. Several dendrites per unit cell. |
| Nagelhout *et al.* (1995) | $Re < 40$ | Square arrays of impermeable fibers. |
| Ghaddar (1995) | $Re < 150$ | Random and regular arrays of cylindrical fibers. Comparison with cell models. Parallel computing. |
| Cai and Berdichevsky (1993) | Small $Re$ | Square, hexagonal and perturbed arrays of cylindrical fibers. Several fibers per unit cell. |
| Bruschke and Advani (1993) | Stokes' | Square and hexagonal arrays of cylindrical fibers. |

area are summarized in Table 9.1. While the studies of Ranganathan *et al.* (1996) and Phelan and Wise (1996) have exploited the advantage of the FEM in handling fluids of complex rheological behavior (they considered the obstructions to be porous bundles of fibers and thus coupled Stokes' flow in the 'empty' regions with Darcy's flow inside the fiber bundles), the works of Bhat *et al.* (1995) and Cai and Berdichevsky (1993) tackled problems of viscous flow in geometries of greater geometrical complexity. Bhat *et al.*(1995) used the FEM to solve the Navier–Stokes equations for flow transverse to regular arrays of columnar dendrites (whose cross-section was approximated as a four-pointed star) as well as through regions of more general (and realistic) shape, whose geometrical outline was obtained by digitizing actual microstructures obtained by directional solidification of Pb–Sn alloys.

Cai and Berdichevsky (1993) extended the application of the finite element method to random arrangements of unidirectional long fibers. They considered a two-dimensional grid of uniform squares (unit cells) and then randomly removed or added one circular particle in each cell. Because of the uniformity of the finite element mesh in each region, only two types of meshes needed to be generated, one type for an empty fluid cell and another for the fluid region in a fiber-containing cell. These cells could be interconnected to generate unidirectional assemblies of randomly placed fibers. The results of Cai and Berdichevsky (1993) show that the random placement of fibers results in a scatter in the predicted effective permeability, which was reduced as the number of fiber cells considered in the simulation increased. Similar scatter in volume-averaged properties due to morphological variations from sample to sample is typical of random systems as will be shown in Section 9.3.1.1 below. Detailed numerical (FEM) investigation of flow through random arrays of fibers has also been reported by Ghaddar (1995), who discussed parallel computation issues and also examined the effect of inertia on permeability and flow patterns.

This chapter is concerned with a method which is very promising in the development of structure-oriented models for the prediction of composite properties, namely the boundary element method (BEM). A critical advantage of the BEM over domain methods is that instead of the entire domain of interest, only its boundaries need to be discretized. Not only does this reduce the dimensionality of the problem by one, but more importantly for problems involving fiber-reinforced composites, the method greatly simplifies grid generation for the associated complex multiconnected domain. Besides geometrical flexibility, the operation count number required by the BEM is smaller than the corresponding number by the FEM (Bettess, 1981). Further, studies by Mukherjee and Morjaria (1984) have shown that for the same level of solution accuracy, the boundary element method requires a coarser mesh than the FEM. Mammoli and Bush (1993) reached the same conclusion in a numerical study of the deformation of particulate MMCs. The BEM is therefore seen as an excellent method for modeling random multiparticle systems, as evidenced by its numerous applications in the fields of suspension and composite mechanics (where it is important to be able to handle large assemblies of particles) which have appeared in the past few years. Some of these applications are listed in Table 9.2. Clearly, the majority of previous BEM simulations in multiparticle systems have dealt with dispersions of spherical or near-spherical particles. Probably only with the exception of the work of Mondy et al. (1991), Ingber and Mondy (1994) and Papathanasiou et al. (1994, 1995), the behavior of fiber-filled solids or fluids has not received much attention in this context. Following an outline of the BEM and a note on its implementation on parallel supercomputers, we will present in this chapter results concerning the application of the BEM for the prediction of 1), the effective stiffness of SFR composites and 2), the prediction of the transverse permeability for viscous flow across arrays of long unidirectional fibers.

*Table 9.2.* Some recent works on the application of the boundary element method in the area of composite (CM) or suspension mechanics (SM)

| Author | Area | Comments |
| --- | --- | --- |
| Li *et al.* (1996) | SM | Shear flow of suspensions of deformable drops incl. interfacial tension. Up to 49 drops |
| Mondy *et al.* (1996) | SM | Motion of agglomerates of spheres and of flake-like particles in Newtonian fluid |
| Phan-Thien and Fan (1995) | CM | Up to 25 rigid particles in elastic matrix |
| Pozrikidis (1995); Zhou and Pozrikidis (1995) | SM | Transient deformation of liquid capsules surrounded by elastic membrane |
| Peirce and Napier (1995) | CM | Large scale models for granular mechanics |
| Mammoli and Bush (1993) | CM | 3D unit cell incl. elastic–plastic deformation |
| Papathanasiou *et al.* (1994, 1995) | CM | 3D assemblies of up to 63 rod-like rigid fibers in and elastic matrix. Stiffness predictions |
| Zhou and Pozrikidis (1994) | SM | Pressure-driven flow of dispersions of liquid-drops |
| Phan-Thien and Tullock (1994) | CM/SM | Sedimentation of spheres through suspensions Elongation of arrays of rigid spheres |
| Ingber and Mondy (1994) | SM | Motion of cylinders in Stokes' flow; Jeffery orbits |
| Li and Ingber (1994) | SM | Migration of spherical particles in tube flow |
| Geller *et al.* (1993) | SM | Linear chains of spheres moving in a quiescent fluid |
| Ilic *et al.* (1992) | SM | Sedimentation of spheres in conduits of various shapes |
| Chan *et al.* (1992) | SM | 3D multiparticle interactions; proximity effect |
| Dingman *et al.* (1992) | SM | 3D particle trajectories in Stokes' flow |
| Phan-Thien *et al.* (1991) | SM | Spheres, cubes and particle clusters in simple shear |
| Mondy *et al.* (1991) | SM | Falling ball rheometry incl. spheres and rods. Up to 40 particles |
| Vincent *et al.* (1991) | SM | Sedimentation of pairs of cubic, tetrahedrical and other non-spherical particles |

## 9.2    Boundary element method

The boundary element formulation for elasticity problems is based on Navier's equations for equilibrium which, in terms of displacement, are given by:

$$\nabla^2 u_i + \frac{1}{1-2v} u_{k,ik} = 0 \tag{1}$$

where $(u_i)$ is the $i$th component of displacement, $v$ is the Poisson ratio and the comma denotes differentiation with respect to the appropriate Cartesian coordinate. To avoid a condition of indeterminacy in Eqn [1], the following identification is made for an incompressible elastic material:

$$P(\mathbf{x}) = -\lim_{v \to 0.5} \frac{1}{(1-2v)} u_{k,k} \tag{2}$$

so, the governing differential equations in the case of incompressible elasticity are given by:

$$\nabla^2 u_i = P_{,i} \qquad u_{i,i} = 0 \qquad x \in \Omega \tag{3}$$

The equations of incompressible elasticity are therefore equivalent to those of Stokes' flow when the displacements and the shear modulus in the former case are replaced by velocities and by the shear viscosity, respectively, in the latter (this is also known as the fluid–solid analogy). Eqn [3] can be recast in an integral form by considering the weighted residual statement of the differential equations with weighting functions given by the fundamental solutions for the displacement $\mathbf{u}^*$ and traction $\mathbf{q}^*$. For three-dimensional problems, these are given by:

$$u_{ij}^*(\mathbf{x}, \mathbf{y}) = \frac{1}{8\pi r}(\delta_{ij} + r_{,i} r_{,j}) \tag{4}$$

and

$$q_{ijk}^*(\mathbf{x}, \mathbf{y}) = \frac{-3}{4\pi r^2} r_{,i} r_{,j} r_{,k} \tag{5}$$

where $r$ is the distance between $\mathbf{x}$ and $\mathbf{y}$ and $\delta_{ij}$ is the Kronecker delta function. Fundamental solutions in the case of compressible elasticity as well as for two-dimensional problems are also available and can be found in standard textbooks (e.g. Brebbia and Dominguez, 1992). Following the introduction of the fundamental solutions, the boundary integral equation (BIE) for the displacement components is given by:

$$c_{ij}(\mathbf{x})u_j(\mathbf{x}) + \int_\Gamma q_{ijk}^*(\mathbf{x}, \mathbf{y})u_k(\mathbf{y})n_j(\mathbf{y})\, d\Gamma = -\int_\Gamma u_{ik}^*(\mathbf{x}, \mathbf{y})t_k(\mathbf{y})\, d\Gamma \tag{6}$$

where $\Gamma$ is the boundary of the domain $\Omega$ and $t_k$ represents the tractions on the surface $\Gamma$. On smooth boundaries, $c_{ij} = 1/2$. In three-dimensional applications, the BIE is discretized by subdividing the boundary of the domain, $\Gamma$, into triangular and quadrilateral boundary elements. A variety of subparametric and isoparametric elements can be used for this purpose. In all cases presented in this chapter, the geometry has been given quadratic approximation whereas the functional approximations for the displacements and tractions have ranged from constant to quadratic. The latter have been shown to offer superior accuracy in the case of particles in close proximity (Chan *et al.*, 1992). Within the $n$th element, the $i$th components of **u** and **t** are therefore approximated by:

$$u_i^{(n)}(x) \cong \sum_{j=1}^{N} \bar{u}_{ij}^{(n)} F_j(x) \qquad t_i^{(n)}(x) \cong \sum_{j=1}^{N} \bar{t}_{ij}^{(n)} F_j(x) \qquad [7]$$

where $\bar{u}_{ij}^{(n)}$ and $\bar{t}_{ij}^{(n)}$ represent the values of the $i$th component of velocity and traction at the $j$th node within the nth element and the $F_j(x)$ are shape functions of the appropriate order. Using these approximations, the BIE can be written in the following discretized form (Brebbia and Dominguez, 1992):

$$c^i u^i + \sum_{n=1}^{Ne} \left\{ \int_{S_n} \mathbf{t}*\mathbf{F} dS \right\} \mathbf{u}^n = \sum_{n=1}^{Ne} \left\{ \int_{S_n} \mathbf{u}*\mathbf{F} dS \right\} \mathbf{t}^n \qquad [8]$$

in which a tensor notation has been used for conciseness and in which $Ne$ indicates the total number of boundary elements and $S$ is the area. Introducing a local coordinate system and applying numerical integration for the integrals in brackets results in:

$$c^i u^i + \sum_{j=1}^{N \cdot Ne} \mathbf{H}^{ij} \mathbf{u}^j = \sum_{j=1}^{N \cdot Ne} \mathbf{G}^{ij} \mathbf{t}^j \qquad [9]$$

where the summations are performed over all nodes in all elements and in which the influence matrices (**H** and **G**) have been introduced (defined as the discrete equivalent of the integrals in brackets in Eqn [8]). By thus collocating the BIE at the area centroid of each element, a system of linear equations is obtained through which the components of traction are related to the components of displacement at the collocation nodes. These equations can be represented symbolically as:

$$[H]\{u\} = [G]\{t\} \qquad [10]$$

where $\{u\}$ and $\{t\}$ are the displacement and traction vectors, respectively, at the collocation nodes.

## 9.2.1 Consideration of dispersed fibers

In typical boundary element applications, either the displacement or the traction is prescribed by the boundary conditions at each collocation node in each coordinate direction and therefore, Eqn [10] is rearranged (assembled) to solve for the remaining boundary unknowns. In applications involving dispersed fibers, the components of the displacement or traction can indeed be specified by the boundary conditions on the fixed portion of the boundary ($\Gamma_f$). However, on the surfaces of the fibers neither tractions nor displacements are known *a priori*. Allowing for some reordering of the vector $\{u\}$ we can rewrite $\{u\} = [\{u^d\}\{u^\tau\}\{u^p\}]^T$, where the superscript $d$ is associated with those components of displacement on $\Gamma_f$ specified by the boundary conditions, the superscript $t$ is associated with those components of the displacement vector on $\Gamma_f$ on which the corresponding traction is specified and the superscript $p$ corresponds to the components of displacement along the surfaces of the particles. Similarly, we can rewrite $\{t\} = [\{t^d\}\{t^t\}\{t^p\}]^T$. The boundary conditions then specify:

$$\{u^d\} = \{u^\delta\} \quad \text{on} \quad \Gamma_d, \text{ and} \tag{11}$$

$$\{t^t\} = \{t^\tau\} \quad \text{on} \quad \Gamma_t, \tag{12}$$

where $\{u^\delta\}$ and $\{t^\tau\}$ are prescribed vectors. Through this, Eqn [10] can be rearranged as:

$$[[H^p][H^d][H^t]]\begin{Bmatrix} \{u^p\} \\ \{u^\delta\} \\ \{u^t\} \end{Bmatrix} = [[G^p][G^d][G^t]]\begin{Bmatrix} \{t^p\} \\ \{t^d\} \\ \{t^\tau\} \end{Bmatrix}. \tag{13}$$

If there are $N$ boundary elements on the fixed boundary $\Gamma_f$ and $M$ boundary elements on the surfaces of the fibers, $\Gamma_i$, then Eqn [13] represents $3N + 3M$ equations in $3N + 6M$ unknowns. In the particular case of rigid fibers, the additional equations required to close the system can be generated by taking into account that the fibers are rigid and at equilibrium. The displacements at the collocation nodes on the surfaces of the fibers are then related to the displacements and rotations at the centroid of the fibers through the kinematic equations

$$u_i = u_i^c + \varepsilon_{ijk}\omega_j x_k \tag{14}$$

where $u_i$ and $u_i^c$ are the components of the displacement vectors on the surface and at the mass centroid of the fiber, respectively. $\omega_j$ represents the angular rotation of the fiber and $\varepsilon_{ijk}$ is the alternating unit tensor. The discretized version of Eqn [14] can be written in the following matrix form:

$$\{u^p\} = [K]\{U\} \tag{15}$$

where $\{U\}$ represents the displacements and rotations of the fibers. Although Eqn [15] provides $3M$ additional equations, it also introduces $6P$ additional unknowns (where $P$ is the number of fibers) associated with the fiber displacements and rotations. To close the system of equations, we impose the equilibrium equations for the fibers. For the $i$th fiber, the equilibrium equations are given by:

$$\int_{\Gamma_i} n \cdot \sigma d\Gamma + b^i = 0 \tag{16}$$

$$\int_{\Gamma_i} (q - q^i) \times (n \cdot \sigma) d\Gamma = 0 \tag{17}$$

where $\sigma$ is the total stress tensor, $b^i$ is the body force acting on the $i$th fiber and $q^i$ is the location of the centroid of the $i$th fiber. These equations can be represented in matrix form in terms of the surface tractions as follows:

$$[\mathbf{M}]\{\mathbf{t}^p\} = \{\mathbf{b}\}. \tag{18}$$

Combining the discretized boundary integral equation Eqn [10]) with the kinematic Eqn [15] and the equilibrium Eqn [18], we close the system of algebraic equations to obtain:

$$\begin{bmatrix} [\mathbf{H}^p][\mathbf{K}] & -[\mathbf{G}^p] & -[\mathbf{G}^d] & [\mathbf{H}^t] \\ 0 & [\mathbf{M}] & 0 & 0 \end{bmatrix} \begin{Bmatrix} \{\mathbf{U}\} \\ \{\mathbf{t}^p\} \\ \{\mathbf{t}^d\} \\ \{\mathbf{u}^t\} \end{Bmatrix} = \begin{Bmatrix} -[\mathbf{H}^d]\{\mathbf{u}^\delta\} + [\mathbf{G}^t]\{\mathbf{t}^\tau\} \\ \{\mathbf{b}\} \end{Bmatrix}. \tag{19}$$

## 9.2.2  Massively parallel implementation

The system of linear equations described by Eqn [19] is solved for the unknown displacements and tractions. When large numbers of fibers are included in a numerical simulation, large computational resources both in terms of CPU effort and, especially, computer memory are required. For example, a numerical simulation with 200 fibers can result in a system of over 20 000 discretized linear equations. This is beyond the capabilities of a single workstation and hence, a parallel BEM implementation on either a parallel supercomputer such as the IBM SP-2, Cray T3D or Intel Paragon or a cluster of networked workstations or PCs is desirable.

The parallel implementation considered here uses MPI (message passing interface) for parallel communication. Data is distributed in a block–block fashion where blocks of rows and columns in the matrix equation are distributed among processors. That is, the main strategy in the current parallel implementation of the boundary element formulation is to have different processors which are responsible for calculating different regions of the matrix Eqn [19]. As a simplified example, consider a dense matrix and an array of processors 'mapped' onto the matrix. The number of rows of processors in this array is denoted by NPROCSR and the number of columns by NPROCSC. The product of NPROCSR and NPROCSC is NP, the total number of processors used. A simple example of a $6 \times 6$ matrix distributed across six processors where NPROCSR $= 3$ and NPROCSC $= 2$ is shown in Fig. 9.1. In this example, processor 3 (P3) would 'own' rows 3 and 4 in columns 4, 5 and 6. Processor 3 would only calculate the coefficients in those rows and columns. In the present scheme, processor 3 would only integrate over the elements containing nodes 4, 5 and 6 for collocation nodes 3 and 4.

From a programming viewpoint this type of block–block data distribution is more complicated than data distributions which would allocate either entire rows or columns to processors. A serial BEM code is typically comprised of three nested loops; an outer loop over collocation nodes, an intermediate loop over elements and an inner loop over Gauss points. If the outer loop is distributed among processors, rows of the matrix are formed in parallel. If the intermediate loop is distributed among processors, columns of the matrix are formed in parallel. Performing the inner loop in parallel results in a fine-grain parallelization which is not very efficient on distributed memory machines. In the current parallel approach, portions of both the outer and intermediate loops are distributed over processors. Despite the additional complexity of the parallel algorithm, the block–block data distribution minimizes interprocessor communication during the solution phase of the algorithm. Interprocessor communication is considered an expense which is to be minimized whenever possible.

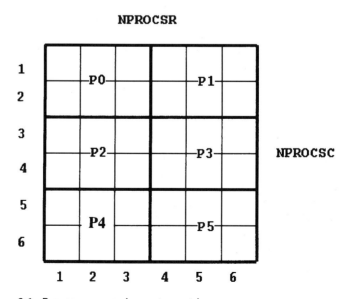

9.1 Processor mapping onto matrix.

The data decomposition shown in Fig. 9.1 must be modified for the current parallel BEM formulation in order to balance the computational load. In particular, Eqn. [19] is rewritten as:

$$
\begin{bmatrix}
[\tilde{\mathbf{H}}^p][\tilde{\mathbf{K}}] & -[\tilde{\mathbf{G}}^p] & -[\tilde{\mathbf{G}}^d] & [\tilde{\mathbf{H}}^t] \\
[\mathbf{H}^p][\mathbf{K}] & -[\mathbf{G}^p] & -[\mathbf{G}^d] & [\mathbf{H}^t] \\
0 & [\mathbf{M}] & 0 & 0
\end{bmatrix}
\begin{Bmatrix}
\{\mathbf{U}\} \\
\{\mathbf{t}^p\} \\
\{\mathbf{t}^d\} \\
\{\mathbf{u}^t\}
\end{Bmatrix}
=
\begin{Bmatrix}
-[\tilde{\mathbf{H}}^d]\{\mathbf{u}^\delta\} + [\tilde{\mathbf{G}}^t]\{\mathbf{t}^\tau\} \\
-[\mathbf{H}^d]\{\mathbf{u}^\delta\} + [\mathbf{G}^t]\{\mathbf{t}^\tau\} \\
\{\mathbf{b}\}
\end{Bmatrix}
$$

[20]

where the matrices with the 'tilde' indicate that they are associated with collocation on the surface $\Gamma_i$ of the embedded fibers while the other matrices (except $[\mathbf{M}]$) are associated with collocation on the outer surface of the system. Before data distribution, the matrix Eqn [20] is first partitioned into nine sub-matrices, as indicated. Similarly, the right-hand vector is partitioned into three sub-vectors.

The parallel program structure of distributing the outer and intermediate loops is modified in order to balance the computational load since the coefficient matrix is not dense and the computational effort for each sub-matrix is not equal. Each sub-matrix is partitioned and assigned to processors in a block–block fashion. A sample parallel data distribution for a system equation on four ($2 \times 2$) processors is shown in Fig 9.2. Notice that each of the nine sub-matrices has been assigned to a $2 \times 2$ processor array. There is some communication

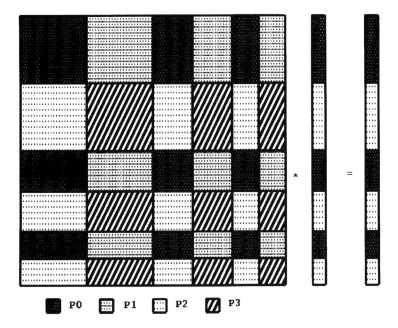

PO    P1    P2    P3

*9.2* Example of parallel data distribution on four ($2 \times 2$) processors.

necessary to form the right-hand side of Eqn [20] as the matrices $[\tilde{\mathbf{H}}^t]$, $[\mathbf{H}^d]$, $[\tilde{\mathbf{G}}^t]$ and $[\mathbf{G}^t]$ are also distributed among the processors. That is, interprocessor communication is necessary to complete the matrix–vector multiplications.

Although the complexity of the algorithm and programming effort is increased, the block–block data distribution is much more efficient compared to more traditional row- and column-wrapped data distributions (Ingber *et al.*, 1994). The LU factorization of the coefficient matrix is computed to solve the linear system. This is followed by a forward substitution and a back substitution. The block–block data distribution minimizes interprocessor communication during the LU factorization. A typical run time for a three-dimensional simulation, involving 100 randomly dispersed individual short fibers in a cylindrical container, resulted in 11 400 linear equations. Using 100 nodes of the Intel Paragon, it took 44 s to assemble the BEM equations, 239 s to factor and solve the equations and 5 s for data input, output and post-processing. The total time of 288 s is reasonable, considering the magnitude of the problem.

## 9.3    Representative results

In the following, the application of the BEM in modeling fiber-reinforced composites is demonstrated by considering the following two problems: first, the prediction of the stiffness of a three-dimensional rod reinforced with up to 200 discrete short fibers arranged randomly and second, Stokes' flow across arrays of aligned long fibers in various configurations or across arrays of fiber clusters. The three-dimensional nature and the sheer size of the first problem necessitates the use of high performance computers. Some results presented in Section 9.3.1 below were obtained on the Cray-YMP vector supercomputer with 512 MB of RAM, while the majority of simulations were run on the Intel Paragon massively parallel supercomputer. The second problem is two-dimensional and thus smaller; a DEC-alpha workstation with 128 MB of memory was used to obtain the results presented in Section 9.3.2 below.

### 9.3.1  Stiffness of an elastic rod reinforced with short fibers

The prediction of the effective stiffness of SFR composites has been the subject of considerable research over more than half of a century. In the preceding chapter, where this subject is dealt with in more detail, it becomes evident that discrepancies between various analytical models are more prominent at low to intermediate fiber aspect ratios, $a_r$, (where $a_r = l/r$, $l$ being the length and $r$ the radius of the fiber) and high fiber volume fractions, $\phi$. In this section, the flexibility afforded by the use of the BEM in conjunction with (parallel or vector) supercomputing is illustrated by studying the (linear-elastic) mechanical response of a cylindrical rod reinforced with up to 200 discrete rigid fibers in the interesting region of small $a_r$ and high $\phi$. In the cases considered the fibers are placed at random locations inside the reinforced rod and with their axes either aligned or at random angles with the axis of the rod (which coincides with the tensile axis). Typical systems, illustrating the difference between the aligned- and random-fiber configurations, are shown in Fig 9.3.

The composite cylinder is subjected to uniform tensile displacement at both ends, simulating a bar tensile test. The tractions $T$ at the edges of the cylinder are computed from the boundary element solution and the force required to cause this displacement is found as:

$$F = \int_{\Gamma} T \cdot ds \qquad [21]$$

where $\Gamma$ represents the area at the edges of the cylinder. Based on this force, the tensile modulus $E_c$ of the composite cylinder can be calculated by straight-forward analysis. The normalized effective modulus of the composite is then found as $E_{eff} = E_c/E_m$, where $E_m$ is the modulus of the matrix material. In these

**(a)**                    **(b)**

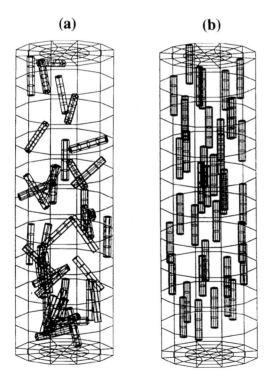

*9.3* Representative composite rods reinforced with rod-like fibers dispersed either randomly (a) or with their axes aligned in the tensile direction (b).

numerical experiments, and in accordance with the conditions in a bar tensile test, the lateral surfaces of the cylinder are considered to be traction free.

### 9.3.1.1 Configurational effects on the effective stiffness

The computations of Papathanasiou *et al.* (1994) and Ingber and Papathanasiou (1997) have indicated that subtle differences in the placement of the reinforcing fibers can result in different values for the predicted effective stiffness of the composite rod. These are manifested as a scatter in the effective moduli of composite rods containing the same numbers of (randomly placed) identical fibers. This is not unexpected, since the details of the spatial distribution of the fibers determine the extent of the interfiber as well as the extent of the fiber–wall interactions, which in turn influence the effective composite stiffness. Figure 9.4 shows the standard deviation of the computed normalized moduli as a function of the fiber volume fraction for fibers of aspect ratios 6, 10 and 20, fully aligned in the tensile direction. These results were obtained on 'samples' containing 100 fibers (Ingber and Papathanasiou, 1997) and were based on 10 or 20 realizations for each of the ($\phi$, $a_r$) combinations considered. As expected, the standard deviation of (and therefore the scatter in) the predicted effective moduli is very small at low fiber concentrations and increases sharply as the fiber volume fraction increases. For fixed $\phi$, the scatter in the predicted effective moduli increases with fiber aspect ratio, reflecting the increased importance of interfiber interactions at higher $a_r$.

### 9.3.1.2 Aligned versus random fibers

The improvement in the tensile modulus and strength afforded by the alignment of fibers in the tensile direction is well known (Guell and Graham, 1996 and

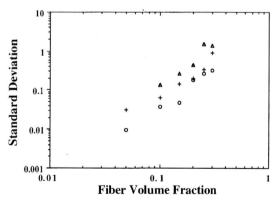

**9.4** Variation of the standard deviation of the dimensionless composite modulus $E_{\text{eff}}$ with fiber volume fraction and fiber aspect ratio. (o) $a_r=6$; (+) $a_r=10$; ($\Delta$) $a_r=20$.

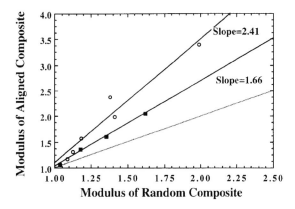

9.5  Normalized tensile modulus of a rod containing aligned fibers plotted against the modulus of an identical composite rod containing randomly oriented fibers. The effect of fiber alignment on the tensile modulus is predicted to increase from 66% for fibers of $a_r = 6$ (■) to 141% for fibers with $a_r = 10$ (○). Broken line has slope of unity.

references therein). This improvement can be better visualized by plotting the results for the $E_{eff}$ of a composite rod containing aligned fibers against the $E_{eff}$ of an otherwise identical (in terms of number and size of fibers) composite rod in which the fibers assume random orientations. This way of plotting the data allows for an immediate quantification of the effect of alignment on the effective stiffness; the fractional increase in stiffness afforded by fiber alignment is simply the slope of the corresponding line minus unity. Figure 9.5 shows numerical results corresponding to fibers of aspect ratio 6 and 10. From the slopes of the corresponding graphs it can be seen that fibers of aspect ratio 6 are predicted to be 66% more efficient in stiffening the composite in the tensile direction when they are aligned with the latter than when they are randomly oriented. From the same figure, the corresponding increase for fibers of aspect ratio 10 is predicted to be 141%. These results were found in reasonable agreement with tensile measurements in carbon-fiber-reinforced epoxy composites (Papathanasiou *et al.*, 1995; Guell and Graham, 1996).

### 9.3.1.3  *Effect of fiber volume fraction and fiber aspect ratio*

Besides the intrinsic properties and the orientation of the fibers, fiber volume fraction $\phi$ and aspect ratio $a_r$ are parameters crucial to the performance of an SFR composite. Models used to predict these dependencies have been presented earlier in this book. In the following, BEM results for the effect of $\phi$ and $a_r$ on stiffness are presented and compared to the predictions of some simple models. In all these cases the reinforcing fibers are aligned in the direction of the tensile stress.

Amongst simple analytical models, the Halpin–Tsai equation is a popular semi-empirical formula frequently used in engineering design for the prediction of the tensile (or longitudinal, $E_L$) modulus of aligned short-fiber composites. The Halpin–Tsai equation reads:

$$E_L = E_m \cdot \frac{(1 + \xi \cdot \eta \cdot \phi)}{(1 - \eta\phi)} \qquad [22]$$

where $\eta$ is a function of the elastic properties of fibers (subscript f) and matrix (subscript m) and of the fiber aspect ratio: $\eta = (A - 1)/(A + \xi)$. Usually, $\xi = a_r = l/r$ and $A = E_f/E_m$. The relationship between the predictions of Eqn [22] and those of other models has been discussed earlier in this book. In relation to experimental measurements, Halpin (1969, 1984) has reported that the predictions of Eqn [22] are in good agreement with experimental data on rubber–nylon short-fiber composites. On the other side, Blumentritt et al. (1974) have reported extensive data on the tensile modulus of aligned short-fiber polymer composites with a number of matrices and a variety of fibers and have shown that their measurements fall consistently below the predictions of the Halpin–Tsai equation. It should always be remembered that agreement or disagreement between theory and experiment should be treated with caution since many of the assumptions underlying the development of analytical models are not met in actual fabricated composite samples. Among the factors that cannot be easily controlled is the quality of the interphase between fiber and matrix (Silvain et al., 1994; Goto and McLean, 1991), the degree of fiber alignment and the uniformity of the spatial configuration as well as of the size of the fibers (Ter Haar and Duszczyk, 1994). Contrary to physical experiments, numerical experimentation is free from such uncertainties and thus ideal for a direct evaluation of the suitability of the Halpin–Tsai equation and of its variants in predicting the stiffness of aligned short-fiber composites.

Figure 9.6 shows typical configurations of short aligned fibers in a composite rod used in the BEM simulations. Figures 9.6(a)–(c) correspond to $\phi = 5\%$ and fiber aspect ratios of 6, 10 and 20 while Fig. 9.6(d) is for $\phi = 5\%$ and $a_r = 10$. Figure 9.7 summarizes the results of a large number of numerical experiments on the composite rods of Fig. 9.6 in the range of fiber volume fractions between 5% and 30%. Because of steric inhibition effects, it can be practically impossible to create statistically random distributions of aligned fibers at higher concentrations (that limit was around 30% in the present studies, see Ingber and Papathanasiou, 1997) and this, along with the need to consider more than one realization at each volume fraction and aspect ratio, determined the limits of the investigated ($\phi$–$a_r$) space. Each numerical data point in Fig. 9.7 is the average of either 10 or 20 individual realizations, each containing 100 discrete fibers.

Also shown in Fig. 9.7 are the 95% confidence intervals at each point. The predictions of the Halpin–Tsai equation are shown as dotted lines. At low volume fractions the Halpin–Tsai equation agrees well with micromechanical

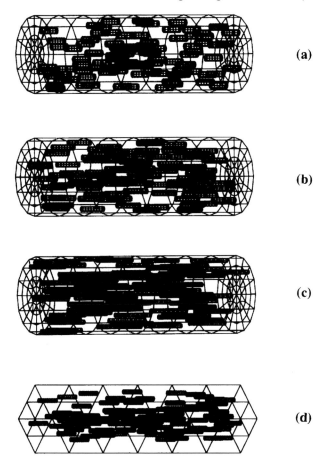

**(a)**

**(b)**

**(c)**

**(d)**

*9.6* Typical random configurations of aligned fibers in a 10 × 30 cylinder (a)–(c) for three aspect ratios $a_r = \{6,10,20\}$ and (d) in a parallepiped for $a_r = 10$. The boundary element mesh on the container walls is only partially shown for clarity.

BEM computations. The critical concentration at which the BEM micromechanical predictions start deviating from those of the Halpin–Tsai equation depends on the fiber aspect ratio: for $a_r = 6$, BEM computations and Eqn [22] give roughly identical results for $\phi$ up to 17%. This critical fiber volume fraction is 10% when $a_r = 10$ and only around 5% when $a_r = 20$. It is intriguing to notice that these concentrations roughly mark the upper limits of the semi-concentrated regime (defined as $(2/a_r)^2 \ll \phi \ll (2/a_r)$; Tucker and Advani, 1994). At higher concentrations, the predictions of the BEM model and those of the Halpin–Tsai equation deviate from each other, with the micromechanical BEM simulations consistently predicting a higher effective tensile modulus.

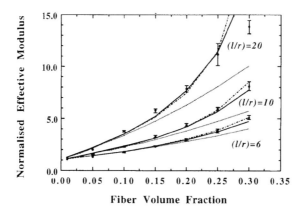

*9.7* Comparison between BEM results for $E_{eff}$ (indicated as data points with error bars showing the 95% confidence intervals), the predictions of the Halpin–Tsai equation (dotted lines) and the predictions of the Nielsen model (solid and broken lines corresponding to the two forms of Eqn [24] for $\psi$). The exponential form of the latter gives the highest predictions (broken lines). $\phi_m$ is taken as 0.6, 0.5 and 0.4 for $a_r=6$, $a_r=10$ and $a_r=20$, respectively. 100 aligned fibers.

One of the obvious reasons for the discrepancy between the predictions of the BEM computations and those of the Halpin–Tsai equation lies in the fact that the latter assumes that fibers can fill up to 100% of the available space. A more robust and useful generalization of the Halpin–Tsai equation, which accounts for the fact that the maximum volume fraction of the reinforcing phase can never reach 100% due to steric interactions between fibers and between fibers and containing surfaces, was proposed by Lewis and Nielsen (1970) (also, Nielsen, 1970,1974). Taking into account the finite amount of the reinforcing phase that can be packed into a fixed volume of composite, they proposed the following expression for the effective longitudinal tensile modulus:

$$E_L = E_m \cdot \frac{(1 + \xi \cdot \eta \cdot \phi)}{(1 - \eta \psi \phi)} \tag{23}$$

where $\eta$ is still calculated as in Eqn [22], and where a new parameter $\psi$ is introduced so that the quantity $\psi \cdot \phi$ represents a reduced volume fraction of the reinforcement with the property of approaching unity as the volume fraction of the reinforcement approaches maximum packing $\phi_m$. Since fiber–fiber interactions become overwhelming as $\phi_m$ is approached, the introduction of the parameter $\psi$ can also be seen as an empirical way of introducing interfiber interactions in the Halpin–Tsai equation. Two frequently used empirical functional forms for $\psi$, which fulfil the above condition as well as the requirement of Einstein behavior at dilute concentrations, are:

$$\psi = 1 + \left[\frac{1 - \phi_m}{\phi_m^2}\right] \cdot \phi \quad \text{and} \quad \psi = \frac{1}{\phi}\left[1 - \exp\left(\frac{-\phi}{1 - (\phi/\phi_m)}\right)\right].$$

$$[24]$$

These yield roughly identical predictions for $\phi$ up to 25% but differ in their asymptotic behavior as $\phi_m$ is approached (Lewis and Nielsen, 1970). Even though, as in the original Halpin–Tsai equation, there are no adjustable parameters in the Lewis–Nielsen model, there is a degree of empiricism in the choice of the function $\psi$ as well as in the value used for the fiber volume fraction at maximum packing $\phi_m$. A comparison between the predictions of this generalized form of the Halpin–Tsai equations and BEM micromechanical results is also shown in Fig 9.7. In this, both models for the function $\psi$ (Eqn [24]) have been used and are shown as solid and broken lines. Evidently, the Lewis and Nielsen modification of the Halpin–Tsai equation is capable of reproducing the results of the micromechanical BEM computations reasonably well. Other empirical modifications to the Halpin–Tsai equation suggest that the parameter $\xi$ in Eqn [22] is written as:

$$\xi^* = \frac{1}{r}\left(1 + \alpha \cdot \phi^\beta\right).$$

$$[25]$$

Ingber and Papathanasiou (1997) have shown that such models are capable of fitting the BEM results perfectly if the 'correct' $\alpha$ and $\beta$ are chosen. However, $\alpha$ and $\beta$ were found to be functions of the fiber aspect ratio; this indicates that little, in terms of useful generalization, can be gained by adopting such modifications.

## 9.3.2   Stokes' flow across arrays of aligned fibers

Analysis of Stokes' flow across assemblies of unidirectional fibers has been the subject of extensive study in recent years. This is not surprising, since reliable knowledge of the manner in which a fluid flows through a fiber preform is essential for the successful application of CAD/CAE (computer aided design and engineering) in the design and optimization of important liquid-molding manufacturing processes such as resin transfer molding, structural reaction injection molding and various metal infiltration processes. By formulating the Stokes' flow problem in the appropriate unit cell, exact or approximate solutions are sought in the corresponding range of porosity (usually low porosity for applications related to melt-infiltration processes and high porosity for applications related to the flow through columnar dendrites (Bhat et al., 1995; Nagelhout et al., 1995). Given such solutions, the volumetric flowrate, $Q$, through the unit cell is determined for a given pressure drop per unit length, $\Delta P/L$ and the effective permeability, $K$, of the medium is found from Darcy's law:

$$Q = \frac{K}{\mu} \cdot \frac{\Delta P}{L} \qquad [26]$$

where $\mu$ is the viscosity of the permeating fluid. Analytical models for the effective transverse permeability of uniform arrays of identical fibers have been developed by applying the lubrication approximation in unit-cells corresponding to square or hexagonal arrays of fibers (Sangani and Acrivos, 1982; Drummond and Tahir, 1984; Bruschke and Advani, 1993; Gebart, 1992). These models have been found in reasonable agreement with experimental data in polymeric (Sadiq et al., 1995) and metallurgical (Bhat et al., 1995) systems at intermediate porosity levels. The unit cell approach has been a useful tool in incorporating the basic fluid-mechanical and geometrical characteristics of a fiber assembly into predictive models for the effective permeability. However, the assumption of macroscopic homogeneity underlying unit-cell models does not allow consideration of microstructural factors, such as the non-uniform placement, size and orientation of the fibers, on the average permeability of the system. This is a significant limitation, since it is known that microstructure plays an important role in the permeability of fiber preforms used in liquid molding processes (Pillai et al., 1993; Pillai and Advani, 1995; Summerscales et al., 1995; Parnas and Phelan, 1991). The use of domain methods to handle more complex problems of flow across arrays of fibers has been outlined in the Introduction (Section 9.1, Table 9.1). The application of the BEM in this direction has been limited. Higdon and Ford (1996) used the spectral BEM developed by Muldowney and Higdon (1995) to solve the equations of Stokes' flow through three-dimensional cubic-lattice networks of cylindrical fibers. Chan et al.(1994) handled the problem of pressure-driven Stokes' flow across fixed beds of fibers, while Ganesan et al. (1992) used the BEM to solve for Stokes' flow parallel to the direction of primary dendrite arms at high levels of porosity, at which experimental measurement of the pertinent permeability fails. Stokes' flow near the interface between a fluid and a porous medium (the latter represented by semi-infinite arrays of cylindrical fibers) has further been analysed by Larson and Higdon (1987) using the BEM.

### 9.3.2.1 Morphological factors

As argued in the Introduction, the main advantage of the BEM over analytical approaches or domain methods is the ability to model complicated morphological features (e.g. random placement or unequal size of fibers) on the effective properties (permeability in this case) of an assembly (Ingber et al. (1994)). To this effect, the influence of morphology on the permeability of unidirectional assemblies of fibers is investigated in this section by using the BEM to solve the equations of Stokes' flow in the following four cases (Papathanasiou and Lee, 1997):

1   Fully random placement of fibers of equal size,
2   Fibers of equal size located in a square array, with a random perturbation in the location of the center of each fiber from the position corresponding to the square array,
3   A square array of fibers in which the fiber radius is randomly distributed around a mean value, and
4   Random perturbation of both the location and size of each fiber in a square array.

These variations are shown schematically in Fig. 9.8. The first case is analogous to a randomly distributed fiber assembly, while cases (2) to (4) could simulate manufacturing-induced modifications in an initially uniformly packed array of fibers. Cases (2) to (4) are also typical to many situations encountered in metallurgy, where the growth of columnar cells or dendrites often forms a near-square array in which both the exact location and the radius of the dendrites varies spatially. BEM computations were carried out in arrays of fibers of the type shown in Fig. 9.8.

By imposing a pressure drop $\Delta P$ across the cell, the flow rate $Q$ through it is calculated numerically and the effective permeability is then found from Eqn

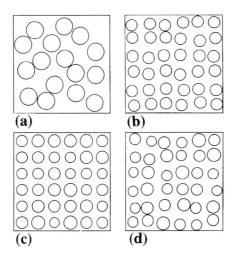

*9.8* Example unit cells for each of the four types of morphological irregularities considered in the BEM simulation of Stokes' flow transverse to arrays of aligned long fibers. (a) Fully random configuration; (b) square packing with perturbed fiber locations; (c) square packing with perturbed fiber radii; (d) square packing with perturbed centers and radii.

[26]. The effective permeability, which has units of square metres, is usually normalized by dividing by the square of the fiber radius $R$:

$$K' = \frac{K}{R^2}.$$                                                          [27]

The mean values of the predicted normalized permeability $K'$ for case (1) are shown in Fig. 9.9 along with error bars of one standard deviation of the log $K'$ values. In Fig. 9.9 the results corresponding to a uniform square array are also shown for comparison. The randomness in the placement of the fibers has no discernible effect at high porosity, while its influence on $K'$ becomes progressively more important as the porosity is reduced. At low porosity values, the effective permeability is strongly dependent on the restriction imposed on the flow by narrow passages formed between fibers, rather than by the drag exerted on each particle by the fluid. This is reflected in a more intense scatter in the predicted $K'$ values and thus bigger error bars at lower porosities.

In reality, it is near impossible to generate a perfectly uniform distribution of fibers. Instead, it is common that the location of the fiber centers may vary about a mean (typically as a result of forces exerted on the fibers during fabrication of the preform or during filling). Similarly, the size of the fibers is not entirely uniform. These microstructural effects are simulated by applying a random variation in the location of each fiber center about the location corresponding to a perfect square array and a similar variation in fiber radius. As in Section 9.3.1, a number of random configurations were generated at each volume fraction. The BEM-calculated mean value of $K'$ for a maximum random variation of 15% and a sample of 10 randomly perturbed microstructures (case 3) at each value of porosity is shown in Fig. 9.10(a), in which the error bars show $\pm 1$ log standard deviation. The scatter in the predicted $K'$ at high porosity is very small; as the

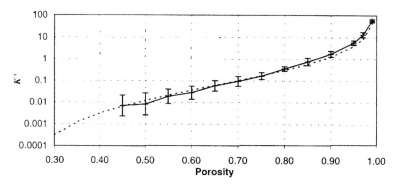

**9.9**  Effect of fully random fiber location on the normalized permeability. The solid line is the mean value with error bars of one standard deviation. The dashed line is the BEM solution for the same number of fibers placed in a perfect square array.

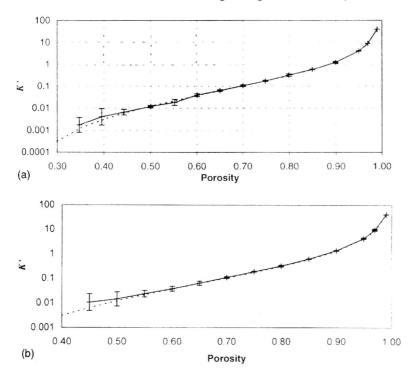

*9.10* (a) Effect of random perturbation of the size of each fiber in a square array on the normalized permeability (case 3) (b) Effect of a random perturbation of the location of each fiber in a square array on the normalized permeability (case 2). The solid line is the mean value with error bars of one log standard deviation. The dashed line is the BEM solution for a perfect square array.

porosity is reduced, the scatter steadily increases, as can be inferred by the magnitude of the error bars. At high porosity, the effective permeability of the perturbed array coincides with the permeability of the unperturbed square array; this is anticipated, since at high porosity the reduction to flow is mainly due to drag on the particles. As the porosity $\varepsilon$ is reduced, and in particular as $\varepsilon < 0.5$, the predicted effective permeability starts deviating upwards from that of the perfect square array. This trend shows the importance of what are sometimes called preferential flow channels which develop during solidification in metallurgical systems (Streat and Weinberg, 1976). The effect of random variation in particle location (case 2) on $K'$, shown in Fig. 9.10(b), appears to be similar to the effect of perturbations in fiber size – a negligible deviation from the result of a perfect square array for high values of porosity, which increases as the porosity decreases.

### 9.3.2.2    *Flow through arrays of fiber clusters*

The presence of fiber clusters (or tows) in which the local porosity is significantly lower than the interstitial (inter-tow) porosity is one of the most troublesome microstructural features of fiber preforms. Indeed, it is now understood that the wide difference in pore size between the inter- and the intra-tow regions is the main source of flow irregularities associated with the mold filling process in resin transfer molding (or other manufacturing processes involving infiltration of molten material into a fibrous preform). It is evident that under a pressure gradient molten material will flow through both, the inter- and the intra-tow space, each of which will contribute to the overall flow resistance of the medium. At least two distinct length scales must therefore be considered in a realistic model for flow through fiber preforms (Lundstrom and Gebart, 1995; Pillai and Advani, 1995): a meso-scale (related to the size of bundles of fibers (tows) comprising the preform) and a micro-scale (associated with the size of the individual filaments comprising each tow). Two porosities can thus be distinguished: a meso-scale porosity (or, inter-tow porosity, $\phi_i$), defined as the voidage between fiber tows in the preform, and a micro-scale porosity (or, intra-tow porosity, $\phi_t$), defined as the voidage between individual filaments inside a tow. In spite of the growing realization of the importance of the inter- and intra-tow length scales, only limited results (theoretical or experimental) exist in quantifying the effect of the interplay between meso- and micro-scale porosities on the permeability of a fiber preform (Summerscales, 1993; Pillai *et al.*, 1993; Pillai and Advani, 1995; Phelan and Wise, 1996). Application of the BEM to this problem has resulted in a novel approach to quantify the effect of the inter- and intra-tow porosities on the apparent transverse permeability of arrays of fiber tows and led to the formulation of appropriate scalings (Papathanasiou, 1996, 1997).

Typical structures associated with square arrays of fiber tows are shown in Fig. 9.11, in which the 'multifiber' unit cell is discernible. Figure 9.12 shows numerically computed trajectories of (massless) tracer particles placed at various locations along the left-hand (inlet) boundary of the unit cell. It is evident from the form of these trajectories that the flow 'senses' the presence of the tow and redistributes itself to take advantage of the low resistance region near the 'top' symmetry line. Weak recirculation regions near the lower symmetry axis appear to persist even for intra-tow porosities as high as 70%. It is also seen that even at intra-tow porosities as high as 50% the tow behaves largely as an impermeable entity, with only a small part of the material flowing through the unit cell travelling through the tow itself. As the internal porosity increases further to 60% and 70%, progressively more of the tracer particles actually pass through the tow.

By calculating the flow rate across the unit cell (obtained under a fixed pressure drop) its permeability can be found through Eqn [26]. The fact that the

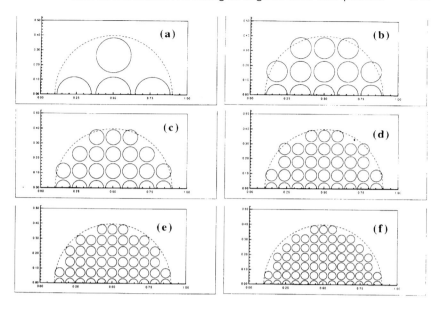

9.11  Typical fiber bundle (tow) configurations containing progressively larger numbers of fibers (a) $N_f=5$; (b) $N_f=21$; (c) $N_f=37$; (d) $N_f=61$; (e) $N_f=89$; (f) $N_f=121$. The (imaginary) outline of the tow is drawn as broken line. The inter-tow porosity is 50%. $N_f$ is the total number of fibers in the tow.

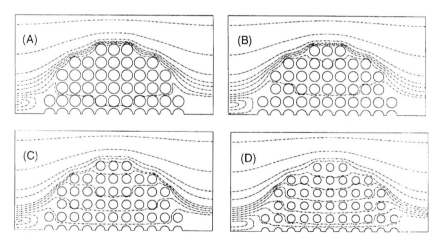

9.12  Typical multifiber unit cells used in the computations of the effective permeability of assemblies of permeable fiber tows, along with the trajectories of tracer particles. Conditions are (A) $\phi_i=0.6$, $\phi_t=0.4$; (B) $\phi_i=0.6$, $\phi_t=0.5$; (C) $\phi_i=0.6$, $\phi_t=0.6$; (D) $\phi_i=0.6$, $\phi_t=0.7$.

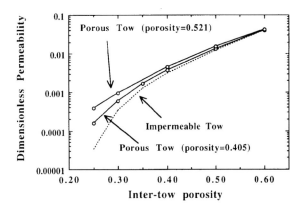

9.13 Effect of the inter-tow porosity on the dimensionless permeability of a square array of permeable tows at two levels of the intra-tow porosity. The curve corresponding to an impermeable tow of equal size is also shown as a broken line.

porous tow behaves as an impermeable entity even at relatively large values of $\phi_t$ is demonstrated in Fig. 9.13, which shows the predicted permeability of arrays of porous and impermeable tows as a function of the inter-tow porosity (Papathanasiou, 1996). It can be seen that for $\phi_i > 0.45$, the permeability of an array of porous tows ($K_p$) becomes practically equal to that of an array of impermeable tows of the same size ($K_s$), even for intra-tow porosity as high as 50%.

The BEM computations of Papathanasiou (1997) have further shown that a power-law relationship exists between the 'intrinsic' permeability $[(K_p/K_s) - 1]$ of an array of porous tows and $\chi$, the effective intra-tow porosity ($\chi = 1 - (1 - \phi_t)/(1 - \phi_{max})$, where $\phi_{max}$ is the porosity at maximum packing). This relationship can be expressed as:

$$K_{in} \equiv \frac{K_p}{K_s} - 1 = \alpha(\phi_i) \cdot \chi^{\beta(\phi_i)} \qquad [28]$$

where the coefficients $\alpha$ and $\beta$ are functions of the inter-tow porosity $\phi_i$ with values reported in Papathanasiou (1997). Equation [28] was found capable of representing the BEM computational results for the dimensionless permeability over a wide range of inter- and intra-tow porosities.

## 9.4    Conclusion

It is now well appreciated that, while domain methods allow for the incorporation of complex physics in a composite-mechanics simulation, the boundary element method is the method of choice when complex microstructural features are to be considered. Besides numerous applications in the field of suspension mechanics (Table 9.2), the suitability of the BEM in considering such features when predicting effective properties of fiber-reinforced composites has been demonstrated in this chapter by presenting results for the effective stiffness of an SFR composite rod. Up to 200 individual fibers, placed randomly in an elastic matrix and with their axes either aligned or at random orientations with the tensile axis, were considered. The geometrical complexity of these microstructures makes them at present intractable to finite element analysis.

Numerical results have shown that aligned-fiber systems are stiffer than randomly oriented ones and that longer fibers are more efficient in stiffening the composite than shorter ones, all other factors being equal. BEM computations were carried out in order to investigate the range of validity of the Halpin–Tsai equation and of its variants. It was found that the former is in agreement with computations for concentrations roughly up to $1/\alpha_r$, while the Lewis–Nielsen modification was in reasonable agreement with computations for the entire range of concentrations considered. The simplicity in mesh generation afforded by the BEM also allowed for a large number of simulations for Stokes' flow across arrays of fiber clusters to be carried out with modest computational resources. These revealed a relationship between the effective permeability of such an array and the characteristic porosities of the inter- and the intra-tow regions.

Currently available computational resources allow for the BEM computations presented in this study to be extended to prediction of the interplay between microstructure and macroscopic deformation in decisively non-homogeneous composite bodies. Figure 9.14 is an example of such a study (carried out on the Intel Paragon), in which the deformation of an elastic rod reinforced with 200 randomly oriented rigid fibers and subjected to a uniform tensile stress is shown. The grey-scale indicates the magnitude of the normal displacement on the side walls of the rod; lighter shade indicates maximum and darker shade minimum deformation. Even though the amount of microstructural detail that can be included in such computations is practically unlimited (the only limit being imposed by the size of the resultant discrete problem), the simulations are restricted to elastic fibers and an elastic matrix.

At present, inclusion of 'complex physics' in the context of the BEM is not straightforward – the method loses its main advantages (ease of mesh generation and computational efficiency) when domain integrals are included. Wearing and Burstow (1994) have shown that problems of elasto–plastic deformation can be successfully tackled by combining the FEM (which handles the usually smaller plastic region) with the BEM (which handles the larger elastic region). Coupling of the FEM with the BEM has already been successful in a range of problems (Hausslercombe, 1996), such as fluid–structure interactions (Duncan *et al.*, 1996, Rajakumar and Ali, 1996), soil–structure interactions (Feng and Owen, 1996) and acoustics (Sgard *et al.*, 1994). This can be a promising way forward in the

*9.14* Boundary element predictions of the normal displacement on the lateral surface of a composite rod reinforced by 200 randomly placed individual fibers. The rod is under (macroscopically homogeneous) tensile stress at the two ends. Due to variations in microstructure, the actual lateral displacement is not spatially uniform and is certainly different from what would have been calculated if some 'average fiber concentration' was assumed. It is seen that rod displacement is highest in regions of low fiber concentration (lighter color).

development of structure-oriented predictive models for fiber-reinforced composites which will combine both physical and microstructural detail.

## References

Bettess, P. (1981) Operation counts for boundary integral and finite element methods, *Internat. J. Numer. Method. Eng.*, **17**, 306–308.

Bhat, M.S., Poirier, D.R. and Heinrich, J.C. (1995) Permeability for cross-flow through columnar dendritic alloys, *Metall. Mater. Trans.*, **26B**, 1049–1056.

Blumentritt, B.F., Vu, B.T. and Cooper, S.L. (1974) The mechanical properties of oriented discontinuous fiber-reinforced thermoplastics: unidirectional fiber orientation, *Polym. Eng. Sci.*, **14**, 633.

Bohm, H.J., Rammerstorfer, F.G., Fischer, F.D. and Siegmund, T. (1994) Microscale arrangement effects on the thermomechanical behavior of advanced 2-phase materials, *J. Eng. Mater. Tech. – Trans. ASME*, **116**(3), 268–273.

Brebbia C.A. and Dominguez, J. (1992) *Boundary Elements: An Introductory Course*, Computational Mechanics Publications. Southampton, UK.

Brockenbrough, J.R., Suresh, S. and Wienecke, H.A. (1991) Deformation of metal matrix composites with continuous fibers: geometrical effects of fiber distribution and shape, *Acta Metall. Mater.*, **39**(5), 735–752.

Bruschke, M.V and Advani, S.G. (1993) Flow of generalised Newtonian fluids across a periodic array of cylinders, *J. Rheol.*, **37**, 479–498.

Cai, Z., and Berdichevsky, A.L. (1993) Numerical simulation on the permeability variations of a fiber assembly, *Polym. Compos.*, **14**(6), 529–539.

Cairns, D.S., Ilcewitz, L.B., Walker, T. and Minguet, P.J. (1995) Fracture scaling parameters of inhomogeneous microstructure in composite structures, *Comp. Sci. Tech.*, **53**(2), 223–231.

Chan, C.Y., Beris, A.N. and Advani, S.G. (1992) 2nd Order boundary element method calculations of hydrodynamic interactions between particles in close proximity, *Internat. J. Numer. Method. Fluids*, **14**(9), 1063–1086.

Chan, C.Y., Beris, A.N. and Advani, S.G. (1994) Analysis of periodic 3D viscous flows using quadratic discrete Galerkin boundary element method, *Internat. J. Numer. Method Fluids*, **18**(10), 953–981.

Dawson, P.R., Needleman, A. and Suresh, S. (1994) Issues in the finite element modelling of polyphase plasticity, *Mater. Sci. Eng. A*, **175**(1–2), 43–48.

Dingman, S.E., Ingber, M.S., Mondy, L.A., Abbott, J.R. and Brenner, H. (1992) Particle-tracking in 3D Stokes flow, *J. Rheol.*, **36**(3), 413–440.

Drummond, J.E. and Tahir, M.I. (1984) Laminar viscous flow through regular arrays of parallel solid cylinders, *Internat. J. Multiphase Flow*, **10**(5), 515–540.

Duncan, J.H., Milligan, C.D. and Zhang, S. (1996) On the interaction between a bubble and a submerged compliant structure, *J. Sound Vibration*, **197**(1), 17–44.

Feng, Y.T. and Owen, D.R.J. (1996) Iterative solution of coupled FE/BE discretisations for plate foundation interaction problems, *Internat. J. Numer. Method Eng.*, **39**(11), 1889–1901.

Finot, M., Shen, Y.L., Needleman, A. and Suresh, S. (1994) Micromechanical modelling of reinforcement fracture in particle-reinforced metal-matrix composites, *Metall. Mater. Trans. A - Phys. Metall. Mater. Sci.*, **25**(11), 2403–2420.

Ganesan, S., Chan, C.L. and Poirier, D.R. (1992) Permeability for flow parallel to primary dendrite arms, *Mater. Sci. Eng. A.*, **151**, 97–105.

Gebart, B.R. (1992) Permeability of unidirectional reinforcements for RTM, *J. Compos. Mater.*, **26**(8), 1100–1133.

Geller, A.S., Mondy, L.A., Rader, D.J. and Ingber, M.S. (1993) Boundary element method calculations of the mobility of non-spherical particles. 1. Linear chains, *J. Aerosol Sci.*, **24**(5), 597–609.

Ghaddar, C.K. (1995) On the permeability of unidirectional fibrous media – a parallel computational approach, *Phys. Fluids*, **7**(11), 2563–2686.

Goto, S. and McLean, M. (1991) Role of interfaces in creep of fiber-reinforced metal-matrix composites - II: Short fibers, *Acta Metall. Mater.*, **39**(2), 165.

Guell, D.C. and Graham, A.L. (1996) Improved mechanical properties in hydro-dynamically aligned, short-fiber composite materials, *J. Compos. Mater.*, **30**(1), 2–12.

Halpin, J.C. (1969) Stiffness and expansion estimates for oriented short fiber composites, *J. Compos. Mater.*, **3**, 723–724.

Halpin, J.C., Ashton, J.E. and Petit, P.H. (1969) *Primer in Composite Materials: Analysis*, Technomic Publishing Co., Westport, CT.

Hausslercombe, U. (1996) Coupling boundary elements and finite elements – a structured approach, *Comput. Method. Appl. Mech. Eng.*, **134**(1–2), 117–134.

Higdon, J.J. L. and Ford, G.D. (1996) Permeability of 3-dimensional models of fibrous porous media, *J. Fluid Mech.*, **308**, 341–361.

Hom, C.L. (1992) Three dimensional finite element analysis of plastic deformation in a whisker-reinforced metal-matrix composite, *J. Mech. Phys. Solids*, **40**(5), 991–1008.

Ilic, V., Tullock, D., Phan-Thien, N. and Graham, A.L. (1992) Translation and rotation of spheres settling in square and circular conduits – experiments and numerical predictions, *Internat. J. Multiphase Flow*, **18**(6), 1061–1075.

Ingber, M.S. and Mondy, L.A. (1994) A numerical study of 3-dimensional Jeffery orbits in shear flow, *J. Rheol.*, **38**(6), 1829–1843.

Ingber, M.S. and Papathanasiou, T.D. (1997) A parallel-supercomputing investigation of the Halpin–Tsai equation in aligned, whisker-reinforced composites, *Internat. J. Numer. Method. Eng.*, in press.

Ingber, M.S., Womble, D.E. and Mondy, L.A. (1994) A parallel boundary element formulation for determining effective properties of heterogeneous media, *Internat. J. Numer. Method. Eng.*, **37**(22), 3905–3920.

Larson, R.E. and Higdon, J.J.L. (1987) Microscopic flow near the surface of two-dimensional porous media. Part 2. Transverse flow, *J. Fluid Mech.*, **178**, 119–136.

Levy, A. and Papazian, J.M. (1991) Elastoplastic finite element analysis of short-fiber-reinforced SiC/Al composites: effects of thermal treatment, *Acta Metall. Mater.*, **39**(10), 2255–2266.

Lewis, T.B. and Nielsen, L.E. (1970) Dynamic mechanical properties of particulate-filled composites, *J. Appl. Polyha. Sci.*, **14**, 1449.

Li, J. and Ingber, M.S. (1994) A numerical study of the lateral migration of spherical particles in Pouseuille flow, *Eng. Anal. with Boundary Elements*, **13**(1), 83–92.

Li, D.S. and Wisnom, M.R. (1994) Finite element micromechanical modelling of unidirectional fiber-reinforced metal matrix composites, *Compos. Sci. Tech.*, **51**(4), 545–563.

Li, X.F., Charles, R. and Pozrikidis, C. (1996) Simple shear flow of suspensions of Liquid drops, *J. Fluid Mech.*, **320**, 395–416.

Lundstrom, T.S. and Gebart, B.R. (1995) Effect of perturbation of fiber architecture on permeability inside fiber tows, *J. Compos. Mater.*, **29**(4), 424–443.

Mammoli, A.A. and Bush, M.B. (1993) A boundary element analysis of metal-matrix composite materials, *Internat. J. Numer. Method. Eng.*, **36**(14), 2415–2433.

Mondy, L.A., Ingber, M.S. and Dingman, S.E. (1991) Boundary element method simulations of a ball falling through quiescent suspensions, *J. Rheol.*, **35**(5), 825–848.

Mondy, L.A., Geller, A.S., Rader, D.J. and Ingber, M. (1996) Boundary element method calculations of the mobility of non-spherical particles. 2. Branched chains and flakes, *J. Aerosol. Sci.*, **27**(4), 537–546.

Mukherjee, S. and Morjaria, M. (1984) On the efficiency and accuracy of the boundary element method and the finite element method, *Internat. J. Numer. Method. Eng.*, **20**, 515–522.

Muldowney, G.P. and Higdon, J.J.L. (1995) A spectral boundary element approach to 3-dimensional Stokes flow, *J. Fluid Mech.*, **298**, 167–192.

Nagelhout, D., Bhat, M.S., Heinrich, J.C. and Poirier, D.R. (1995) Permeability for flow normal to a sparse array of fibers, *Mater. Sci. Eng. A*, **191**(1–2), 203–208.

Nakamura, T. and Suresh, S. (1993) Effects of thermal residual stresses and fiber packing on deformation of metal-matrix composites, *Acta Metall. Mater.*, **41**(6), 1665–1681.

Nielsen, L.E. (1970) Generalised equation for the elastic moduli of composite materials, *J. Appl. Phys.*, **41**, 4626–4627.

Nielsen, L.E. (1974) *Mechanical Properties of Polymers and Composites*, Marcel Dekker, New York, Vol. 2.

Pandey, P. and Sundaresan, M.J. (1996) Strain distributions and over-stressed volumes in unidirectional composites with multiple fiber breaks, *J. Thermoplastic Compos. Mater.*, **9**(2), 199–215.

Papathanasiou, T.D. (1996) A structure-oriented micromechanical model for viscous flow through square arrays of fiber clusters, *Compos. Sci. Tech.*, **56**, 1055–1069.

Papathanasiou, T.D. (1997) On the effective permeability of square arrays of permeable fiber tows, *Internat. J. Multiphase Flow*, **23**(1) 81–92.

Papathanasiou, T.D. and Lee, P.D. (1997) Morphological effects on the transverse permeability of arrays of aligned fibers, *Polym. Compos.*, in press.

Papathanasiou, T.D., Ingber, M.S., Mondy, L.A. and Graham, A.L. (1994) The effective modulus of fiber-reinforced composites, *J. Compos. Mater.*, **28**, 288–304.

Papathanasiou, T.D., Ingber, M.S. and Guell, D.C. (1995) Stiffness enhancement in aligned short-fiber composites: a computational and experimental investigation, *Compos. Sci. Tech.*, **54**, 1–9.

Parnas, R.S. and Phelan, F.R. Jr. (1991) The effect of heterogeneous porous media on mold filling in resin transfer molding, *SAMPE Quart.*, **22**(2), 53–60.

Peirce, A.P. and Napier, J.A.L. (1995) Spectral multipole method for efficient solution of large-scale boundary element models in elastostatics, *Internat. J. Numer. Method. Eng.*, **38**(23), 4009–4034.

Phan-Thien, N. and Fan, X.J. (1995) Traction-based completed adjoint double-layer boundary element method in elasticity, *Comput. Mech.*, **16**(5), 360–367.

Phan-Thien, N. and Tullock, D. (1994) Completed double-layer boundary element method in elasticity and Stokes flow: distributed computing through PVM, *Comput. Mech.*, **14**(4), 370–383.

Phan-Thien, N., Tran-Cong, T. and Graham, A.L. (1991) Shear flow of periodic arrays of particle clusters: a boundary-element method, *J. Fluid Mech.*, **228**, 275–293.

Phelan, F.R. and Wise, G. (1996) Analysis of transverse flow in aligned fibrous porous media, *Compos. Part A*, **27**(1), 25–33.

Pillai, K.M and Advani, S.G. (1995) Numerical and analytical study to estimate the effect of two length scales upon the permeability of a fibrous porous medium, *Transport in Porous Media*, **21**(1), 1–17.

Pillai, K.M., Luce, T.L., Bruschke, M.V., Parnas, R.S. and Advani, S.G. (1993) Modelling the heterogeneities present in preforms during mold filling in RTM, *25th International SAMPE Technical Conference*, Volume 25, 279–293.

Pozrikidis, C. (1995) Finite deformation of liquid capsules enclosed by elastic membranes in simple shear flow, *J. Fluid Mech.*, **297**, 123–152.

Rajakumar, C. and Li, A.A. (1996) BEM-FEM coupled eigenanalysis of fluid structure systems, *Internat. J. Numer. Method. Eng.*, **39**(10), 1625–1634.

Rammerstorfer, F.G., Fischer, F.D., Bohm, H.J. and Daves, W. (1992) Computational micromechanics of multiphase materials, *Comput. Struct.*, **44**(1–2), 453–458.

Ranganathan, S., Phelan, F.R. and Advani, S.G. (1996) A generalised model for the transverse fluid permeability in unidirectional fibrous media, *Polym. Compos.*, **17**(2), 222–230.

Reiter, T.J., Bohm, H.J., Krach, W., Pleschberger, M. and Rammerstorfer, F.G (1994) Some applications of the finite element method in biomechanical stress analysis, *Int. J. Comput. Appl. Tech.*, **7**(3–6), 233–241.

Sadiq, T.A.K., Advani, S.G. and Parnas, R.S. (1995) Experimental investigation of transverse flow through aligned cylinders, *Internat. J. Multiphase Flow*, **21**(5), 755–774.

Sangani, A.S. and Acrivos, A. (1982) Slow flow past periodic arrays of cylinders with application to heat transfer, *Internat. J. Multiphase Flow*, **8**(3), 193–206.

Sgard, F., Atalla, N. and Nicolas, J. (1994) Coupled FEM-BEM approach for mean flow effects on vibroacoustic behavior of planar structures, *AIAA J.*, **32**(12), 2351–2358.

Shen, Y.L., Finot, M. Needleman, A. and Suresh, S. (1995) Effective plastic response of 2-phase composites, *Acta Metall. Mater.*, **43**(4), 1701–1722.

Silvain, J.F., Petitcorps, Y., Sellier, Bonniau, P. and Heim, V. (1994) Elastic moduli, thermal expansion and microstructure of copper-matrix composites reinforced by continuous graphite fibers, *Composites*, **25**(7), 570–574.

Sorensen, N., Needleman, A. and Tvergaard,V. (1992) Three-dimensional analysis of creep in metal matrix composites, *Mater. Sci. Eng. A*, **158**, 129–137.

Steinkopff, T. and Sautter, M. (1995) Simulating the elastoplastic behavior of multiphase materials by advanced finite element techniques. 1 A rezoning technique and the multiphase element method, *Comput. Mat. Sci.*, **4**(1), 10–14.

Streat, N. and Weinberg, F. (1976) Interdendritic fluid flow in a lead–tin alloy, *Met. Trans.* **7B**, 417–423.

Summerscales, J. (1993) A model for the effect of fiber clustering on the flow rate in resin transfer moulding, *Compos. Manuf.*, **4**(1), 27–31.

Summerscales, J., Griffin, P.R., Grove, S.M. and Guild, F.J. (1995) Quantitative microstructural examination of RTM fabrics designed for enhanced flow, *Compos. Struct.*, **32**(1–4) 519–529.

Ter Haar, J.H. and Duszczyk, J. (1994) Mechanical properties and microstructure of a P/M aluminium matrix composite with δ-alumina fibers and their relation to extrusion, *J. Mater. Sci.*, **29**, 1011–1024.

Tucker III, C.L. and Advani, S.G. (1994) Processing of short-fiber systems, in *Flow and Rheology in Polymer Composites Manufacturing* ed. S.G. Advani, Elsevier, Amsterdam.

Vincent, J., Phan-Thien, N., and Tran-Cong, T. (1991) Sedimentation of multiple particles of arbitrary shape, *J. Rheol.*, **35**(1), 17–27.

Wearing, J.L. and Burstow, M.C. (1994) Elastoplastic analysis using a coupled BEM/FEM technique, *Eng. Anal. with Boundary Elements*, **14**(1), 39–49.

Weissenbek, E. and Rammerstorfer, F.G. (1993) Influence of the fiber arrangement on the mechanical and thermo-mechanical behavior of short fiber reinforced MMCs, *Acta Metall. Mater.*, **41**(10), 2833–2843.

Wulf, J., Steinkopff, T. and Fischmeister, H.F. (1996) FE simulation of crack paths in the real microstructure of an Al/SiC composite, *Acta Mater.*, **44**(5), 1765–1779.

Zhou, H. and Pozrikidis, C. (1994) Pressure-driven flow of suspensions of liquid drops, *Phys. Fluids*, **6**(1), 80–94.

Zhou, H. and Pozrikidis, C. (1995) Deformation of liquid capsules with incompressible interfaces in simple shear flow, *J. Fluid Mech.*, **283**, 175–200.

# Index